S0-BDP-262

Intermediate Algebra

Concepts Through Applications
Class Test Edition, volume I

MARK CLARK
Palomar College

BROOKS/COLE
CENGAGE Learning™

Australia • Brazil • Japan • Korea • Mexico • Singapore • Spain • United Kingdom • United States

BROOKS/COLE
CENGAGE Learning™

Intermediate Algebra : Concepts through Applications Class Test Edition, vol. 1
Mark Clark

Executive Editor: Charlie Van Wagner

Development Editor: Donald Gecewicz

Assistant Editor: Shaun Williams

Editorial Assistant: Mary De La Cruz

Marketing Manager: Joe Rogove

Marketing Assistant: Angela Kim

Project Manager, Editorial Production: Cheryll Linthicum

Creative Director: Rob Hugel

Art Director: Vernon T. Boes

Photo Researcher: Terri Wright

Copy Editor: Barbara Willette

Illustrators: Hearthside, ICC India

Print Buyer: Barbara Britton

Permissions Editor: Stephanie Lee

Production Service: Hearthside Publishing Services

Text Designer: Geri Davis

Cover Designer: Roger Knox

Cover Image: Photonica

Compositor: ICC Macmillan Inc.

© 2009, 2006 Brooks/Cole, Cengage Learning

ALL RIGHTS RESERVED. No part of this work covered by the copyright herein may be reproduced, transmitted, stored, or used in any form or by any means graphic, electronic, or mechanical, including but not limited to photocopying, recording, scanning, digitizing, taping, Web distribution, information networks, or information storage and retrieval systems, except as permitted under Section 107 or 108 of the 1976 United States Copyright Act, without the prior written permission of the publisher.

For product information and technology assistance, contact us at **Cengage Learning Customer & Sales Support, 1-800-354-9706.**

For permission to use material from this text or product, submit all requests online at **www.cengage.com/permissions**
Further permissions questions can be e-mailed to **permissionrequest@cengage.com**

Student Edition:

ISBN-13: 978-0-495-82842-6

ISBN-10: 0-495-82842-4

Brooks/Cole Cengage Learning
10 Davis Drive
Belmont, CA 94002-3098
USA

Cengage Learning is a leading provider of customized learning solutions with office locations around the globe, including Singapore, the United Kingdom, Australia, Mexico, Brazil, and Japan. Locate your local office at **international.cengage.com/region**.

Cengage Learning products are represented in Canada by Nelson Education, Ltd.

For your course and learning solutions, visit **academic.cengage.com**

Purchase any of our products at your local college store or at our preferred online store **www.ichapters.com**

Printed in the United States of America
1 2 3 4 5 6 7 12 11 10 09 08

Contents

3 EXPONENTS, POLYNOMIALS AND FUNCTIONS

4 QUADRATIC FUNCTIONS

Linear Functions

- Solve a linear equation and explain the meaning of the solution in context.

- Create a scatterplot from data.

- Understand and recognize model breakdown in a context.

- Find and understand intercepts and their meaning in a context.

- Understand the meaning of a function and function notation.

- Use the vertical line test to verify a function.

- Find and understand the domain and range of a linear function.

- Find the slope of a linear equation and explain its meaning in a context.

- Graph a linear function by hand.

- Find and use a linear model by hand.

© Jeff Greenberg/Photo Edit

he United States Census Bureau found that the average square feet of floor area in new one-family houses in 1980 was 1740 square feet. In 2003, the average square feet of floor area increased to 2330 square feet. In this chapter, we will discuss how to use linear models to analyze trends in real life data. One of the chapter projects will ask you to investigate the costs associated with installing new flooring in a home.

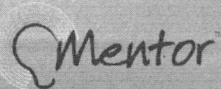

Solving Linear Equations

Equations can be used to represent many things in life. One of the uses of algebra is to solve equations for an unknown quantity, or variable. In this section you will learn how to solve linear equations for a missing variable and to write a complete solution. Complete sentence answers that give the units and the meaning of the solution you found will provide the reader with a clear understanding of that solution.

EXAMPLE  SOLVING APPLICATIONS AND PROVIDING COMPLETE SOLUTIONS

U-Haul charges $19.95 for the day and $0.79 per mile driven to rent a 10-foot truck. The total cost to rent a 10-foot truck for the day can be represented by the equation:

$$U = 19.95 + 0.79m$$

where U is the total cost in dollars to rent a 10-foot truck from U-Haul for the day, and m is the number of miles the truck is driven.

a. Determine how much it will cost to rent a 10-foot truck from U-Haul and drive it 75 miles.

b. Determine the number of miles you can travel for a total cost of $175.00.

Solution

a. Because the number of miles driven was given, you can replace the variable m in the equation with the number 75 and solve for the missing variable U as follows:

$$U = 19.95 + 0.79m$$
$$U = 19.95 + 0.79(75)$$
$$U = 19.95 + 59.25$$
$$U = 79.20$$

This solution indicates that renting a 10-foot truck from U-Haul and driving it 75 miles would cost you $79.20.

b. Because the total cost of $175.00 is given in the statement, we can substitute 175.00 for the variable U and solve for the missing variable m.

$$U = 19.95 + 0.79m$$
$$175.00 = 19.95 + 0.79m$$

$\underline{-19.95 \quad -19.95}$ \qquad Subtract 19.95 from both sides.

$$155.05 = 0.79m$$
$$\frac{155.05}{0.79} = \frac{0.79m}{0.79}$$ \qquad Divide both sides by 0.79.

$$196.266 \approx m$$

Concept Connection

What is the difference between the (=) symbol and the (≈) symbol?

In mathematics we use these symbols and others to show a relationship between two quantities or between two expressions.

The equal sign (=) is used when two quantities or expressions are equal and thus exactly the same.

The approximation symbol (≈) is used to show that two quantities or expressions are approximately the same. We will use the approximation symbol whenever a quantity is rounded.

How are we going to round?

In general we will round values to at least one additional decimal place than the given numbers in the problem. We will follow this rule unless otherwise stated in the problem.

Skill Connection

In solving an equation for a variable, the goal is to isolate the variable (get it by itself) on one side of the equation and simplify the other side. When you have more than one term on the same side of the equation where the variable is, start by undoing any addition or subtraction. After addition and subtraction you can undo multiplication and division using the opposite operation.

SC-Example 1: Solve

$$5z + 16 = 3$$

Solution:

Once again the variable z is not isolated on a side of the equation, so we need to eliminate the 16 and 5. The 16 is being added, so we should eliminate it first. Because it is being added, we will subtract 16 from both sides.

$$\begin{array}{r} 5z + 16 = 3 \\ \underline{-16 \quad -16} \\ 5z = -13 \end{array}$$

$$\frac{5z}{5} = \frac{-13}{5}$$

$$z = -\frac{13}{5}$$

Once the 16 was moved, we divided both sides by 5 to finish isolating the variable. Some instructors will want this solution in a mixed number or decimal.

$$z = -2\frac{3}{5} \quad \text{or} \quad z = -2.6$$

Since U-haul would charge for a full mile for the 0.266 we will round down to 196 miles so we can stay within the budget of $175. We can check this solution by substituting $m = 196$ to be sure U will equal $175.

$$U = 19.95 + 0.79(196)$$
$$U = 19.95 + 154.84$$
$$U = 174.79$$

This solution indicates that for a cost of $175.00 you can rent a 10-foot truck from U-Haul for a day and drive it 196 miles.

Example 1 uses an algebraic equation to represent the cost of the rental in terms of the mileage.

EXAMPLE PRACTICE PROBLEM

A team of engineers is trying to pump down the pressure in the vacuum chamber shown in the figure. They know that the following equation represents the pressure in the chamber.

$$P = 35 - 0.07s$$

where P is the pressure in pounds per square inch (psi) of the vacuum chamber and s is the time in seconds.

Photo Courtesy of Mill Lane Engineering, Lowell, MA

a. What will the pressure be after 150 seconds?

b. When will the pressure inside the chamber be 1 psi?

Many problems in this book will investigate applications that involve money and business. We must understand some business terms to understand these problems and communicate clearly the meaning of the solutions. The three main concepts in business are **revenue, cost,** and **profit.**

A simple definition of revenue is the total amount of money that is brought into the business through sales. For example if a pizza place sells 10 pizzas for $12 each, their revenue would be 10 pizzas · $12 per pizza = $120. Revenue is often calculated as the price times the quantity sold. The revenue for a business can never be a negative number.

Cost is defined as the amount of money paid out for expenses. Expenses often are categorized in two ways: fixed costs and variable costs. The same pizza place would probably have fixed costs such as rent, utilities, and perhaps salaries, and it would have variable costs of supplies and food ingredients per pizza made. The cost for the business would be the fixed costs and the variable costs added together.

The profit for a business is the revenue minus the cost. If this pizza place had $100 in cost when making the 10 pizzas, they would have a profit of $120 − $100 = $20. Although a business cannot have a negative revenue, profit can be negative. When profit is negative we sometimes call that a loss but it is still understood to be a negative profit.

A business is often interested in what its **break-even point** is. That is when the revenue from a product is the same as the cost. The break-even point is also when the profit is zero. The point when profit changes from negative to positive is important to a company who is considering a new product and needs to know how many need to be produced or sold for the company to start making a profit.

These definitions should help you to understand some of the examples and exercises throughout this book.

DEFINITIONS

Revenue: The amount of money brought into a business through sales. Revenue is often calculated as

$$revenue = price \cdot quantity\ sold$$

Costs: The amount of money spent by a business to create and/or sell a product. Costs usually include both fixed costs and variable costs. Fixed costs are the same each month or year and variable costs change consistently depending on the number of items produced and/or sold.

$$cost = fixed\ cost + variable\ cost$$

$$Or\ \ cost = fixed\ cost + marginal\ cost \cdot quantity\ sold$$

Profit: The amount of money left after all costs.

$$profit = revenue - cost$$

Break-even point: A company breaks even when their revenue equals their cost or when their profit is zero.

$$revenue = cost$$

$$profit = 0$$

EXAMPLE **2** WRITING AN EQUATION AND PROVIDING COMPLETE SOLUTIONS

You are in charge of creating and purchasing T-shirts for a local summer camp. After calling a local silk-screening company, you find that to purchase 100 or more T-shirts, there will be a $150 setup fee and a $5 per T-shirt charge.

a. Write an equation for the total cost, C, of making t T-shirts.

b. How much would 300 T-shirts cost?

c. How many T-shirts can you purchase with a budget of $1500?

Solution

a. First we should determine the variables given, since we want to find the total cost and the number of T-shirts let

C = Total cost of T-shirts, in dollars

t = Number of T-shirts produced

Because each T-shirt will cost \$5, we need to multiply 5 by the number of T-shirts produced t. Then we need to add the \$150 setup fee to get the total.

$$C = 5t + 150$$

b. The number of T-shirts is given, so we can substitute 300 for t and simplify to find the total cost C.

$$C = 5t + 150$$
$$C = 5(300) + 150$$
$$C = 1500 + 150$$
$$C = 1650$$

The total cost to produce 300 T-shirts will be \$1650.

c. The total budget represents the total cost of the T-shirts, so we can substitute 1500 for C and solve for the number of T-shirts, t.

$$C = 5t + 150$$
$$1500 = 5t + 150$$

$$\underline{-150 \qquad -150} \qquad \text{Subtract 150 from both sides.}$$

$$1350 = 5t$$

$$\frac{1350}{5} = \frac{5t}{5} \qquad \text{Divide both sides by 5.}$$

$$270 = t$$

We can check this solution by substituting 270 for t and making sure the cost is \$1500.

$$C = 5(270) + 150$$
$$C = 1350 + 150$$
$$C = 1500$$

For a budget of \$1500 you can purchase 270 T-shirts.

EXAMPLE ② PRACTICE PROBLEM

Rockon, a small-town rock band, wants to produce a CD before their next summer concert series. They have looked into a local recording studio and found that it will cost them \$1500 to produce the master recording and then an additional \$1.50 to make each CD up to 500.

a. Write an equation for the total cost, C, in dollars, of producing n CDs.

b. How much will it cost Rockon to make 250 CDs?

c. If Rockon has \$2000 to produce CDs, how many can they order?

In both of the previous examples it is important to pay attention to the definition of each variable. The definitions of the variables help you to determine which variable value was given and which variable you need to solve for. In some questions you will need to define your own variables. Use meaningful variable names to make it easy to remember what they represent. For example,

- t = time in years
- h = hours after 12 noon
- p = population of San Diego (in thousands)
- P = profit of IBM (in millions of dollars)
- S = Salary (in dollars per hour)

Units or how the quantity is measured, are very important in communicating what a variable represents. The meaning of $P = 100$ is very different if profit for IBM is measured in dollars and not millions of dollars. The same for $S = 6.5$. If S represents your salary for your first job out of college, it would be great if S were measured in millions of dollars per year and not dollars per hour. As you can see, units can make a large difference in the meaning of a quantity. When defining variables, be sure to give units.

When solving an equation that represents something in an application you should always check that the solution you find is a reasonable one for the situation. Use the following concept investigation to practice determining which solutions might be reasonable and which would not make sense in the situation given.

CONCEPT INVESTIGATION 1

IS THAT A REASONABLE SOLUTION?

In each part choose the value that seems the most reasonable for the given situation. Explain why the other given value(s) do not make sense in that situation.

a. If P is the population of the United States in millions of people, which of the following is a reasonable value for P?

   i. $P = -120$

   ii. $P = 300$

   iii. $P = 5,248,000,000$

b. If H is the height of an airplane's flight path in feet, which of the following is a reasonable value for H?

   i. $H = -2000$

   ii. $H = 3,500,000$

   iii. $H = 25,000$

c. If P is the annual profit in dollars of a new flower shop the first year it opens, which of the following is a reasonable value for P?

   i. $P = -40,000$

   ii. $P = 50,000$

   iii. $P = 3,000,000$

EXAMPLE   **3** BUSINESS APPLICATIONS

A small bicycle company produces high-tech bikes for international race teams. The company has fixed costs of $5000 per month for rent, salary, and utilities. For every bike they produce, it costs them $755 in materials and other expenses related to that bike. The company can sell each bike for an average price of $1995 but can produce a maximum of only 20 bikes per month.

Photo Provided by Trek Bicycle Corporation. All Rights Reserved.

a. Find an equation for the monthly cost of producing b bikes.

b. How much does it cost the bicycle company to produce 20 bikes in a month?

c. Find an equation for the monthly revenue from selling b bikes.

d. How much revenue will the bicycle company make if they sell 10 bikes in a month?

e. Find an equation for the monthly profit the company makes if they produce and sell b bikes. (You can assume that they will sell all the bikes they make.)

f. What is the profit of producing and selling 15 bikes in a month?

g. How many bikes does the company have to produce and sell in a month to make $15,000 profit?

h. How many bikes does the company have to produce and sell in a month to make $30,000 profit?

Solution

a. First define the variables in the problem.

b = The number of bikes produced each month. (Remember that a maximum of 20 bikes can be produced each month.)

C = The monthly cost, in dollars, to produce b bikes.

Each bike cost $755 for materials and other expenses, so we need to multiply b by 755, and then the fixed costs need to be added on, to get the total monthly cost. This gives the following equation.

$$C = 755b + 5000$$

b. The number of bikes produced is given, so we can substitute 20 for b and simplify the right side to find C.

$$C = 755b + 5000$$
$$C = 755(20) + 5000$$
$$C = 20100$$

A monthly production of 20 bikes will result in a total monthly cost of $20,100.

c.   First define the variables in the problem. Recall that *b* was already defined in part a.

b = The number of bikes produced each month.

R = The monthly revenue, in dollars, from selling *b* bikes.

The bicycle company can sell each bike for an average price of $1995, so the revenue can be calculated by using the equation

$$R = 1995b$$

d.   The number of bikes is given so we can substitute 10 for *b*.

$$R = 1995b$$
$$R = 1995(10)$$
$$R = 19950$$

The total monthly revenue from selling 10 bikes is $19,950.

e.   Profit is calculated by taking the revenue and subtracting any costs incurred.

b = The number of bikes produced each month.

P = The monthly profit, in dollars, from producing and selling *b* bikes.

Thus we can use the equations for revenue and cost we found earlier.

$$P = R - C$$
$$P = (1995b) - (755b + 5000) \qquad \textbf{Substitute for } R \textbf{ and } C.$$

This profit equation can be simplified by distributing the negative and combining like terms.

$$P = (1995b) - (755b + 5000)$$
$$P = 1995b - 755b - 5000 \qquad \textbf{Distribute the negative sign.}$$
$$P = 1240b - 5000 \qquad \textbf{Combine like terms.}$$

f.   The number of bikes is given, so we can substitute 15 for *b*.

$$P = 1240(15) - 5000$$
$$P = 13600$$

The monthly profit from producing and selling 15 bikes is $13,600.

g.   The amount of profit desired is given, so we can substitute 15000 for *P*.

$$P = 1240b - 5000$$
$$15000 = 1240b - 5000$$

$$\underline{+ 5000 \qquad\qquad + 5000} \qquad \textbf{Add 5000 to both sides.}$$
$$20000 = 1240b$$
$$\frac{20000}{1240} = \frac{1240b}{1240} \qquad \textbf{Divide both sides by 1240.}$$
$$16.129 \approx b$$

Because we found a decimal answer, we need to compare the profits for the whole bikes represented on both sides of this decimal.

$$P = 1240(16) - 5000$$
$$P = 14840$$

$$P = 1240(17) - 5000$$
$$P = 16080$$

If we want to make at least $15,000 profit for the month, we will need to produce at least 17 bikes. We round up, since producing 16 bikes would not quite make $15,000 profit.

h. The amount of profit desired is given, so we can substitute 30000 for *P*.

$$P = 1240b - 5000$$
$$30000 = 1240b - 5000$$

$$\underline{+\ 5000 \qquad\qquad +5000}\qquad\qquad \textbf{Add 5000 to both sides.}$$

$$35000 = 1240b$$

$$\frac{35000}{1240} = \frac{1240b}{1240} \qquad\qquad \textbf{Divide both sides by 1240.}$$

$$28.226 \approx b$$

We can check this solution by substituting in 28.226 for *b* and confirming that it gives a profit of $30,000

$$P = 1240(28.226) - 5000$$
$$P = 35000.24 - 5000 \qquad \textbf{The solution was rounded so the check}$$
$$P = 30000.24 \qquad\qquad \textbf{is not exact but close enough.}$$

The algebra once again came out with a decimal answer, so we would need to round to the whole number of bikes that will produce the desired profit. That would give us 29 bikes produced in a month. However, this level of production is not possible, since the problem stated that the company could produce a maximum of only 20 bikes per month. Therefore the correct answer is that the company cannot make $30,000 profit in a month with its current production capacity.

Example 3 shows that you need to check each answer to determine whether it is a reasonable solution or not. Many times this is something that requires only some common sense; other times a restriction that is stated in the problem will need to be considered.

EXAMPLE ③ PRACTICE PROBLEM

A local chiropractor has a small office where she cares for patients. She has $8000 in fixed costs each month that cover her rent, basic salaries, equipment and utilities. For each patient she sees she has an average additional cost of about $15. The chiropractor charges her patients, or their insurance company, $80 for a visit.

a. Find an equation for the total monthly cost when *v* patient visits are done in a month.

b. What is the total monthly cost if the chiropractor has 100 patients visit during a month.

c. Find an equation for the monthly revenue when *v* patients visit a month.

d. What is the revenue if the chiropractor only sees 75 patients in a month?

e. Find an equation for the monthly profit the chiropractor makes if she has v patient visits in a month.

f. What is the monthly profit when 150 patients visit in a month?

g. How many patient visits does this chiropractor need to have in a month in order for her profit to be $5000.00?

When solving equations that involve fractions, you could work with the fractions or eliminate the fractions at the beginning of the problem by multiplying both sides of the equation by the common denominator and then finish solving the equation as you would any other equation.

EXAMPLE 4 SOLVING EQUATIONS

Solve the following equations.

a. $\frac{2}{3}x + \frac{5}{6} = 7$

b. $\frac{1}{4}(x + 5) = \frac{1}{2}x - 6$

c. $4t - 2(3.4t + 7) = 5t - 17.3$

Solution

a. In this case, to eliminate the fractions, we can multiply both sides of the equation by the common denominator 6 to clear the fractions and then continue solving

$$\frac{2}{3}x + \frac{5}{6} = 7$$

$$6\left(\frac{2}{3}x + \frac{5}{6}\right) = 6(7)$$
Multiply both sides by the common denominator 6.

$$6\left(\frac{2}{3}x\right) + 6\left(\frac{5}{6}\right) = 6(7)$$

$$\overset{2}{\cancel{6}}\left(\frac{2}{\cancel{3}}x\right) + \cancel{6}\left(\frac{5}{\cancel{6}}\right) = 6(7)$$
Reduce to eliminate the fractions.

$$4x + 5 = 42$$
Finish solving.

$$\underline{-5 \quad -5}$$

$$4x = 37$$

$$\frac{4x}{4} = \frac{37}{4}$$

$$x = \frac{37}{4}$$

We need to check this solution.

$$\frac{2}{3}\left(\frac{37}{4}\right) + \frac{5}{6} = 7 \qquad \text{Check the solution.}$$

$$\frac{37}{6} + \frac{5}{6} = 7$$

$$\frac{42}{6} = 7$$

$$7 = 7 \qquad \text{The solution works.}$$

Therefore the solution is $x = \frac{37}{4} = 9\frac{1}{4} = 9.25$. This solution can be written

as an improper fraction, a mixed number or a decimal.

b. For this equation we should distribute the $\frac{1}{4}$ first and then we can eliminate the
fractions by multiplying both sides of the equation by the common denominator
4.

$$\frac{1}{4}(x + 5) = \frac{1}{2}x - 6$$

$$\frac{1}{4}x + \frac{1}{4}(5) = \frac{1}{2}x - 6 \qquad \text{Distribute the 1/4.}$$

$$\frac{1}{4}x + \frac{5}{4} = \frac{1}{2}x - 6$$

$$4\left(\frac{1}{4}x + \frac{5}{4}\right) = 4\left(\frac{1}{2}x - 6\right) \qquad \begin{array}{l}\text{Multiply both sides of the equation by} \\ \text{the common denominator 4.}\end{array}$$

$$4\left(\frac{1}{4}x\right) + 4\left(\frac{5}{4}\right) = 4\left(\frac{1}{2}x\right) - 4(6) \qquad \text{Distribute the 4 through both sides.}$$

$$\cancel{4}\left(\frac{1}{\cancel{4}}x\right) + \cancel{4}\left(\frac{5}{\cancel{4}}\right) = \overset{2}{\cancel{4}}\left(\frac{1}{\cancel{2}}x\right) - 4(6) \qquad \text{Reduce and multiply.}$$

$$x + 5 = 2x - 24$$

$$\underline{ -2x \qquad -2x } \qquad \begin{array}{l}\text{Subtract } 2x \text{ from both sides of the} \\ \text{equation to get the variable terms} \\ \text{together.}\end{array}$$

$$-x + 5 = -24$$

$$\underline{ -5 \qquad -5 } \qquad \begin{array}{l}\text{Subtract 5 from both sides of the} \\ \text{equation to get the variable term} \\ \text{isolated.}\end{array}$$

$$-x = -29$$

$$-1(-x) = -1(-29) \qquad \text{Multiply by -1 to make } x \text{ positive.}$$

$$x = 29$$

$$\frac{1}{4}(29 + 5) = \frac{1}{2}(29) - 6 \qquad \text{Check the solution.}$$

$$\frac{1}{4}(34) = \frac{29}{2} - \frac{12}{2}$$

$$\frac{17}{2} = \frac{17}{2}$$

c.   To start solving, we need to distribute the negative 2 through the parentheses.

$$4t - 2(3.4t + 7) = 5t - 17.3$$ Distribute the 2.

$$4t - 6.8t - 14 = 5t - 17.3$$
          Combine like terms.
$$-2.8t - 14 = 5t - 17.3$$

$$\begin{array}{r} -2.8t - 14 = 5t - 17.3 \\ + 2.8t \qquad\quad + 2.8t \\ \hline -14 = 7.8t - 17.3 \end{array}$$ Add 2.8*t* to both sides of the equation to get the variable terms together.

$$\begin{array}{r} + 17.3 \qquad\quad +17.3 \\ \hline 3.3 = 7.8t \end{array}$$ Add 17.3 to both sides of the equation to isolate the variable term.

$$\frac{3.3}{7.8} = \frac{7.8t}{7.8}$$ Divide both sides of the equation by 7.8 to isolate *t*.

$$0.423 \approx t$$

Check the solution.

$$4(0.423) - 2(3.4(0.423) + 7) = 5(0.423) - 17.3$$
$$1.692 - 2(8.4382) = 2.115 - 17.3$$
$$-15.1844 \approx 15.185$$

Because of rounding, the two sides will not be exactly the same but are very close.

EXAMPLE ④ PRACTICE PROBLEM

Solve the following equations.

a. $\frac{2}{5}(x - 6) = \frac{1}{4}x + 15$

b. $3 + 2.5(3x + 4) = 1.75x - 8$

Equations that contain more than one variable are called literal equations. Formulas are literal equations that are used to express relationships among physical quantitites. We use literal equations and formulas in many areas of our lives. For example, $D = rt$ is a formula to calculate the distance traveled when you are given the rate at which you are traveling and for how long (time) you traveled that rate. This formula can be rearranged (solved) for one of the other variables to make it easier to use to find the rate if you know the distance and time or to find the time if you know the distance and rate. Solving literal equations and formulas for other variables works the same way that other solving does except that many of the calculations will not be able to be completed until values for the variables are known.

EXAMPLE 5 SOLVING LITERAL EQUATIONS FOR A DIFFERENT VARIABLE

Solve the following literal equations for the variable indicated.

a. $D = rt$ for *r*.

b.   Distance in free fall: $D = \frac{1}{2}Gt^2$ for *G*.

c. $y = mx + b$ for *m*.

Solution

a. Because r is being multiplied by t, we need to divide both sides by t.

$$D = rt$$

$$\frac{D}{t} = \frac{rt}{t} \qquad \text{Divide both sides by } t.$$

$$\frac{D}{t} = r$$

b. We want to solve for G.

$$D = \frac{1}{2}Gt^2$$

$$2D = 2\left(\frac{1}{2}Gt^2\right) \qquad \text{Multiply both sides by 2 to undo the } 1/2.$$

$$2D = Gt^2$$

$$\frac{2D}{t^2} = \frac{Gt^2}{t^2} \qquad \text{Divide both sides by } t \text{ squared.}$$

$$\frac{2D}{t^2} = G$$

c. We want to solve for m, so we will need to isolate the term with m and then divide both sides by x.

$$y = mx + b$$

$$\underline{-b \qquad\qquad -b}$$

$$y - b = mx \qquad \text{Subtract } b \text{ from both sides to isolate the } mx \text{ term. Because } y \text{ and } b \text{ are not the same variable, we cannot subtract them but must leave the left side as } y - b.$$

$$\frac{y - b}{x} = \frac{mx}{x} \qquad \text{Divide both sides by } x \text{ to isolate the } m.$$

$$\frac{y - b}{x} = m$$

EXAMPLE ⑤ PRACTICE PROBLEM

Solve the following for the indicated variable.

a. Velocity in free fall: $V = Gt$ for t.

b. Velocity: $v = v_o + at$ for a. (Note: v_o is the initial velocity.)

1.1   Exercises

For Exercises 1 through 10, solve each equation.

1. $2x + 10 = 40$

2. $3x + 14 = 35$

3. $-4t + 8 = -32$

4. $-7m + 20 = 48$

5. $2.5x + 7.5 = 32.5$

6. $3.4x - 8.2 = 15.6$

7. $20 = 5.2x - 0.8$

8. $45 = -3.6c + 189$

9. $0.05(x - 200) = 240$

10. $0.03(n - 500) = 108$

11. During the first day of training on the job a new candy maker gets faster at making candies. The number of candies a new employee can produce during an hour can be represented by $C = 10h + 20$ candies, where h is the number of hours of training.

    a. Find the number of candies a new employee can produce in an hour after 1 hour of training.

    b. Find the number of candies a new employee can produce in an hour after 4 hours of training.

    c. How many hours of training must an employee do before they can produce 150 candies an hour?

12. The number of students who are enrolled in math classes at a local college can be represented by $E = -17w + 600$, where E represents the math class enrollment at the college w weeks after the start of the fall semester.

    a. Find the total enrollment in math classes at the college at the beginning of the fall semester. (*Hint:* Since the semester is just starting, $w = 0$.)

    b. In what week will the total enrollment be 430 students?

    c. What will the total enrollment be in math classes after 8 weeks?

13. The number of homicides, N, of 15- to 19-year-olds in the United States t years after 1990 can be represented by the equation $N = -315.9t + 4809.8$

    Source: Based on data from Statistical Abstract 2001.

    a. Find the number of homicides of 15- to 19-year-olds in the United States in 1992. (1992 is 2 years after 1990, so $t = 2$.)

    b. Find the number of homicides of 15- to19-year-olds in the United States in 2002.

    c. When was the number of homicides 7337?

14. The gasoline prices in Southern California can increase very quickly during the summer months. The equation $p = 2.399 + 0.03w$ represents the gasoline prices p in dollars per gallon w weeks after the beginning of summer.

    a. What does gasoline cost after 5 weeks of summer?

    b. When will gasoline cost $2.759 per gallon?

15. $P = 1.5t - 300$ represents the profit in dollars from selling t printed T-shirts.

    a. How much profit will you earn if you sell 100 printed T-shirts?

    b. How much profit will you earn if you sell 400 printed T-shirts?

    c. How many printed T-shirts must you sell to make $1000 in profit?

16. $P = 5.5b - 500.5$ represents the profit in dollars from selling b books.

    a. How much profit will you earn if you sell 75 books?

    b. How much profit will you earn if you sell 200 books?

    c. How many books must you sell to make $3600 in profit?

17. Salespeople often work for commissions on the sales that they make for the company. As a new salesperson at a local technology company you are told that you will receive an 8% commission on all sales you make after the first $1000. Your pay can be repre-

sented by $p = 0.08(s - 1000)$ dollars, where s is the amount of sales you make in dollars.

a. How much will you earn from $2000 in sales?

b. How much will you earn from $50,000 in sales?

c. If you need at least $500 per week to pay your bills, what sales do you have to make per week?

18. At a new job selling high end clothing to women you earn 6% commission on all sales you make after the first $500. Your pay can be represented by $p = 0.06(s - 500)$ dollars, where s is the amount of sales you make in dollars.

a. How much will you earn from $2000 in sales?

b. How much will you earn from $5,000 in sales?

c. If you need at least $450 per week to pay your bills, what sales do you have to make per week?

19. After calling U-Haul truck rental, you decided to compare their prices to Budget truck rental. Budget charges $29.95 for the day and $0.55 per mile driven to rent a 15-foot moving truck.

Source: Budget.com.

a. Let B be the cost of renting a 15-foot moving truck from Budget for a day and driving the truck m miles. Find an equation for the cost of renting from Budget.

b. How much would it cost to rent a 15-foot truck from Budget if you were to drive it 75 miles?

c. How many miles could you drive the truck if you could only pay $100 for the rental?

20. A local cellular phone company has a pay-as-you-talk plan that costs $10 per month and $0.20 per minute you talk on the phone.

a. Write an equation for the total monthly cost C of this plan if you talk for m minutes.

b. Use your equation to determine the total monthly cost of this plan if you talk for 200 minutes.

c. How many minutes did you talk on the phone if your bill for June was $37?

21. If you are a salesperson and are guaranteed $250 per week plus a 7% commission on all your sales, find the following.

a. Write an equation for your pay per week, P, if you make s dollars of sales.

b. What will your pay be if you have sales of $2000?

c. How many dollars of sales do you need to make to have a weekly pay of $650?

22. If you are a salesperson and are guaranteed $300 per week plus a 5% commission on all your sales, find the following.

a. Write an equation for your pay per week, P, if you make s dollars of sales.

b. What will your pay be if you have sales of $4000?

c. How many dollars of sales do you need to make to have a weekly pay of $750?

23. You are planning a trip to Las Vegas and need to calculate your expected costs for the trip. You found that you can take a tour bus trip for up to 7 days, and it will cost you $125 for the round trip. You figure that you can stay at a hotel and eat for about $100 per day.

a. Write an equation for the total cost of this trip depending on the number of days you stay. (We will ignore the gambling budget.)

b. How much will it cost for a 3-day trip?

c. If you have $700 and want to gamble half of it, how many days can you stay in Las Vegas, assuming that you do not win any money?

24. Your family is planning a trip to Orlando Florida to visit the amusement parks and you need to budget for your expected costs. You find round trip flights for your 4 person family that total $1000. You expect the hotel, food and admissions to cost about $400 per day.

a. Write an equation for the total cost of this trip depending on the number of days you stay.

b. How much will it cost for a 5-day trip?

c. If your family can afford to spend $3500 on this trip how many days can you stay in Orlando?

25. A professional photographer has several costs involved in taking pictures at an event such as a wedding. The film costs about $3.39 per roll and it costs about $15 per roll to develop the proofs. The photographer also has to pay salaries for the day of $500.

© Ludovic Maisant/CORBIS

a. Find an equation for the total cost to shoot a wedding depending on the number of rolls of film used.

b. How much will it cost if the photographer shoots 15 rolls of film?

c. How many rolls of film can the photographer shoot if the total cost cannot exceed a budget of $750?

26. The photographer from the previous problem charges her clients a $45 fee per roll of film plus a flat fee of $400 for the wedding.

a. Find an equation for the total revenue for shooting the wedding depending on the number of rolls of film used.

b. How much will the photographer charge the client for a wedding at which she shot 15 rolls of film?

c. Find an equation for the profit made by the photographer depending on the number of rolls of film used.

d. How much profit will the photographer make on a wedding at which she shot 15 rolls of film?

e. How many rolls of film must the photographer shoot to break even? (Breaking even means that profit = 0.)

27. The Squeaky Clean Window Cleaning Company has several costs included in cleaning windows for a business. The materials and cleaning solutions cost about $1.50 per window. Insurance and salaries for the day will cost about $230.

a. Find an equation for the total cost to clean windows for a day depending on the number of windows cleaned.

b. How much will it cost if the company cleans 60 windows?

c. How many windows can the company clean if the total cost cannot exceed a budget of $450?

28. The Squeaky Clean Window Cleaning Company from the previous problem charges companies $7 per window cleaned plus a travel charge of $50.

a. Find an equation for the total revenue for cleaning windows at a business depending on the number of windows cleaned.

b. How much will the Squeaky Clean Window Cleaning Company charge a business to clean 20 windows?

c. Find an equation for the profit made by the Squeaky Clean Window Cleaning Company depending on the number of windows cleaned.

d. How much profit will the company make from cleaning 40 windows for a business?

e. How many windows must the company clean in a day to break even? (Breaking even means that profit = 0.)

29. Enviro-Safe Pest Management charges new clients $150 for an in-home inspection and initial treatment for ants. Monthly pre-planned treatments cost $38.

a. Find an equation for the total cost for pest management from Enviro-Safe Pest Management depending on the number of months your house is treated.

b. If your house has an initial treatment and then is treated for 11 more months, how much will Enviro-Safe charge you?

30. A small manufacturer of golf clubs is concerned about monthly costs. The workshop costs $23,250 per month to run in addition to the $145 in materials per set of irons produced.

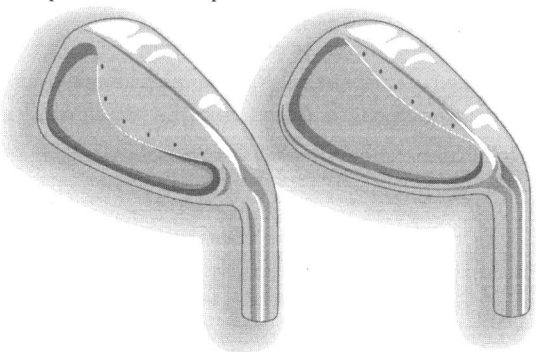

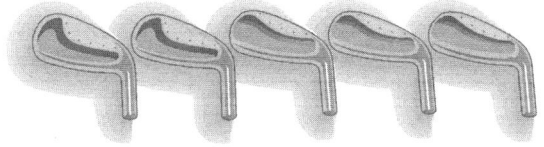

   a. Find an equation for the monthly costs of this club manufacturer.

   b. What are the monthly costs for this company if they make 100 sets of irons?

   c. How many sets of irons does this manufacturer need to produce for their costs to be $20,000?

   d. If this company wants to break even making 100 sets of irons per month, what should they charge for each set? (To break even the company needs the revenue to equal cost. Use the cost from part b and the fact that revenue can be calculated as the price times quantity.)

31. The population of the United States during the 1990s can be estimated by the equation $P = 2.57t + 249.78$, where P is the population in millions t years since 1990.

Source: Based on data from Statistical Abstract 2001.

   a. What was the population of the United States in 1993?

   b. When was the population of the United States 270,000,000?

   c. When did the population of the United States reach 300 million?

32. The percent P of companies that are still in business t years after the fifth year in operation can be represented by the equation

$$P = -3t + 50$$

   a. What percent of companies are still in business after 1 year in operation?

   b. What percent of companies are still in business after 25 years in operation?

   c. After how many years are there only 35% of companies still in business?

For Exercises 33 through 54, solve each equation.

33. $5x + 60 = 2x + 90$

34. $6x + 20 = 9x + 5$

35. $1.25d - 3.4 = -2.3(5d + 4)$

36. $3.7m - 4.6 = -1.8(6m + 8)$

37. $\frac{2}{5}d + 6 = 14$

38. $\frac{3}{4}x - 17 = 20$

39. $3(c + 5) - 21 = 107$

40. $5k + 7 = 2(6k - 14) + 56$

41. $-3x - 6 = 14 + 8x$

42. $5r - 9 = 18r + 2$

43. $1.7d + 5.7 = 29.7 + 5d$

44. $2.1m + 3.4 = 7.2 - 9.4m$

45. $\frac{1}{3}m + \frac{4}{3} = 4$

46. $\frac{1}{2}x + \frac{3}{2} = 5$

47. $\frac{3}{7}(2z - 5) = \frac{4}{7}(-3z + 9)$

48. $\frac{2}{5}(3r - 8) = \frac{3}{5}(-4r + 6)$

49. $-3(2v + 9) - 3(3v - 7) = 4v + 6(2v - 8)$

50. $4(2x + 7) - 6(4x - 8) = 12x + 3(4x - 9)$

51. $\frac{5}{7}d - \frac{3}{10} = \frac{4}{7}d + 4$

52. $\frac{3}{8}p - \frac{4}{9} = \frac{5}{8}p + 7$

53. $-\dfrac{8}{9}(3t + 5) = \dfrac{2}{3}t - 12$

54. $-\dfrac{2}{7}(4x + 2) = \dfrac{3}{28}x - 15$

For Exercises 55 through 72, solve for the indicated variable.

55. Force: $F = ma$ for a.

56. Weight (newtons): $W = mg$ for m.

57. Impulse: $J = Ft$ for F.

58. $P = 10h$ for h.

59. Angular acceleration: $\omega = \omega_o + \alpha t$ for α (Note that ω is the Greek symbol "omega" and α is the Greek symbol "alpha.")

60. $y = mx + b$ for b.

61. Rotational kinetic energy (J): $K = \dfrac{1}{2}I\omega^2$ for I.

62. Elastic potential energy (J): $U = \dfrac{1}{2}kx^2$ for k.

63. Kinetic energy (J): $K = \dfrac{1}{2}mv^2$ for m.

64. $y = \dfrac{1}{2}xz^2$ for x.

65. $ax + by = c$ for y.

66. $2x - y = z$ for x.

67. $ax + 5 = y$ for x.

68. $4m + n = p$ for m.

69. $b = 2c + 3d$ for c.

70. $x = 3y + 5z$ for y.

71. $5x^2 + 3y = z$ for y.

72. $4a - 5b^2 = c$ for a.

Using Data to Create Scatterplots

Data are collected in many ways and places around us. Grocery stores collect data on how and what we buy in their store. Scientists collect data during experiments to study different characteristics of the subject. Historians collect data to study trends in past events. The Centers for Disease Control and Prevention (CDC) collect data on cases of infectious diseases to determine when an outbreak has occurred or how dangerous a virus might be. Governments collect population data to predict future trends in public funds and needs for services.

After data are collected, the observations have to be organized and presented in a useful form. Often data are organized in a table.

Year	Number of Deaths
1994	3532
1995	3262
1996	2894
1997	2601
1998	2283
1999	2093

Source: Centers for Disease Control and Prevention.

This table alone does not give enough information about the context of the situation to be of much help. The number of deaths in this table could represent many different things. Some possibilities may include the following:

- Number of deaths due to jaywalking
- Number of deaths caused by poisoning
- Number of deaths of bears living in Canada
- Number of homicides of 15- to 19-year-olds in the United States

These data were collected by the Centers for Disease Control and Prevention and actually do represent the number of homicides of 15- to 19-year-olds in the United States.   We can see that as the years go by the number of deaths is decreasing.

USING DATA TO CREATE SCATTERPLOTS

Although a table is useful to display data, it sometimes is better to graph the data, giving a more visual picture of the situation. To create a graph, we first define the variables involved and determine which variable depends on the other. In cases such as renting a U-Haul truck and driving it m miles for the day, it is clear that the cost U depends on how many miles you drive the truck. In many cases it is not as clear which variable depends on the other. You may find that it is often easier to determine which variable is independent rather than dependent. In the number of deaths data given in the table we could define the following variables:

t = Year

N = Number of Homicides of 15- to 19-year-olds in the United States

Say What?

The following words or phrases all mean the same thing. We will discuss some of these in later sections.

Independent Variable:

* independent variable
* input variable
* input
* domain value
* usually *x*

Dependent Variable:

* dependent variable
* output variable
* output
* range value
* usually *y*

scatterplot:

* scatter diagram
* scattergram
* statplot (Texas Instruments term)

Since the year does not depend on the number of homicides committed we would call t the **independent variable** (input variable), thus making N the **dependent variable** (output variable). Someone might say that N depends on t. It is important to note that the number of homicides is not caused by what year it is; the number of homicides depends on what year you are wanting to discuss.

Now that we know which variable depends on the other, we can build a **scatterplot.** The independent variable is usually placed on the horizontal axis and the dependent variable on the vertical axis. It is best to label each axis with at least the units for each variable being represented.

EXAMPLE   1   CREATING A SCATTERPLOT

Create a scatterplot of the data given in the table.

Year	Number of Deaths
1994	3532
1995	3262
1996	2894
1997	2601
1998	2283
1999	2093

Source: Centers for Disease Control and Prevention.

Solution

First we will use the variables as defined above.

t = Year

N = Number of Homicides of 15- to 19-year-olds in the United States

Since t is the independent variable, it will be on the horizontal axis, and the dependent variable N will be on the vertical axis. The next thing to be decided is the **scale** (spacing) for each axis. Please note that the scale does not need to be the same on the horizontal and vertical axis. For any one axis, though, the scale must remain equal. The scale remaining equal on an axis means that every space of the same size represents the same number of units. For the horizontal axis, representing the year, the scale can be 1 and start at about 1990; for the vertical axis, representing the number of deaths, the scale could be 200 starting at about 2000. This results in the following graph.

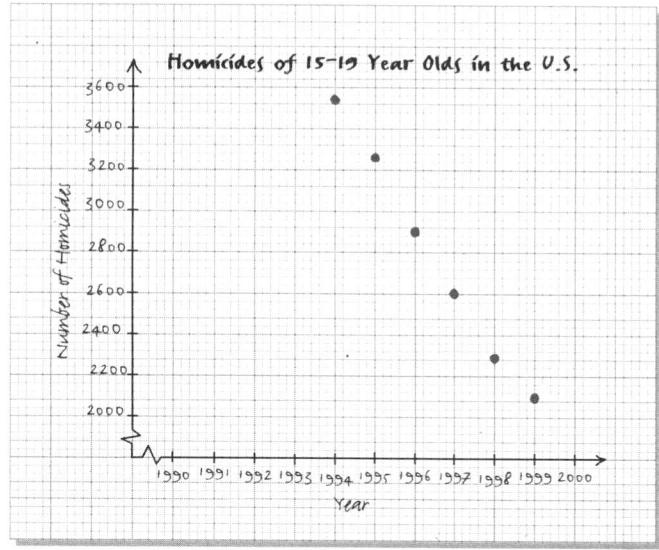

You should notice that both axes in this case do not start at 0, so a zigzag pattern is placed at the beginning of the axis to show a jump in the numbering. This is a valid way of making the graph easier to create, but it does distort how the relationship looks. With this scale, the decrease in homicides appears very steep. Using a different scale would cause this same decrease to appear much less drastic. By adjusting the definitions of the variables and the data, you can get an accurate graph that is not distorted. One option for such changes would be

t = Time in years since 1990

N = Number of homicides of 15- to 19-year-olds in the U.S. (in thousands)

Adjusting the data allows for smaller labels on the axes and often a more readable graph. The years have been represented with the year 1990 being year 0. You can choose any year you would like for the base year in a problem. This base year only gives you a place to start counting from; it does not determine the starting year of the data. Choosing a "nice" year such as 1990 or 1980 makes it easier to figure out what each number represents. For example, for a base year 1990, $t = 7$ represents 1997; for a base year 1980, 1997 would have been represented by $t = 17$.

Be careful that once you define your variables, you stay consistent with the values you use for each variable.

To adjust the years to time in years since 1990 we will subtract 1990 from each year. For example $1994 - 1990 = 4$. 1994 is 4 years since 1990 so $t = 4$ would represent 1994.

To change the number of homicides data into thousands we will multiply each number by

$$\frac{1\,\text{thousand}}{1000}$$

This is basically dividing each number by 1000 and adding the word thousand to the end of the number. For example

$$3532 \cdot \frac{1\,\text{thousand}}{1000} = 3.532 \text{ thousand}$$

Changing the years and number of homicides will result in the following adjusted data table:

t	N
4	3.532
5	3.262
6	2.894
7	2.601
8	2.283
9	2.093

The above change in data will result in the following graph. The previous graph was reasonable, but the following graph is easier to read and will be easier to use in the future. The rate of decline in homicides does not look as steep as in the previous graph.

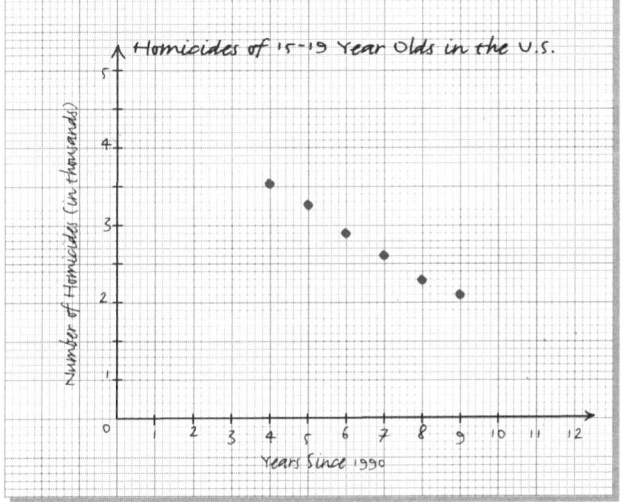

On this graph you should notice that although the data for the years starts at $t = 4$ we still must include 1,2, and 3 on the horizontal axis. You should not skip numbers on the horizontal axis if you can avoid it.

EXAMPLE ① PRACTICE PROBLEM

The state of Arizona's population for various years is given in the table.

Year	Population of Arizona
2000	5,165,993
2001	5,295,929
2002	5,438,159
2003	5,577,784
2004	5,739,879
2005	5,939,292
2006	6,166,318*

*Estimate.
Source: U.S. Census Bureau.

a. Define variables for these data.

b. Adjust the data.

c. Create a scatterplot.

GRAPHICAL MODELS

The data shown thus far are considered to be **linearly related,** because it generally falls along the path of a straight line. In the concept investigation below we will draw an "eyeball best fit" line through the data points on the scatterplots. This is best done with a small clear ruler. Choose a line that comes as close to all the plotted points as possible. The points that miss the line should be equally spread out above and below the line. This will allow for each point that the line misses to be balanced out by another point missed by the line.

CONCEPT INVESTIGATION I

WHAT IS AN "EYEBALL BEST FIT" LINE?

In parts a, b, and c choose which graph has the best "eyeball best fit" line for the data. Remember you want the line to be as close as possible to ALL the data points. For each set of data describe what makes the line you chose the "eyeball best fit" line.

a.

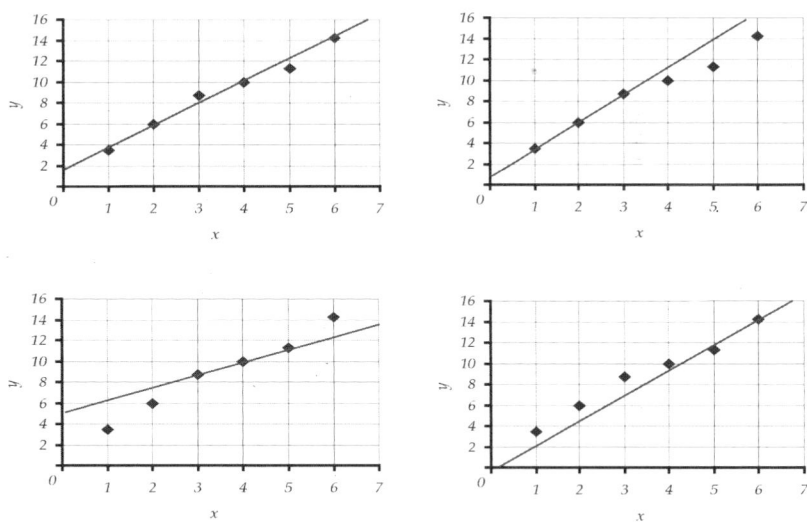

b.

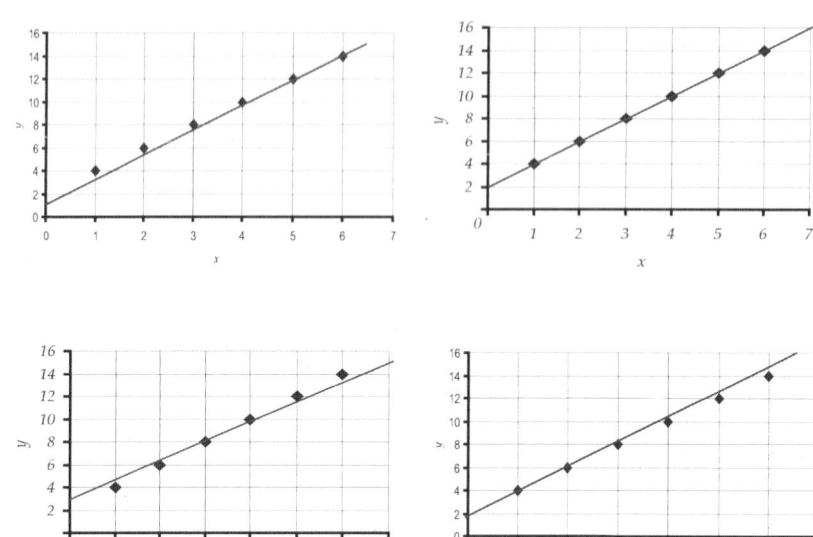

c.

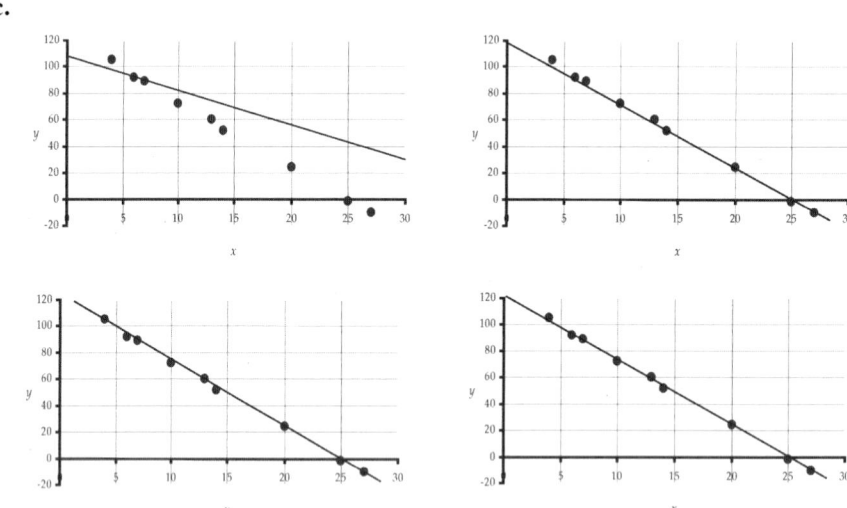

For parts d, e, and f draw lines that pass through each of the two points listed and decide which line would make a good "eyeball best fit" line for the data.

d.

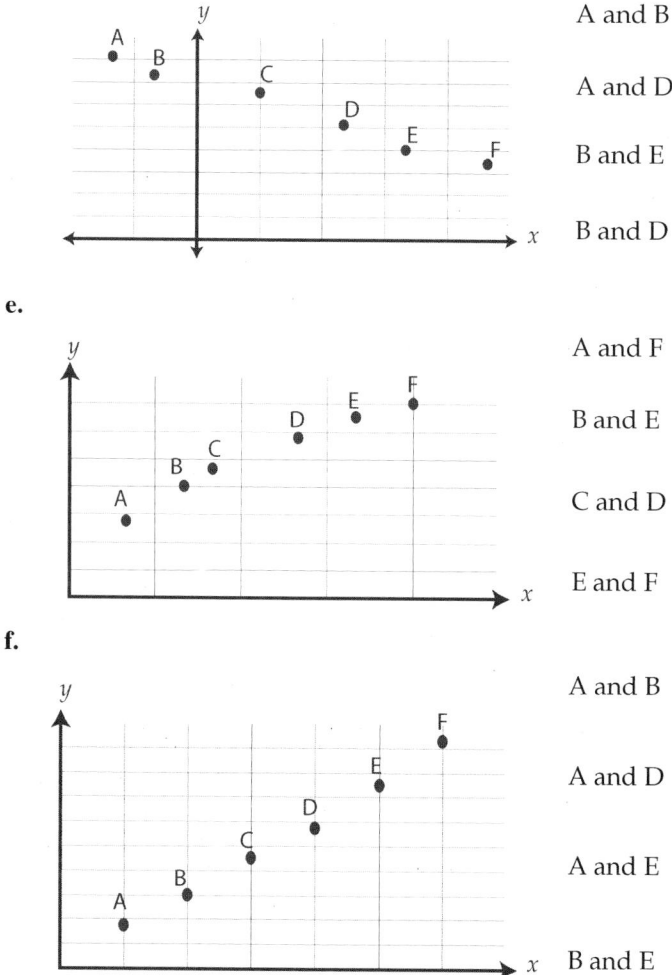

A and B

A and D

B and E

B and D

e.

A and F

B and E

C and D

E and F

f.

A and B

A and D

A and E

B and E

Remember that the "eyeball best fit" line will not necessarily hit any points on the graph but will be as close to ALL the points as possible. This line can be considered a **graphical model** of the data. A graphical model is used to gain additional information about the situation described by the data. Models may also be used to make predictions beyond the given data.

EXAMPLE ▣ 2 **DRAWING "EYEBALL BEST FIT" LINES**

a. Using the scatterplot of the data from example 1 draw an "eyeball best fit" line through the data.

b. Using your "eyeball best fit" line, make a prediction for the number of homicides of 15- to 19-year-olds in 1993.

Solution

a. See the following graph.

t	N
4	3.532
5	3.262
6	2.894
7	2.601
8	2.283
9	2.093

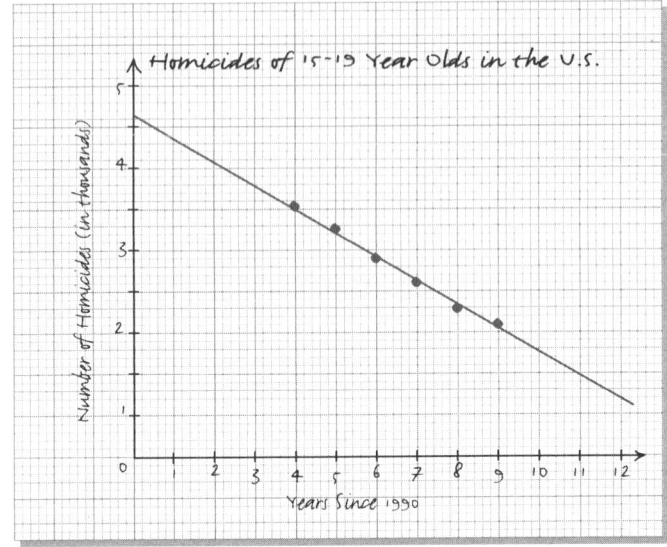

b. We are asked to estimate the number of homicides in 1993. The year 1993 is represented by $t = 3$ which is on the horizontal axis. Start at the 3 on the horizontal axis and look up to the line. The output value for the line when $t = 3$ is about 3.8 (slightly less than 4). Therefore, the graph indicates that in 1993 the number of homicides was approximately 3.8 thousand.

EXAMPLE ② PRACTICE PROBLEM

Draw an "eyeball best fit" line for the population of Arizona data you graphed in Problem 1.

INTERCEPTS, DOMAIN, AND RANGE

The points that a graph crosses the axes are called **intercepts.** The **vertical intercept** or y-intercept is the point where the graph crosses the vertical axis. A vertical intercept will always have an input(x) value of zero. The **horizontal intercept** or x-intercept is the point where the graph crosses the horizontal axis. A horizontal intercept will always have an output(y) value of zero.

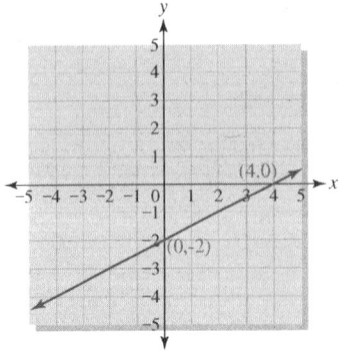

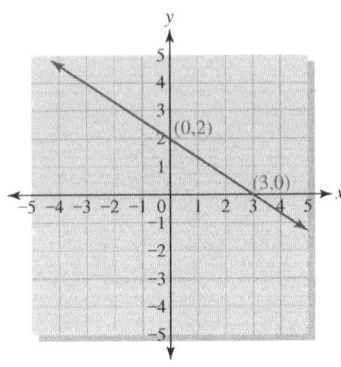

Say What?

The following words or phrases all mean the same thing.

Vertical Intercept:
- y-intercept
- Can take on the name of the variable representing the vertical axis, such as N-intercept in Example 2.

Horizontal Intercept:
- x-intercept
- Can take on the name of the variable representing the horizontal axis, such as t-intercept in Example 2.

DEFINITIONS

Vertical intercept: The point where the graph crosses the vertical axis. This will always occur when the input variable is zero. Vertical intercepts are written as an ordered pair $(0, k)$ Where k is a real number.

$$(0, 5) \qquad (0, 23) \qquad (0, -6)$$

Horizontal intercept: The point where the graph crosses the horizontal axis. This will always occur when the output variable is zero. Horizontal intercepts are written as an ordered pair $(k, 0)$ Where k is a real number.

$$(2, 0) \qquad (3.7, 0) \qquad (-4, 0)$$

In the homicides graph from example 2 the vertical intercept or N-intercept is approximately at the point $(0, 4.75)$. This point means that when $t = 0$, $N = 4.75$. $t = 0$ can be translated as zero years since 1990, so 1990, and $N = 4.75$ represents 4,750 homicides. Together $t = 0$ and $N = 4.75$ result in the statement: "In 1990 there were about 4750 homicides of 15- to 19-year-olds in the United States."

The horizontal intercept or t-intercept for the homicides graph cannot be seen, since the line does not extend to hit or cross the t-axis. To see the t-intercept, you will need to continue the values on the horizontal axis and extend the line until it reaches the t-axis. Determining intercepts is one reason to make your graph extend past the data given in the problem. If the graph were to reach the horizontal axis, N would be zero, meaning no homicides of 15-19 year olds. Having no homicides, sadly, will probably never happen so the horizontal intercept would not make sense in the situation.

EXAMPLE 3 READING A GRAPH

Use the graph to answer the following questions.

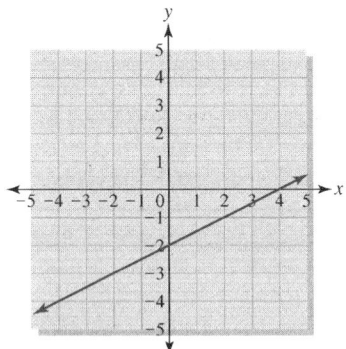

a. Estimate the vertical intercept.

b. Estimate the horizontal intercept.

c. Estimate the input value that makes the output of this graph equal -1.

d. Estimate the input value that makes the output of this graph equal -3.5.

e. Estimate the output value of this graph when the input value is -2.

Solution

a. This graph crosses the vertical axis (y-axis), at about -2 so the vertical intercept is $(0, -2)$.

b. This graph crosses the horizontal axis (x-axis), at about 4 so the horizontal intercept is $(4, 0)$.

c. First we need to locate the output -1 on the vertical axis and trace over until we find the line.

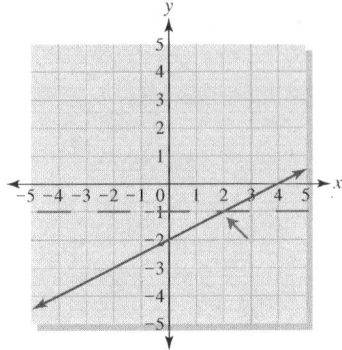

The line has an output of -1 when the input, x, is equal to 2.

d. We locate the output of $y = -3.5$, on the vertical axis, and trace over to the line.

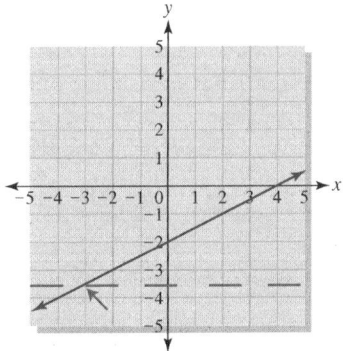

The line has an output of $y = -3.5$ when the input is $x = -3$

e. We locate the input value of $x = -2$ on the horizontal axis and trace down to the line.

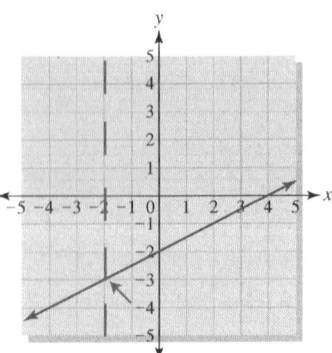

The line has an output of $y = -3$ when the input is $x = -2$.

EXAMPLE ③ PRACTICE PROBLEM

Use the graph to answer the following questions.

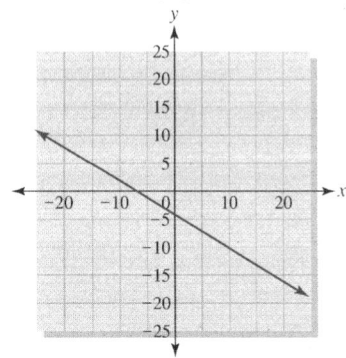

a. Estimate the y-intercept.

b. Estimate the x-intercept.

c. Estimate the value of x that results in $y = -10$.

d. Estimate the value of y when $x = -15$.

The values of the independent variable (inputs) that result in reasonable values for the dependent variable (outputs) are considered the **domain** of the model. The resulting outputs from a given domain are called the **range** of the model. We will use interval notation or inequalities when stating the domain and range of a model. For a refresher on interval notation see the review in appendix A. In using an "eyeball best fit" line that you have drawn on a scatterplot, the domain and range will be estimated according to the graph and what you believe is reasonable. This means that every student's domain and range may be different but still equally correct.

Say What?

Domain:

- Input values for the model.
- Values for the Independent variable.
- x-values.

Range:

- Output values for the model.
- Values for the dependent variable.
- y-values.

DEFINITIONS

Domain: The set of values for the independent variable that results in reasonable output values with no model breakdown. A domain should be written in interval notation or using inequality symbols.

Example: [2, 15] $2 \leq x \leq 15$

Range: A set of values for the dependent variable resulting from the given domain values. The outputs that come from the given domain's input values. A range should be written in interval notation or using inequality symbols.

Example: [5, 48] $5 \leq x \leq 48$

Because a model is often meant to be used to extrapolate, that is, to predict a future or past value, the domain should extend beyond the data whenever reasonable. The main thing that you will want to avoid is model breakdown. **Model breakdown** is when a domain value results in an output that does not make sense in the context of the situation or makes an equation undefined mathematically. An example of model breakdown in the number of homicides example would be a negative number of homicides. Another example of model breakdown would be a percentage of people or things that was over 100% or a negative percentage.

DEFINITION

Model breakdown: When input values give you outputs that do not make sense in the context of the problem.

EXAMPLE 4 DOMAIN AND RANGE OF A MODEL

Determine a reasonable domain and range for the homicide model found in example 2.

Solution

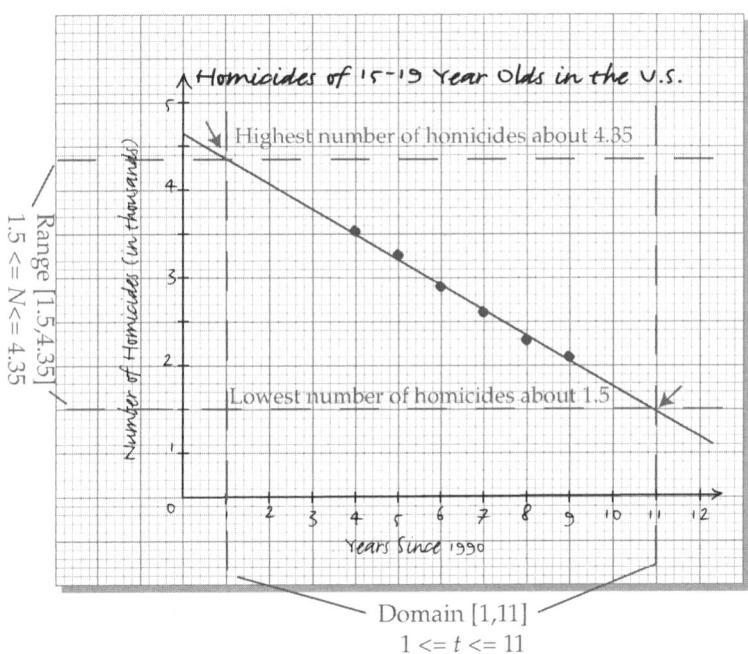

Since the model fits the data pretty well we should be able to extend our domain beyond the data values. How far we spread out from the data is somewhat arbitrary. Because there is no obvious model breakdown shown on the graph, such as negative numbers of homicides, we will be able to extend the input values somewhat in both directions. The data set starts in year 4 (1994) and goes to year 9 (1999), so in this situation we could choose a domain of $1 \leq t \leq 11$.

Looking at the graph the lowest number of homicides predicted during the years within the domain is about 1.5 thousand in year 11 (2001). The highest number of homicides predicted during the years within the domain seems to be about 4.35 thousand in year 1 (1991). This gives us a range of $1.5 \leq N \leq 4.35$.

EXAMPLE 4 PRACTICE PROBLEM

Determine a reasonable domain and range for the graphical model you drew in example 2 practice problem.

EXAMPLE   5   INTERPRETING INTERCEPTS, DOMAIN AND RANGE

The number of Home Depot stores has been increasing on a very consistent pace. The bar graph gives the number of Home Depot stores for each year.

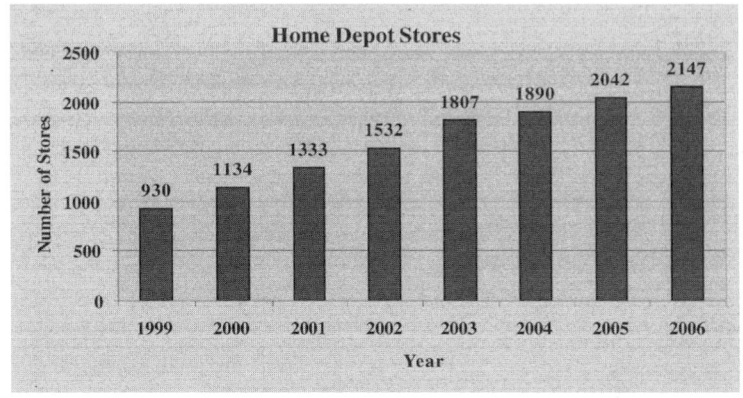

Source: Home Depot Annual Reports www.Homedepot.com.

a. Create a scatterplot for these data and draw an "eyeball best fit" line through the data.

b. Determine the vertical-intercept for this model and explain its meaning in this context.

c. Find a reasonable domain and range for this model.

d. According to your graphical model, how many Home Depot stores were there in 2009?

Solution

a. First define the variables.

H = Number of Home Depot stores.

t = Time in years since 1995.

We use the year 1995 because it is an easy year to count from and will make the input values smaller, and thus easier to graph. The outputs are already reasonable, so no adjustment is needed.

t	H
4	930
5	1134
6	1333
7	1532
8	1807
9	1890
10	2042
11	2147

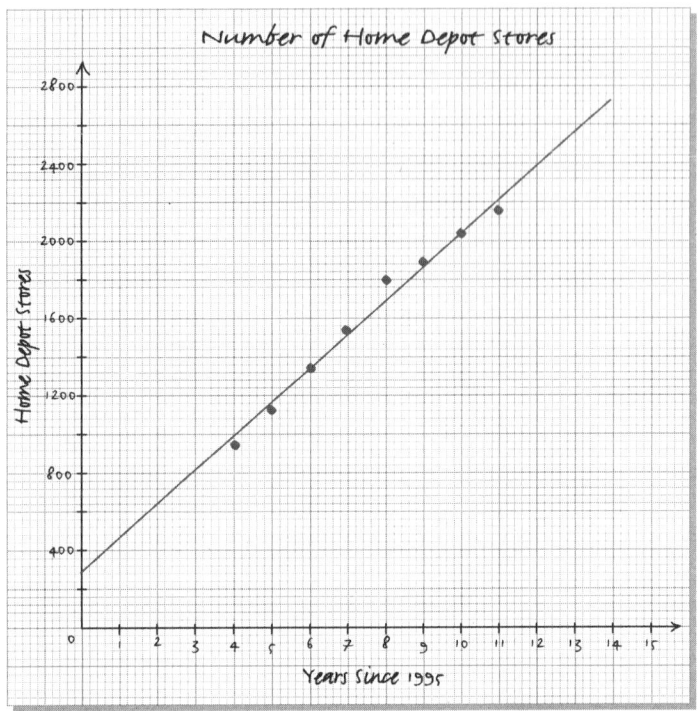

b. The vertical-intercept for this model is when the line crosses the *H*-axis at approximately (0 , 300), as shown in the graph. The 0 represents the year 1995, and the 300 represents number of Home Depot stores. Together, this means that in early 1995 Home Depot had approximately 300 stores.

c. Because this model does a reasonable job of representing these data, we should be able to extend the domain beyond the given data. One possible domain for this model would be [2, 13] or $2 \le t \le 13$, which would represent the years 1997 to 2008. The range that results from that model would be the number of stores that the model predicts for Home Depot between the years 1997 and 2008.

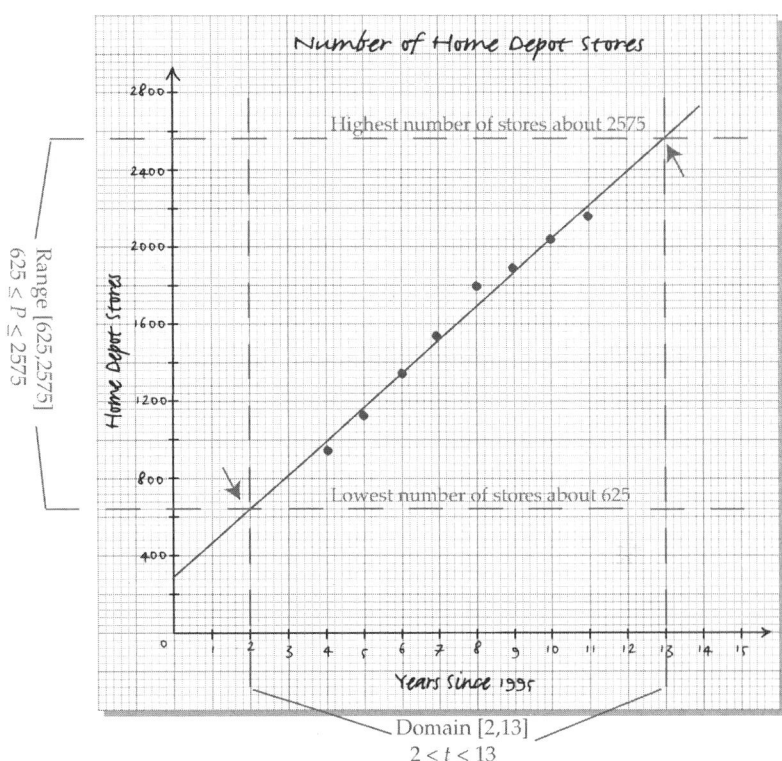

From the graph we can estimate that the least number of stores during those years is about 625 and the most stores is about 2575. This results in a range of [625, 2575] or $625 \leq H \leq 2575$.

d. 2009 would be represented by $t = 14$, and according to the graph, the model is at approximately 2750 when $t = 14$. Thus, in 2009, there were approximately 2750 Home Depot stores.

Domain and range are essential parts of any model and help the user of that model to know when it is appropriate to use that model versus when it is not. Be sure to pay special attention in the coming sections to any domain and range questions.

1.2 Exercises

For exercises 1 through 4 determine which two points an "eyeball best fit" line for the data would pass through. (You may want to use a clear ruler to help you draw lines with.)

1.

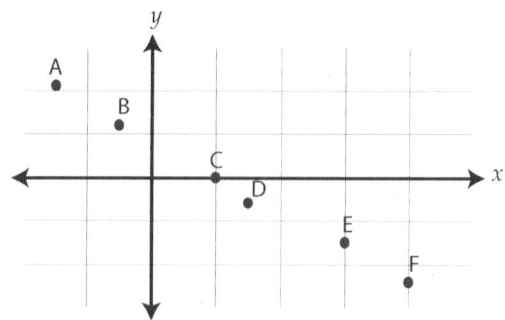

2.

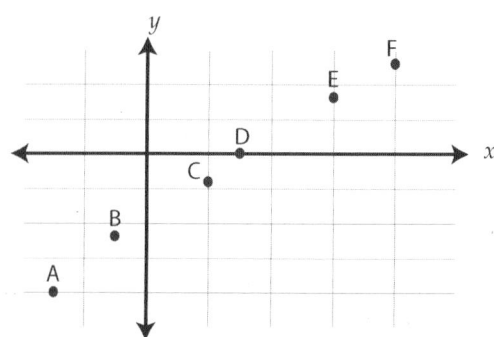

3.

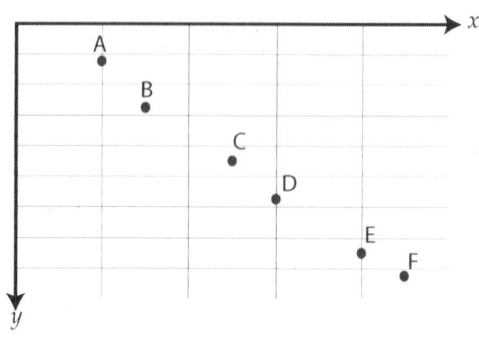

4.

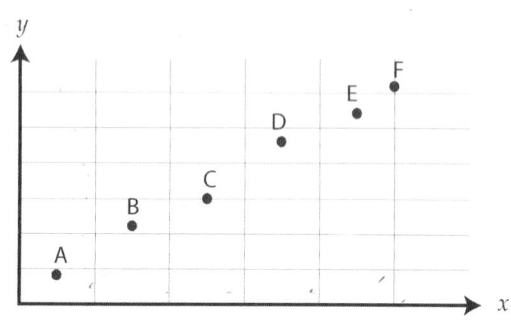

5. Quiksilver, Inc., the makers of the Quiksilver and Roxy clothing lines, has had steadily increasing profits over the past several years. The profits for Quicksilver, Inc. are given in the bar graph.

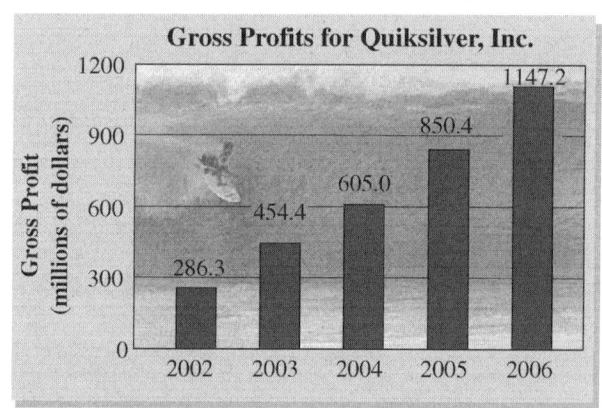

Source: CBS.Marketwatch.com.

a. Define the variables for this problem. Identify which is the independent and dependent variable. Adjust the data if needed.

b. Create a scatterplot and draw an "eyeball best fit" line through the data.

c. Using your graphical model, estimate the profit for Quiksilver, Inc. in 2010.

d. What are a reasonable domain and range for your graphical model? Write your answer using inequalities.

6. Home Depot's net sales are given in the bar graph

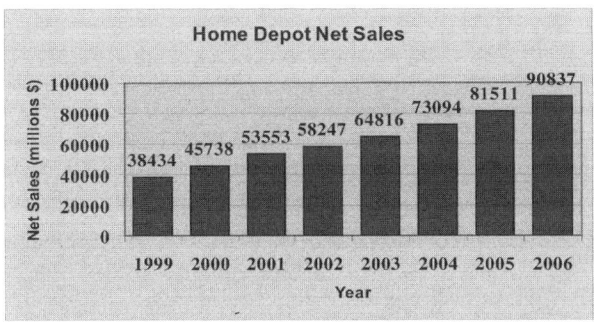

Source: Home Depot annual reports. www.homedepot.com.

a. Define the variables for this problem. Identify which is the independent and dependent variable. Adjust the data if needed.

b. Create a scatterplot and draw an "eyeball best fit" line through the data.

c. Using your graphical model estimate the net sales for Home Depot in 2009.

d. What are a reasonable domain and range for your graphical model? Write your answer using inequalities.

7. The population of the United States is given in the bar graph.

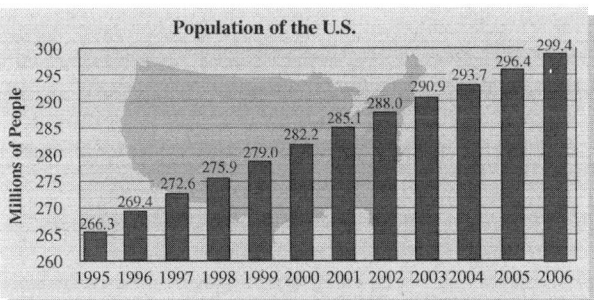

Source: U.S. Census Bureau.

a. Define the variables for this problem. Identify which is the independent and dependent variable. Adjust the data if needed.

b. Create a scatterplot and draw an "eyeball best fit" line through the data.

c. Using your graphical model estimate the population of the United States in 2009.

d. What are a reasonable domain and range for your graphical model? Write your answer in interval notation.

e. What is the vertical intercept for your model, and what does it mean in this context?

8. The population of Florida is given in the bar graph.

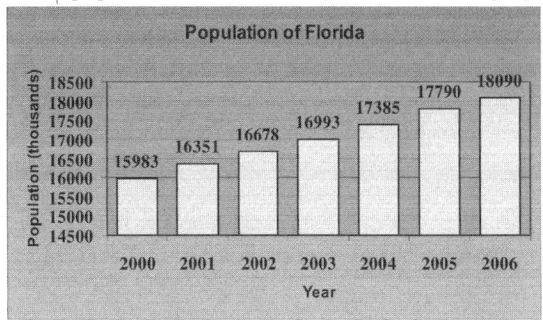

Source: U.S. Census Bureau.

a. Define the variables for this problem. Identify which is the independent and dependent variable.

b. Adjust the data if needed.

c. Create a scatterplot and draw an "eyeball best fit" line through the data.

d. Using your graphical model estimate the population of Florida in 2015.

e. What are a reasonable domain and range for your graphical model? Write your answer in interval notation.

f. What is the vertical intercept for your model, and what does it mean in this context?

9. The number of deaths of women, induced by illegal drugs in the United states is given in the table.

Year	2000	2001	2002	2003	2004
Deaths	6583	7452	9306	10297	11349

Source: Centers for Disease Control and Prevention.

a. Create a graphical model for these data. (Remember to define variables.)

b. Using your graphical model, estimate the number of drug-induced deaths of females in the United States in 2007.

c. Use your graphical model to estimate in what year the number of deaths was 4000.

d. What are a reasonable domain and range for your graphical model? Write your answer using inequalities.

10. The death rate (per 100,000 people) for heart disease in the United States for various years is given below as:

Year	Death Rate
1999	266.5
2000	257.6
2001	247.8
2002	240.8
2003	232.3
2004	217.5

Source: Centers for Disease Control and Prevention.

a. Find a graphical model for these data.

b. According to your model, when will the death rate be approximately 200 deaths per 100,000 people?

c. What are a reasonable domain and range for your model?

d. What does your model predict the death rate to be in 1995?

e. What is the vertical intercept for your model, and what does it mean in this context?

11. The expected number of years someone will live is given below:

Age	Years
0 (at birth)	77.5
10	68.2
20	58.4
30	48.9
40	39.5
50	30.6
60	22.2

Source: www.census.gov. Statistical Abstract 2007.

a. Find a graphical model for these data.

b. According to your graphical model, what is the expected number of years a 45 year old person would live?

c. What are a reasonable domain and range for your model?

d. What does your graphical model predict the number of years a 90 year old person should expect to live?

e. What is the vertical intercept for your model, and what does it mean in this context?

12. The amount teachers are paid per year is typically determined by their years of experience. The salaries for several different teachers at one school district is given in the table.

Years of Experience	Salary (dollars)
0	49,900
2	52,400
5	62,100
8	67,000
10	74,300
12	79,200

a. Find a graphical model for these data.

b. According to your graphical model, what is the expected salary if a teacher has 7 years of experience?

c. What are a reasonable domain and range for your model?

d. How many years of experience does a teacher need in order to earn $60,000?

e. What is the vertical intercept for your model, and what does it mean in this context?

In Exercises 13 through 20, use the graph to answer the questions. Answers may vary.

13.

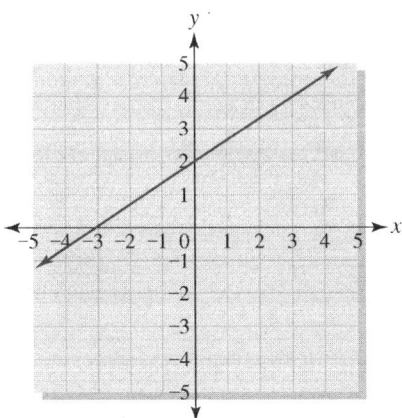

a. Estimate the y-intercept.

b. Estimate the x-intercept.

c. Estimate the value of x that results in $y = 4$.

d. Estimate the value of y when $x = 1$.

14.

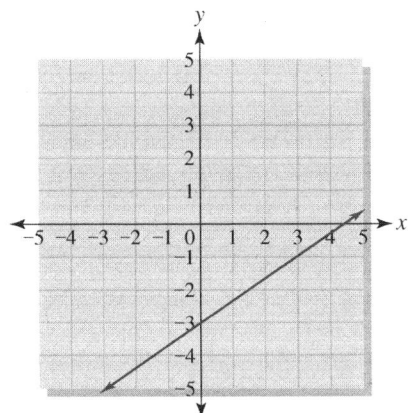

a. Estimate the y-intercept.

b. Estimate the x-intercept.

c. Estimate the value of x that results in $y = -2$.

d. Estimate the value of y when $x = -3$.

15.

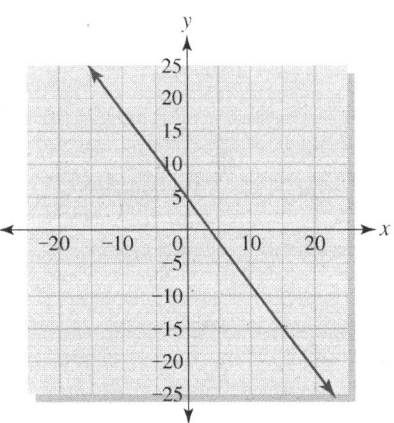

a. Estimate the vertical intercept.

b. Estimate the horizontal intercept.

c. Estimate the input value that makes the output of this graph equal 18.

d. Estimate the input value that makes the output of this graph equal -15.

e. Estimate the output value of this graph when the input value is 10.

16.

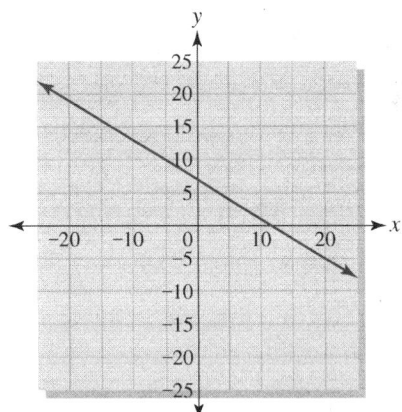

a. Estimate the vertical intercept.

b. Estimate the horizontal intercept.

c. Estimate the input value that makes the output of this graph equal 15.

d. Estimate the input value that makes the output of this graph equal -2.

e. Estimate the output value of this graph when the input value is -10.

17.

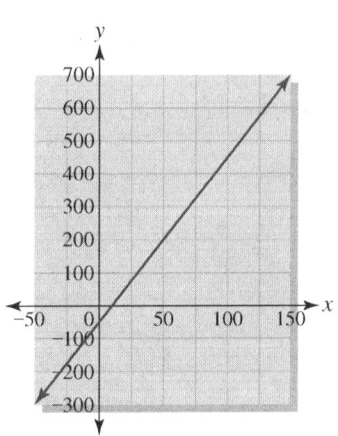

a. Estimate the *y*-intercept.
b. Estimate the *x*-intercept.
c. Estimate the value of *x* that results in $y = -5$.
d. Estimate the value of *x* that results in $y = -10$.
e. Estimate the value of *y* when $x = 20$.

18. .

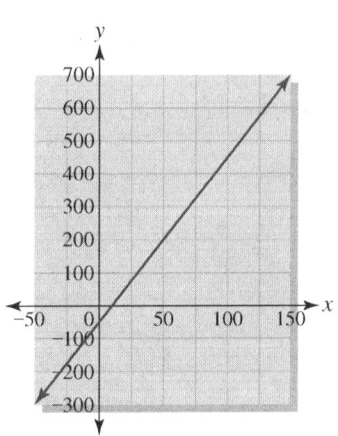

a. Estimate the *x*-intercept.
b. Estimate the *y*-intercept.
c. Estimate the value of *x* that results in $y = 500$.
d. Estimate the value of *x* that results in $y = -200$.
e. Estimate the value of *y* when $x = 100$.

19.

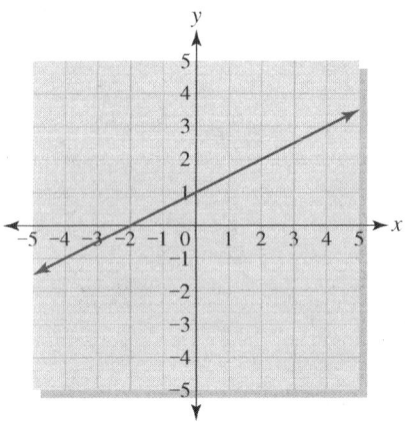

a. Estimate the vertical intercept.
b. Estimate the horizontal intercept.
c. Estimate the input value that makes the output of this graph equal 1.5.
d. Estimate the input value that makes the output of this graph equal -1.
e. Estimate the output value of this graph when the input value is 3.

20.

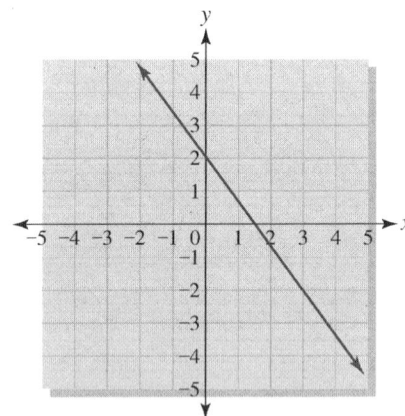

a. Estimate the vertical intercept.
b. Estimate the horizontal intercept.
c. Estimate the input value that makes the output of this graph equal 4.
d. Estimate the input value that makes the output of this graph equal -2.
e. Estimate the output value of this graph when the input value is 4.

1.3 Introduction to Graphing

INTRODUCTION TO GRAPHING EQUATIONS

In Section 1.2, we created scatterplots of data by hand and drew an "eyeball best fit" line on that scatterplot to give us an approximate graphical model for the data and situation we were considering. Although the graphical model is useful to estimate some values, it is often easier to find values for a model if we have an algebraic equation for it. In this section, we will investigate the different characteristics of a linear equation and its graph.

EXAMPLE 1 GRAPHING EQUATIONS BY PLOTTING POINTS

Graph the equations by creating a table of values and plotting the points.

a. $y = 3x - 8$

b. $x = -2y + 10$

c. $y = x^2 + 3$

Solution

a. We begin by finding ordered pairs that satisfy the equation. Because y is already isolated in this equation, it is easier to choose values for x and find the y values that go with them.

x	$y = 3x - 8$	(x,y)
-2	$y = 3(-2) - 8 = -14$	$(-2, -14)$
0	$y = 3(0) - 8 = -8$	$(0, -8)$
2	$y = 3(2) - 8 = -2$	$(2, -2)$
3	$y = 3(3) - 8 = 1$	$(3, 1)$

Now we can plot these points.

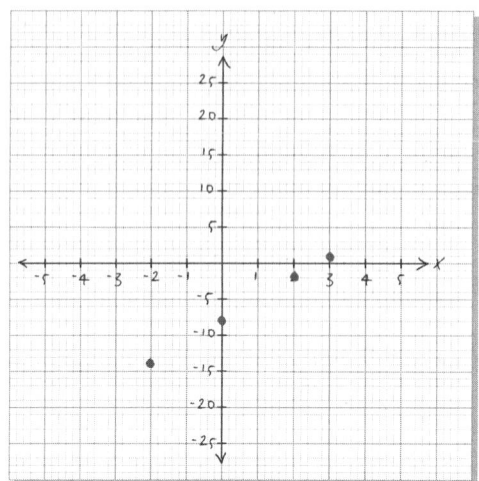

Using Your TI Graphing Calculator

We can use our graphing calculators to do many things with equations. Graphing calculators will consider x the independent (input) variable and y the dependent (output) variable.

When entering an equation in the Y= screen you will need to have the dependent variable isolated so that the equation can be entered in as $y = \text{expression}$.

To enter the equation
$$y = 2x + 5$$

you will press [Y=] and then in the Y1 = line you will enter $2x + 5$.

Use the [X,T,θ,n] button for the variable x.

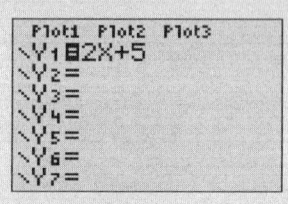

We can see that all of these points lie on a straight line. Now we can draw a line through the points. Since this equation would have an infinite number of points that satisfy it, we will extend the line beyond the points we drew and put arrows on the ends of the line to indicate that the line continues in both directions infinitely. The line itself actually represents all of the possible combinations of x and y that satisfy this equation.

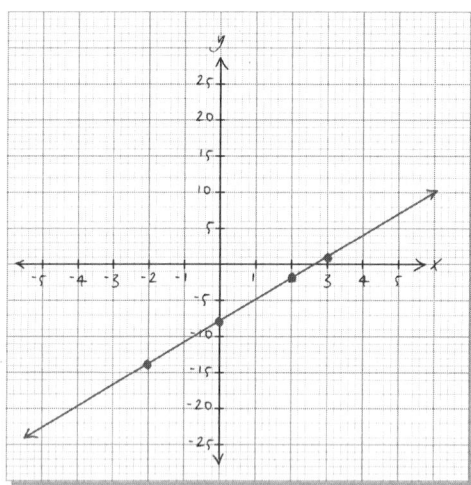

b.   Because x is already isolated in this equation, it would be easier to choose values for y and find values for x.

$x = -2y + 10$	y	(x, y)
$x = -2(-1) + 10 = 12$	-1	$(12, -1)$
$x = -2(0) + 10 = 10$	0	$(10, 0)$
$x = -2(2) + 10 = 6$	2	$(6, 2)$
$x = -2(4) + 10 = 2$	4	$(2, 4)$

Now we can plot these points and draw a straight line through them.

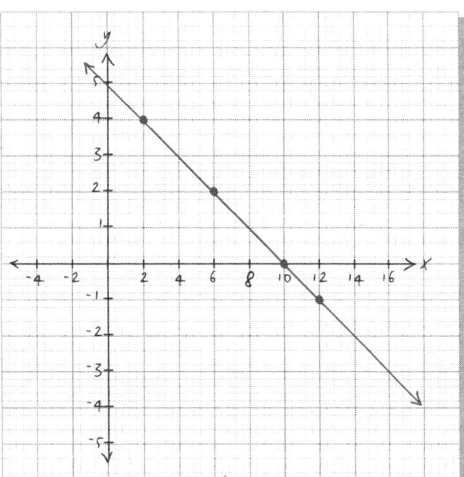

Using Your TI Graphing Calculator

Using the TABLE Feature:

First you need to enter the function into the Y= screen. To set up the table, you need to press [2nd] [TBLSET] above the WINDOW button.

The only things you need to be sure are set correctly are the Indpnt and Depend settings.

Indpnt should be set to **Ask** Depend should be set to **Auto.**

These settings will allow you to enter a value for the input variable, and the calculator will automatically calculate the related output.

Now press [2nd] [TABLE] above the GRAPH button and enter the values of x you want.

When the Indpnt setting is set to Auto, the table will automatically fill with input values based on the TblStart setting and the ΔTbl (change in table) setting. This is used when you need an incremental list.

For more on this feature see the calculator guide in the appendix.

c. This equation is a little more complicated than the others because the variable x has an exponent of 2, so we will want to find more points to plot to help us see what shape the graph has. We will choose values of x and find the values of y.

x	$y = x^2 + 3$	(x, y)
-2	$y = (-2)^2 + 3 = 7$	$(-2, 7)$
-1	$y = (-1)^2 + 3 = 4$	$(-1, 4)$
0	$y = (0)^2 + 3 = 3$	$(0, 3)$
1	$y = (1)^2 + 3 = 4$	$(1, 4)$
2	$y = (2)^2 + 3 = 7$	$(2, 7)$
3	$y = (3)^2 + 3 = 12$	$(3, 12)$

Now we can plot the points and draw a smooth curve through them.

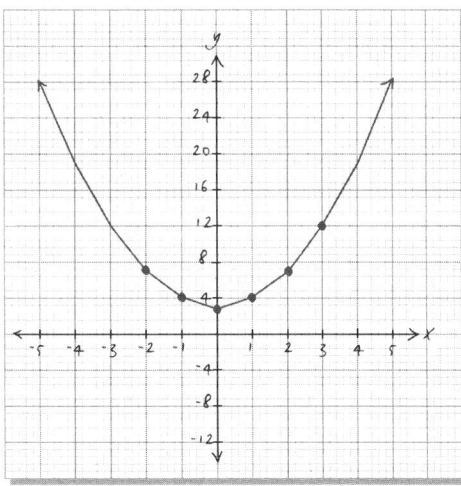

This graph is more complicated than the lines in the first two parts of this example. By graphing more points, we get a better idea of the shape of the graph. In general, plotting 3 points will be enough for graphing a line, but 5 or more are needed to graph more complicated equations. We will study this type of equation more in chapter 4.

EXAMPLE ① PRACTICE PROBLEM

Graph the equations by creating a table of values and plotting the points.

a. $y = 2x - 6$

b. $y = x^2 - 8$

Graphing equations by plotting points and then connecting them with a smooth line or curve can be used with most equations. This technique is the most basic method to graph an equation and can be used in applications as well as in basic algebra problems.

EXAMPLE 2 GRAPHING EQUATIONS BY PLOTTING POINTS

In section 1.1, we were given the equation $U = 19.95 + 0.79m$ where U is the cost in dollars to rent a 10ft truck from U-Haul when it is driven m miles.

a. Create a table of points that satisfy this equation.

b. Create a graph for the equation using your points. Remember to label your graph with units.

Solution

a. Since we are investigating an equation about renting a truck and driving it m miles, we should choose values of m that would make sense for miles driven. We will actually start with zero miles just to remind us of how much the rental costs even if we don't happen to go anywhere.

m	$U = 19.95 + 0.79m$	(m, U)
0	$U = 19.95 + 0.79(0) = 19.95$	$(0, 19.95)$
50	$U = 19.95 + 0.79(50) = 59.45$	$(50, 59.45)$
100	$U = 19.95 + 0.79(100) = 98.95$	$(100, 98.95)$
200	$U = 19.95 + 0.79(200) = 177.95$	$(200, 177.95)$
300	$U = 19.95 + 0.79(300) = 256.95$	$(300, 256.95)$

b. Now we can plot the points. Since the cost depends on the miles driven, the cost will be the dependent variable and be plotted on the vertical axis and the miles driven is the independent variable and will be plotted on the horizontal axis. We will need to choose a scale for each axis that will allow for all the points to be plotted. We could choose a scale of 20 for the vertical axis and a scale of 25 for the horizontal axis. These scales will allow us to see all the points in our table as well as a little beyond them.

Using Your TI Graphing Calculator

The table feature of the calculator can be used to calculate several values of an equation very quickly. First you must enter the equation you are working with into the Y= screen.

Now go to table using 2nd GRAPH and enter in the input values you want to evaluate the equation for.

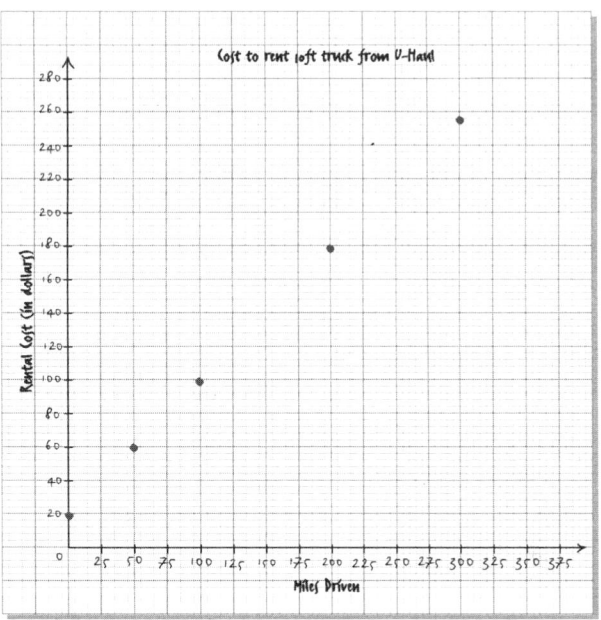

Now that we can see that all the points lie on the same line, we can draw that line and extend it farther out to represent more possible combinations of miles and costs that satisfy this equation.

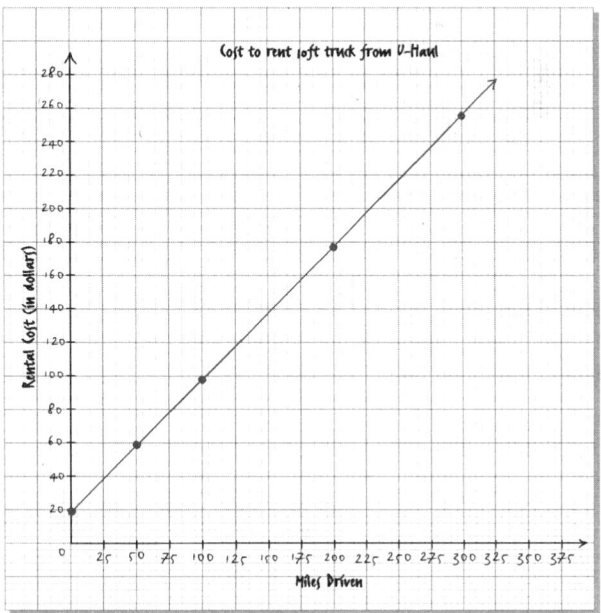

SLOPE-INTERCEPT FORM OF A LINE

Most equations can be graphed by finding points and plotting them. Graphing calculators use this technique to graph an equation that you enter. When you graph by hand, an understanding of the basic characteristics of the graph of the equation can make graphing it quicker and more accurate. In the concept investigation below, we will examine some of the characteristics of the graph of a line and how they are related to the equation of a line.

 CONCEPT INVESTIGATION I WHAT DIRECTION IS THAT LINE GOING IN? ·······························

Use your graphing calculator to examine the following.
Start by setting up your calculator by doing the following steps.

i. Change the window to a standard window. (Press [zoom], [6](ZStandard).)

ii. Clear all equations from the Y= screen. (Press [Y=], [CLEAR].)

Now your calculator is ready to graph equations. The Y= screen is where equations will be put into the calculator to graph them or evaluate them at input values. We are now going to graph several simple equations to investigate how the graph of an equation for a line reacts to changes in the equation. (Note that your calculator uses y as the dependent (output) variable and x as the independent (input) variable.)

a. Graph the following equations on a standard window. Enter each equation in its own row (Y1, Y2, Y3, . . .).

(Note: To enter an x, you use the $[X, T, \theta, n]$ button next to the green [ALPHA] button.)

i. $y = x$ **ii.** $y = 2x$
iii. $y = 5x$ **iv.** $y = 8x$

In your own words, describe what the coefficient (number in front) of x does to the graph. Remember to read graphs from left to right.

b. Now graph the following equations that have negative coefficients.

i. $y = -x$ **ii.** $y = -2x$
iii. $y = -5x$ **iv.** $y = -8x$

In your own words, describe what a negative coefficient of x does to the graph.

Say What?

Coefficient: The number in front of a variable expression.

For example:
$$-7x \qquad 5x^2$$

-7 is the coefficient for x.

5 is the coefficient of x^2.

Remember that a variable that is by itself (x) has a coefficient of 1.

Using Your TI Graphing Calculator

In entering fractions in the calculator, it is often best to use parentheses.

$$y = (1/5)x$$

On many graphing calculators, parentheses are needed in almost all situations. In some calculators when entering

$$1/5x$$

the calculator will interpret this as

$$\frac{1}{5x}$$

instead of

$$\frac{1}{5}x$$

To be sure the calculator does what you intended using parenthesis is a good idea.

The TI-83 does not need them in some situations, but in other situations they are required. To keep confusion down, one option is to use parentheses around every fraction. Extra parentheses do not usually create a problem, but not having them where they are needed can cause miscalculations.

c. Graph the following equations with coefficients that are between zero and one.

i. $y = x$ **ii.** $y = \frac{1}{5}x$ **iii.** $y = \frac{1}{2}x$

iv. $y = \frac{2}{3}x$ **v.** $y = 0.9x$

In your own words, describe what a coefficient of x between 0 and 1 does to the graph.

• ◆

These three sets of graphs demonstrate the **slope** of a line. Slope can be described in several ways and is basically the direction or steepness of the line. The graph of a line will have the same steepness (direction) over the entire graph. When considering slope or direction of a graph, always look from left to right. Here are some ways of thinking about slope of a line.

* The steepness of a line
* The direction in which a line is traveling (left to right)
* How fast something is changing

In mathematics, we calculate the slope of a line as a ratio (fraction) of the vertical change and horizontal change. This ratio must stay constant no matter where on the line you are. Change in mathematics is the difference between two quantities or variables and is calculated by subtraction. This concept results in some of the following ways to remember how to calculate the slope of a line; m is often the letter that is used to represent the slope of a line.

* $m = \dfrac{\text{rise}}{\text{run}}$ The vertical rise divided by the horizontal run.

* $m = \dfrac{\text{change in } y}{\text{change in } x} = \dfrac{y_2 - y_1}{x_2 - x_1}$ where (x_1, y_1) and (x_2, y_2) are two distinct points on the line

All of these descriptions can be used to remember the idea of slope and how to calculate it. If you use one of the word descriptions, remember that change in mathematics is most often measured by using subtraction. When interpreting the slope of a model, remember that slope is a measurement of how fast something is changing.

In particular, the slope measures the increase or decrease in the output variable for a unit change in the input variable. This is often stated as the increase or decrease in y for a unit change in x. It is important to remember that all graphs in mathematics are read from left to right. Therefore an increase in y will cause the line to go up from left to right and will result in a positive slope. A decrease in y will cause the line to go down from left to right and result in a negative slope.

Say What?

Slope:

- slope
- rate of change
- *m* in the slope-intercept form of a line.

When asked for the slope of the line $y = 4x + 9$, be sure to give only the constant 4 as the slope. The variable *x* is <u>NOT</u> part of the slope.

DEFINITION

Slope:

- The ratio of the vertical change and horizontal change of a line.
- The increase or decrease in **y** for a unit change in **x**.
- For a line going through the two distinct points (x_1, y_1) and (x_2, y_2),

$$\text{slope} = \frac{\text{rise}}{\text{run}} = \frac{\text{change in } y}{\text{change in } x} = \frac{y_2 - y_1}{x_2 - x_1}$$

- Slope represents the amount the output variable is changing for every unit change in the input variable.

EXAMPLE **3** FINDING SLOPE FROM THE GRAPH OF A LINE

Use the graph to estimate the slope of the line and determine if the line is increasing or decreasing.

a.

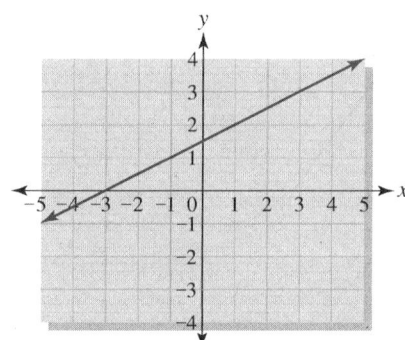

b.

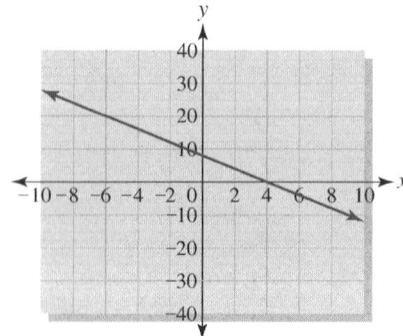

Solution

a. On reading the graph, it appears that the points $(1, 2)$ and $(3, 3)$ lie on the line.

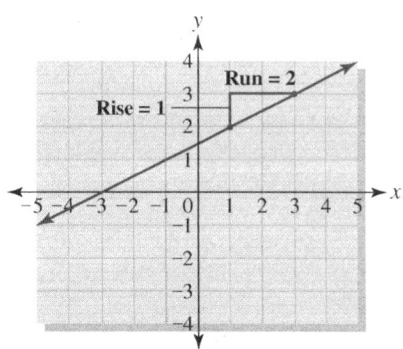

Reading the graph from left to right and using these two points, we can see that the rise is 1 when the run is 2. This gives us a slope of

$$m = \frac{1}{2}$$

If we pick another set of points on this line, we should get the same slope. Because this line is going up from left to right, it confirms that the slope should be positive. We say this line is increasing.

b. On reading this graph, it appears that the points $(-6, 20)$ and $(4, 0)$ lie on the line.

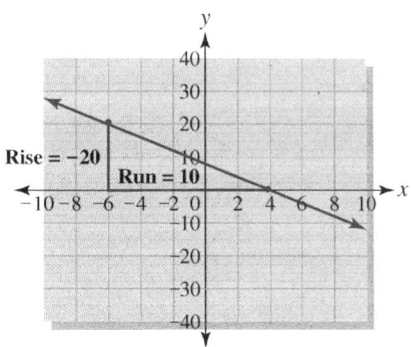

Reading the graph from left to right and using these two points, we can see that the rise is -20 when the run is 10. This gives us a slope of

$$m = \frac{-20}{10} = -2$$

This could also have been calculated using the formula for slope resulting in

$$m = \frac{20 - 0}{-6 - 4} = \frac{20}{-10} = -2$$

Because this line is going down from left to right, it confirms that the slope should be negative. We say that this line is decreasing.

EXAMPLE ③ PRACTICE PROBLEM

Use the graph to estimate the slope of the line and determine if the line is increasing or decreasing.

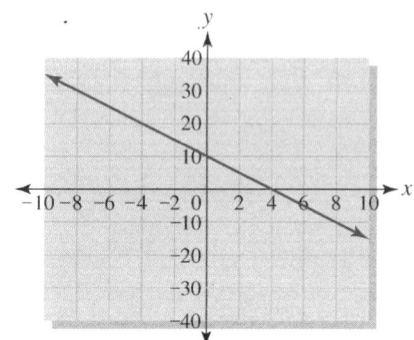

EXAMPLE ④ FINDING SLOPE FROM A TABLE

Find the slope of the line passing through the points given in the table.

x	y
−3	29
0	20
2	14
11	−13
14	−22

Solution

If we calculate the change in the values given in the table we can find the slope.

x	y	Slope
−3	29	$\dfrac{20-29}{0-(-3)} = \dfrac{-9}{3} = -3$
0	20	$\dfrac{14-20}{2-0} = \dfrac{-6}{2} = -3$
2	14	$\dfrac{-13-14}{11-2} = \dfrac{-27}{9} = -3$
11	−13	$\dfrac{-22-(-13)}{14-11} = \dfrac{-9}{3} = -3$
14	−22	

All of the slope calculations are equal to -3 and thus confirm that these points would all lie on a line with slope -3.

EXAMPLE (4) PRACTICE PROBLEM

Find the slope of the line passing through the points given in the table.

x	y
0	140
5	100
10	60
15	20
20	−20

As you can see in Example 4, if the points in a table all lie on the same line, each pair of points will have the same slope. Therefore, you can calculate the slope by selecting **any** two points on the line. We can determine if points given in a table all lie on a line if the slopes remain constant between all the points. If the slope between two points is not the same as that of the other pairs of points, then it does not lie on the same line.

EXAMPLE [5] DETERMINING IF GIVEN POINTS LIE ON A LINE

Determine if the points given in the table all lie on a line.

a.

x	y
1	-15
3	-5
5	5
7	15

b.

x	y
-3	1
0	2
3	3
9	8

Solution

a. To determine if the points all lie on a line, we need to calculate the slope between each pair of points and see if the slopes remain constant.

x	y	slope
1	-15	$\dfrac{-5-(-15)}{3-1} = \dfrac{10}{2} = 5$
3	-5	$\dfrac{5-(-5)}{5-3} = \dfrac{10}{2} = 5$
5	5	$\dfrac{15-5}{7-5} = \dfrac{10}{2} = 5$
7	15	

All of the slopes are equal to 5 so these points lie on the same line. We can confirm this by plotting the points and drawing a line.

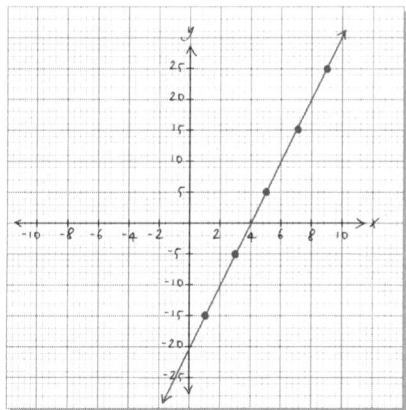

b. Again we will check to see if the slopes remain constant.

x	y	Slope
-3	1	$\dfrac{2-1}{0-(-3)} = \dfrac{1}{3}$
0	2	
3	3	$\dfrac{3-2}{3-1} = \dfrac{1}{3}$
9	8	$\dfrac{8-3}{9-3} = \dfrac{5}{6}$

The slopes do not remain constant so these points do not all lie on the same line. We can confirm this by plotting the points and trying to draw a line through them.

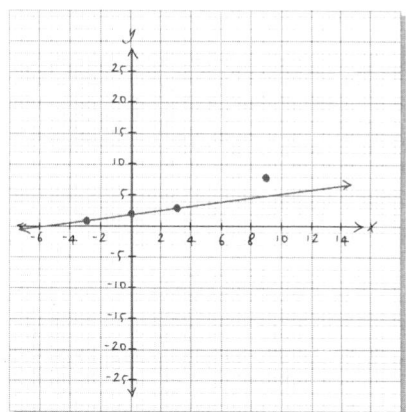

We can see that the last point does not lie on the same line as the first three.

EXAMPLE ⑤ PRACTICE PROBLEM

Determine if the points given in the table all lie on a line.

a.

x	y
1	-5
4	1
7	7
10	13

b.

x	y
-5	8
0	6
5	4
10	7

Next we will use the concept investigation to investigate another important characteristic of linear equations.

CONCEPT INVESTIGATION 2 WHAT'S MOVING THAT GRAPH? •••

Start by clearing the equations from the Y= screen.

a. Graph the following equations on the same calculator screen.

    i. $y = x$ **ii.** $y = x + 1$ **iii.** $y = x + 2$

    iv. $y = x + 3$ **v.** $y = x + 7$

In your own words, describe what the constant term does to the graph.

b. Graph the following equations on the same calculator screen.

    i. $y = x$ **ii.** $y = x - 1$ **iii.** $y = x - 2$

    iv. $y = x - 3$ **v.** $y = x - 7$

In your own words, describe what a negative constant term does to the graph.

••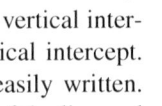

The constant term of a linear equation in this form represents the vertical intercept of the linear graph. In mathematics, we let *b* represent the vertical intercept. With the slope and vertical intercept, an equation of a line can be easily written. Every line can be described with two pieces of information: the slope of the line and a point on the line. Both of these are needed, since a slope does not tell you where the line is and a point does not tell you where to go.

An equation for a line can be written in many forms, but the **slope-intercept form** is the most common and most useful form. On a graphing calculator the slope-intercept form is the easiest to use. The slope-intercept form of an equation of a line is represented by the equation $y = mx + b$. In this equation, *m* represents the slope of the line and *b* represents the output of the *y*-intercept (vertical intercept).

The variables *x* and *y* are the independent and dependent variables, respectively. Any equation that can be simplified into the form $y = mx + b$ is the equation of a line.

DEFINITION

Linear Equation: An equation is linear if it has a constant rate of change (the slope is constant). That is, for every unit change in the input the output has a constant amount of change.

Slope-Intercept Form of a Line:

$$y = mx + b$$

Slope: The increase, or decrease in the output variable for a unit change in the input variable. In the slope-intercept form of a line, slope is represented by *m*.

Vertical Intercept: The point where the line crosses the vertical axis. In the slope-intercept form of a line, the vertical intercept is $(0, b)$. This is often called the *y*-intercept.

EXAMPLE **6** DETERMINING SLOPE AND Y-INTER-CEPT OF LINEAR EQUATIONS

Find the slope and y-intercept of the following lines.

a. $y = 2x + 5$ **b.** $y = \frac{3}{2}x - \frac{7}{2}$

c. $2x + 5y = 20$

Solution

a. This equation is in slope-intercept form, so the slope is 2, and the y-intercept is $(0, 5)$.

b. This equation is in slope-intercept form, so the slope is $\frac{3}{2}$, and the y-intercept is $\left(0, -\frac{7}{2}\right)$.

c. This equation should be put into slope-intercept form before we try to read the slope and intercept.

$$2x + 5y = 20$$
$$\underline{-2x \qquad\quad -2x} \qquad \text{Isolate } y.$$
$$5y = -2x + 20$$
$$\frac{5y}{5} = \frac{-2x + 20}{5}$$
$$y = -\frac{2}{5}x + 4$$

Therefore, the slope is $-\frac{2}{5}$, and the y-intercept is $(0, 4)$.

EXAMPLE **6** PRACTICE PROBLEM

Find the slope and y-intercept of the following lines.

a. $y = -3x + 7$ **b.** $y = \frac{2}{7}x - 8$

c. $3x - 4y = 15$

THE MEANING OF SLOPE IN AN APPLICATION

Understanding the meaning of the slope of a line is an important part of understanding the role slope plays in the context of an application. The slope is the rate of change of the line and tells us how much the output variable changes for a unit change in the input variable. The idea of a rate of change is easier to understand in the context of an application.

In section 1.1, we found the equation $C = 38m + 150$ for the cost in dollars for pest management from Enviro-Safe Pest Management when m months of services are provided. This equation is in slope-intercept form, so the slope is the coefficient of the input variable.

$$38 = \frac{38}{1} = \frac{\$38}{1\,\text{month}}$$

This slope means that the cost for pest management is increasing by 38 dollars per month of service. Since the output variable represents cost and the slope is positive, the cost is increasing for each month.

In the U-Haul example, $U = 0.59m + 19.95$, the slope was

$$0.59 = \frac{0.59}{1} = \frac{\$0.59}{1\,\text{mile}}$$

The cost for renting a 10-foot truck from U-Haul for the day increases by 0.59 dollar, 59 cents, per mile driven.

If the variables are defined, writing this sentence is just a matter of putting the pieces together properly. Always remember that slope is the output change per input change so you will state the meaning correctly.

EXAMPLE ⑦ FINDING AND INTERPRETING SLOPE

Find the slope and give its meaning in the context of the given model.

a. Let $P = 115b - 4000$ be the profit in dollars Bicycles Galore makes from selling b bikes.

b. Let $S = -0.76a + 33.04$ be the percentage of women a years old who smoked cigarettes during their pregnancy.

Source: U.S. National Center for Health Statistics, National Vital Statistics Reports.

Solution

a. The slope of this model is

$$115 = \frac{115}{1} = \frac{\$115}{1\,\text{bike}}$$

The output variable is P profit in dollars, and the input variable is b bikes sold. Therefore the slope means: The profit for Bicycles Galore in increasing by $115 per bike sold.

b. The slope of this model is

$$-0.76 = \frac{-0.76}{1} = \frac{-0.76\,\text{percentage points}}{1\,\text{year of age}}$$

The output variable is S percent of women who smoked cigarettes during their pregnancy, and the input variable is a years old. Therefore the slope means: The percent of women who smoke cigarettes during their pregnancy is decreasing by 0.76 percentage points per year of age. In other words, the older the women get, the fewer of them smoke during pregnancy.

Note that when a model is measuring a percentage, the slope will measure the number of percentage points that the output is changing, not the percent itself.

EXAMPLE ⑦ PRACTICE PROBLEM

Find the slope and give its meaning in the context of the given model.

a. The pressure inside a vacuum chamber can be represented by $P = 35 - 0.07s$ where P is the pressure in pounds per square inch (psi) of the vacuum chamber after being pumped down for s seconds.

Say What?

PERCENT OR PERCENTAGE POINTS?

When working with percentages, you may encounter some confusion between a percent change and a percentage points change.

If the percentage of people who smoke is currently 40% and it increases to 45%, that is a 5 percentage point increase. It is not a 5% increase. A 5% increase would only be to 42%-- the 2 being 5% of 40.

When working with how much a percentage is changing (slope), it will be a percentage point increase or decrease.

b. The cost for making tacos at a local street stand can be represented by $C = 0.55t + 140.00$, where C is the cost in dollars to make tacos at the local street stand when t tacos are made.

GRAPHING LINES USING SLOPE AND INTERCEPT

Using the information gained from the slope-intercept form of a line, we can graph linear equations. Use the vertical intercept as a starting point on the line, and then use the slope as the direction from left to right that you should follow to find additional points on the line. Remember that slope is the rise over the run that the line takes from one point to another. Once you have two or three points, you can draw the entire line by connecting these points and extending the line in both directions. In most cases, you will want to graph the line so that both the vertical and horizontal intercepts are shown.

EXAMPLE 8

Sketch the graph of the following lines. Label the vertical intercept.

a. $y = \frac{2}{3}x - 2$

b. $y = 2x + 5$

c. $y = -3x + 2$

Solution

a. The vertical intercept is $(0, -2)$, and the slope is $\frac{2}{3}$, so the rise is 2 when the run is 3. Working from left to right and using the vertical intercept and slope, we get the following graph

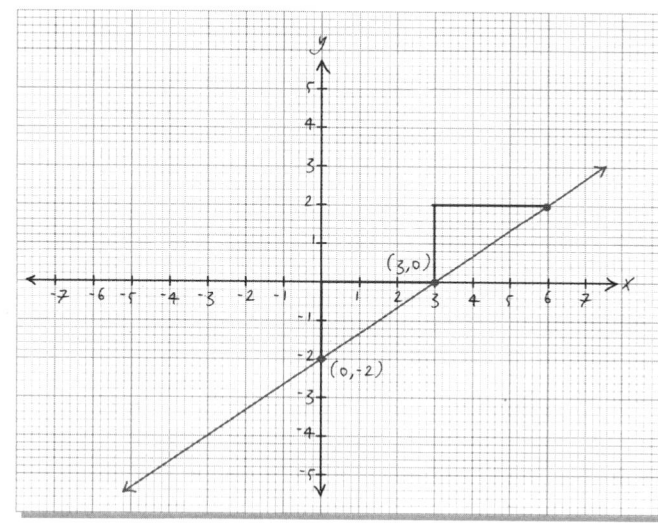

Starting at $(0, -2)$ and going up 2 and over 3, we get to $(3, 0)$. Doing this again gets us to $(6, 2)$. Using these three points, we connect the points and extend the line in both directions.

b. The y-intercept is (0, 5), and the slope is $2 = \dfrac{2}{1}$, so the rise is 2 when the run is 1. Working from left to right and using the vertical intercept and slope, we get the following graph.

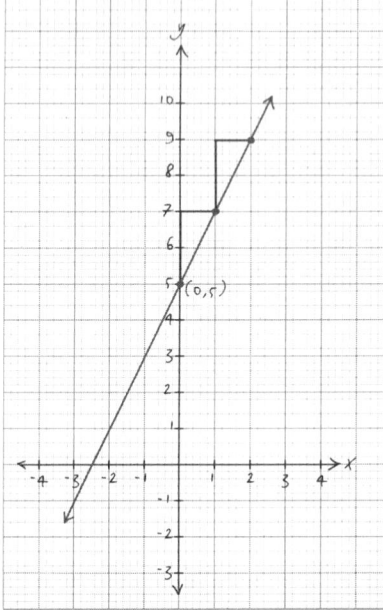

Starting at (0, 5) and going up 2 and over 1, we get to (1, 7). Doing this again gets us to (2, 9). Using these three points, we connect the points and extend the line in both directions.

c. The y-intercept is (0, 2), and the slope is $-3 = \dfrac{-3}{1}$, so the rise is -3 when the run is 1. Because the rise is negative, the graph will actually go downward when looking from left to right. Using this information, we get the following graph

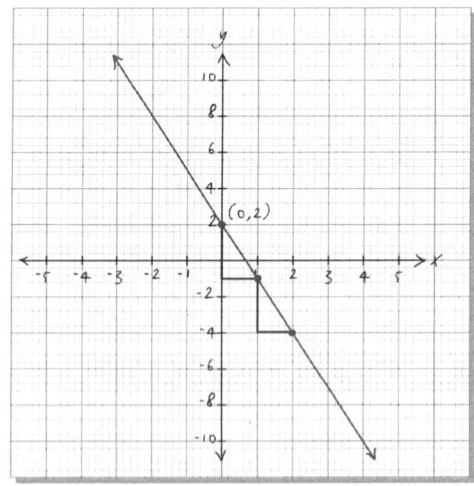

EXAMPLE ⑧ PRACTICE PROBLEM

Sketch the graph of the following lines. Label the vertical intercept.

a. $y = 4x - 5$

b. $y = -\dfrac{3}{2}x + 6$

1.3   Exercises

1. Use the graph to find the following.

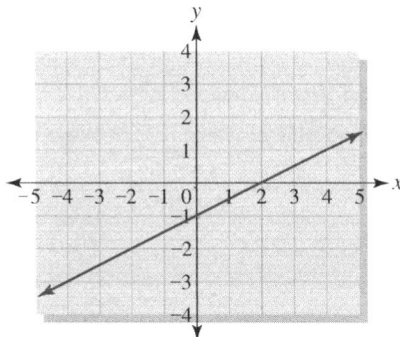

   a. Find the slope of the line.

   b. Is the line increasing or decreasing?

   c. Estimate the vertical intercept.

   d. Estimate the horizontal intercept.

2. Use the graph to find the following.

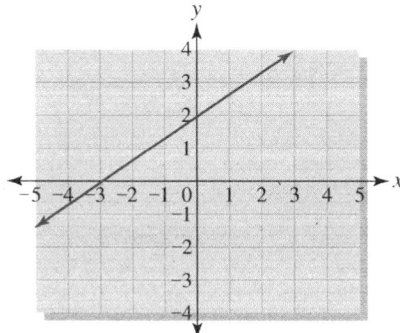

   a. Find the slope of the line.

   b. Is the line increasing or decreasing?

   c. Estimate the vertical intercept.

   d. Estimate the horizontal intercept.

3. Use the graph to find the following.

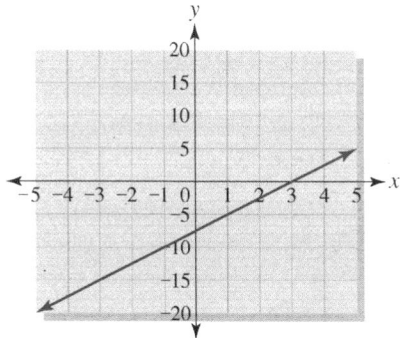

   a. Find the slope of the line.

   b. Is the line increasing or decreasing?

   c. Estimate the y-intercept

   d. Estimate the x-intercept

4. Use the graph to find the following.

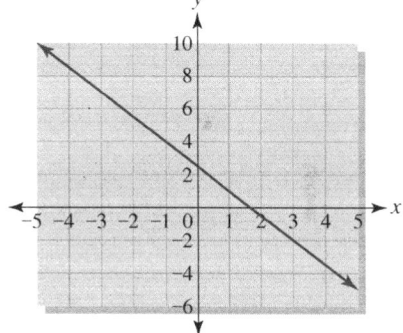

   a. Find the slope of the line.

   b. Is the line increasing or decreasing?

   c. Estimate the y-intercept

   d. Estimate the x-intercept

5. Use the graph to find the following.

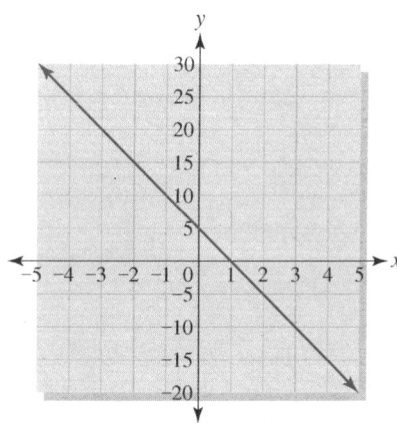

a. Find the slope of the line.

b. Is the line increasing or decreasing?

c. Estimate the vertical intercept.

d. Estimate the horizontal intercept.

6. Use the graph to find the following.

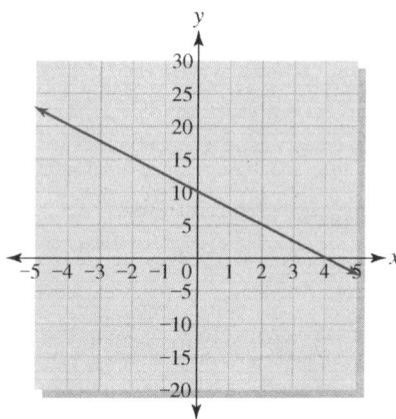

a. Find the slope of the line.

b. Is the line increasing or decreasing?

c. Estimate the vertical intercept.

d. Estimate the horizontal intercept.

For exercises 7 through 22 graph the equations by plotting points. For linear equation use at least 3 points, for more complicated equations use at least 5 points.

7. $y = 2x + 3$

8. $y = 3x - 8$

9. $x = 5y + 4$

10. $x = -2y - 7$

11. $y = x^2 + 2$

12. $y = x^2 - 4$

13. $y = \frac{2}{3}x + 6$

14. $y = \frac{1}{4}x - 3$

15. $x = \frac{2}{3}y - 4$

16. $x = \frac{3}{4}y + 1$

17. $y = 0.5x - 3$

18. $y = -0.4x + 5$

19. $x = -1.5y + 7$

20. $x = -2.5x + 10$

21. $y = -2x^2 + 15$

22. $y = -1.5x^2 + 8$

23. In section 1.1 exercises, we investigated the equation $B = 0.55m + 29.95$, which represented the cost in dollars for renting a 10-ft truck from Budget and driving it m miles.

a. Create a table of points that satisfy this equation.

b. Create a graph for the equation using your points. Remember to label your graph with units.

24. One type of investment that institutions and very wealthy people use are called hedge funds. These funds typically give a high return on the money invested. That return comes at a price. A typical hedge fund charges 20 percent on the earnings of the fund. If someone invested 5 million dollars, their fees could be calculated as $F = 0.20e + 100000$, where F represented the fee in dollars charged on the investment that had e dollars worth of earnings.

Source: Equation based on information from abcnews.go.com May 26, 2006 Hundreds of Millions for Top Hedge Funders, by Bob Jamieson.

a. Create a table of points that satisfy this equation. (Note: earnings for this type of investment can easily be $500,000 or more.)

b. Create a graph for the equation using your points. Remember to label your graph with units.

25. A sales clerk at a high end clothing store earns 4% commission on all of her sales in addition to her $100 base weekly salary.

 a. Write an equation for the sales clerk's salary per week when she sells s dollars of merchandise during the week.

 b. Create a table of points that satisfy the equation you found in part a.

 c. Create a graph for the equation using your points. Remember to label your graph with units.

26. A sales clerk at a computer store earns 3% commission on all of his sales in addition to his $200 base weekly salary.

 a. Write an equation for the sales clerk's salary per week when he sells s dollars of merchandise during the week.

 b. Create a table of points that satisfy the equation you found in part a.

 c. Create a graph for the equation using your points. Remember to label your graph with units.

27. Western Washington University charges each resident undergraduate student taking up to ten units $137 per unit in tuition plus $208 in other fees.

 a. Write an equation for the total tuition and fees that Western Washington University charges its resident undergraduate students who take u units.

 b. Create a table of points that satisfy the equation you found in part a.

 c. Create a graph for the equation using your points. Remember to label your graph with units.

28. A local specialty delivery service charges a fee of $5 for each delivery plus 25 cents for each mile over 10 miles.

 a. Write an equation for the cost in dollars for a delivery of m miles over 10 miles. (Note a 15 mile delivery would be represented by $m = 5$.)

 b. Create a table of points that satisfy the equation you found in part a.

 c. Create a graph for the equation using your points. Remember to label your graph with units.

29. The percent P of companies that are still in business t years after the fifth year in operation can be represented by the equation. $P = -3t + 50$.

 a. Create a table of points that satisfy this equation.

 b. Create a graph for the equation using your points. Remember to label your graph with units.

30. The number E of golf events on the Champions Tour t years after 2000 can be represented by the equation. $P = -2t + 39$.

Source: Model derived from data found in the USA Today 5-25-2005

 a. Create a table of points that satisfy this equation.

 b. Create a graph for the equation using your points. Remember to label your graph with units.

For exercises 31 through 36 find the slope of the line passing through the points given in the table.

31.

x	y
0	7
2	11
4	15
6	19

32.

x	y
−4	−18
0	−4
4	10
8	24

33.

x	y
−2	29
0	15
2	1
4	−13

34.

x	y
−3	14.25
0	9
3	3.75
6	−1.5

35.

x	y
2	−18
5	−6
8	6
11	18

36.

x	y
4	7
9	−5.5
14	−18
19	−30.5

For exercises 37 through 40 determine if the points given in the table all lie on a line.

37.

x	y
-2	-14
0	-6
2	2
4	10

38.

x	y
0	5
1	2
2	-1
3	-4

39.

x	y
-6	12
-3	10
0	8
3	5

40.

x	y
-8	-8
-2	-3.5
0	-2
10	7.5

For Exercises 41 through 54 determine the slope and y-intercept of the given linear equations.

41. $y = 3x + 5$

42. $y = 7x + 12$

43. $y = -4x + 8$

44. $y = -9x + 25$

45. $y = \frac{1}{2}x + 5$

46. $y = \frac{2}{3}x + 43$

47. $y = 0.4x - 7.2$

48. $y = 2.3x - 8.4$

49. $y = \frac{4}{3}x + \frac{7}{5}$

50. $y = -\frac{6}{5}x + \frac{4}{9}$

51. $2x + y = 20$

52. $5x + y = 11$

53. $4x - 2y = 20$

54. $6x - 3y = 18$

55. Using the application found in Exercise 23, give the slope of the equation, and explain its meaning in the context of the situation.

56. Using the application found in Exercise 24, give the slope of the equation, and explain its meaning in the context of the situation.

57. Using the application found in Exercise 25, give the slope of the equation, and explain its meaning in the context of the situation.

58. Using the application found in Exercise 26, give the slope of the equation, and explain its meaning in the context of the situation.

59. Using the application found in Exercise 27, give the slope of the equation, and explain its meaning in the context of the situation.

60. Using the application found in Exercise 28, give the slope of the equation, and explain its meaning in the context of the situation.

61. Using the application found in Exercise 29, give the slope of the equation, and explain its meaning in the context of the situation.

62. Using the application found in Exercise 30, give the slope of the equation, and explain its meaning in the context of the situation.

For Exercises 63 through 80, sketch the graph on graph paper and label the vertical intercept. Round the values of the intercepts to two decimal places if needed.

63. $y = 2x - 7$

64. $y = 3x - 12$

65. $y = \frac{4}{5}x - 6$

66. $y = \frac{2}{3}x + 4$

67. $y = -2x + 5$

68. $y = -4x - 3$

69. $y = -\frac{7}{5}x + 11$

70. $y = -\frac{3}{4}x + 2$

71. $y = 0.5x + 4$

72. $y = 0.75x - 8$

73. $y = -0.25x + 2$

74. $y = -0.3x + 6$

75. $y = -\frac{2}{5}x + \frac{3}{5}$

76. $y = -\frac{1}{4}x + \frac{9}{4}$

77. $y = \frac{1}{4}x - \frac{3}{4}$

78. $y = \frac{1}{8}x - \frac{5}{8}$

79. $y = -0.75x + 2.5$

80. $y = -0.4x + 3.6$

THE GENERAL FORM OF LINES

Linear equations can take another form, called the general form. The general form is written with both variables on one side of the equal sign and the constant term on the other side.

$$Ax + By = C$$

Notice that the variables x and y both have exponents of 1. The constants A, B and C are integers and A should be positive. To have the constants be integers means that to put a linear equation into general form you will remove any fractions by multiplying both sides of the equation by the common denominator. All linear equations can be put into general form.

Say What?

WHAT IS AN INTEGER?

Integers are the whole numbers and their opposites. So they include all positive and negative whole numbers and zero.

. . . -3, -2, -1, 0, 1, 2, 3, . . .

EXAMPLE REWRITING EQUATIONS IN GENERAL FORM

Rewrite the following equations in general form.

a. $y = -4x + 12$

b. $y = \dfrac{2}{3}x + \dfrac{2}{5}$

Solution

a. To put the equation into general form, we need to bring the x-term to the left side of the equation.

$$y = -4x + 12$$
$$\underline{+4x \qquad +4x}$$
$$4x + y = 12$$

b. First, we will eliminate the fractions by multiplying both sides of the equation by the common denominator, and then we will get the x-term to the left side of the equation.

EXAMPLE ▮ PRACTICE PROBLEM

Rewrite the following equations in general form.

a. $y = 2x + 10$

b. $y = \dfrac{1}{2}x - \dfrac{3}{7}$

FINDING INTERCEPTS AND THEIR MEANING

Recall that the intercepts of a graph are the points where the graph crosses the vertical, y, and horizontal, x, axes.

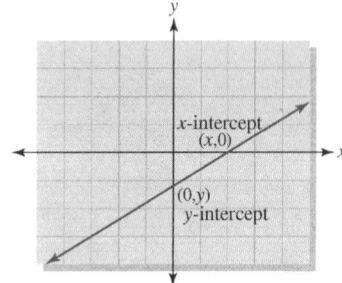

$$y = \frac{2}{3}x + \frac{2}{5}$$

$$15(y) = 15\left(\frac{2}{3}x + \frac{2}{5}\right)$$ **Multiply by 15 to eliminate the fractions.**

$$15y = 10x + 6$$
$$\underline{-10x \quad -10x}$$ **Move the x term to the left side of the equation.**

$$-10x + 15y = 6$$
$$-1(-10x + 15y) = -1(6)$$ **Multiply by -1 to make the coefficient of x positive.**
$$10x - 15y = -6$$

Explaining the meaning of the intercepts of a line is an important part of understanding what role they play in the context of an application. The intercepts of a line are points on the graph so they represent combinations of input and output values that satisfy the equation.

Both parts of each intercept will have a meaning and should be explained in the context of the problem. In the U-Haul example from section 1.1, the vertical intercept, U-intercept, is (0, 19.95) and means that if you rented a truck from U-Haul and drove it zero miles it would cost $19.95 for the day. It may not make much sense to rent a truck and not drive it, but this is still the meaning of the vertical intercept. The horizontal intercept, m-intercept, is $(-33.81, 0)$ and means that if you rented a truck from U-Haul and drove it negative 33.81 miles the truck would not cost you anything. This is definitely model breakdown since you cannot drive a truck a negative number of miles. In the world of mathematics, both of these intercepts exist but they may not make much sense in the context of the application. Always explain what both numbers in the intercept mean in the context, and remember to state if an intercept is a case of model breakdown.

To find an intercept from an equation, set the "other" variable equal to zero and solve. What this means is that if you are looking for a y-intercept you make $x = 0$ and solve for y. If you want an x-intercept you make the $y = 0$ and solve for x. In an application such as the U-Haul equation, $U = 0.79m + 19.95$, to find the U-intercept, you make $m = 0$ and solve for U. To find the m-intercept, you make $U = 0$ and solve for m.

FINDING THE INTERCEPTS OF AN EQUATION

Set the "other" variable equal to zero and solve.
    To find the y-intercept, set $x = 0$ and solve for y.
    To find the x-intercept, set $y = 0$ and solve for x.

EXAMPLE 2 FINDING AND INTERPRETING INTER-CEPTS

Find the vertical and horizontal intercepts, and give their meaning in the context of the given model.

a. The number of students who are enrolled in math classes at a local college can be represented by $E = -17w + 600$, where E represents the math class enrollment at the college w weeks after the start of the fall semester.

b. The number of homicides, N, of 15- to 19-year-olds in the United States t years after 1990 can be represented by the equation $N = -315.9t + 4809.8$.

Solution

a. The equation is in slope-intercept form so the vertical intercept, E-intercept, is $(0, 600)$. We can also find this intercept by setting $w = 0$ and solving for E.
$$E = -17(0) + 600$$
$$E = 600$$

The vertical intercept means that during week zero, the beginning of the fall semester, there are 600 students enrolled in math classes at the college.

The horizontal intercept, w-intercept, can be found by substituting zero for E and solving for w.

$$E = -17w + 600 \qquad \text{Substitute zero for } E \text{ and solve for } w.$$
$$0 = -17w + 600$$
$$\underline{-600 \qquad\qquad -600} \qquad \text{Subtract 600 from both sides.}$$
$$-600 = -17w$$
$$\frac{-600}{-17} = w \qquad \text{Divide both sides by -17.}$$
$$35.29 \approx w$$

Therefore the horizontal intercept is $(35.29, 0)$ and means that about 35 weeks after the start of the fall semester there will be no students enrolled in math classes at the college. This is model breakdown since a semester would not last 35 weeks and some students love math and would never drop their math class.

b. The equation is in slope-intercept form so the vertical intercept is $(0, 4809.8)$ and means that in 1990 there were about 4810 homicides of 15- to 19-year-olds in the United States. In this case, the zero would be years after 1990, and zero years after 1990 would be 1990. Also, you should round the intercept values to something that is reasonable to the situation. 4809.8 homicides has 4809 plus 0.8 of a homicide; 0.8 of a homicide does not make sense, so you should round the intercept to 4810.

The horizontal intercept can be found by substituting zero for N and solving for t.

$$N = -315.9t + 4809.8$$

Substitute zero for N and solve for t.

$$0 = -315.9t + 4809.8$$

$$\underline{-4809.8 \qquad\qquad -4809.8}$$

Subtract 4809.8 from both sides.

$$-4809.8 = -315.9t$$

$$\frac{-4809.8}{-315.9} = t$$

Divide both sides by -315.9

$$15.23 \approx t$$

Therefore, the horizontal intercept is (15.23, 0) and means that in about 2005 there were no homicides of 15-to 19-year-olds in the United States. This must be model breakdown because there were definitely homicides of 15-to 19-year-olds in the United States in 2005. Note the 15.23 was rounded and represented the year and the zero meant that there were no homicides.

EXAMPLE 2 PRACTICE PROBLEM

Find the vertical and horizontal intercepts, and give their meaning in the context of the given model.

a. The pressure inside a vacuum chamber can be represented by $P = 35 - 0.07s$ where P is the pressure in pounds per square inch (psi) of the vacuum chamber after being pumped down for s seconds.

b. The cost for making tacos at a local stand can be represented by $C = 0.55t + 140.00$ where C is the cost in dollars to make tacos at the neighborhood stand when t tacos are made.

With an equation in the general form, solving for both intercepts is simple, since it is easy to solve for x or y if the other variable is zero.

EXAMPLE 3 FINDING INTERCEPTS FROM THE GENERAL FORM

Find the horizontal and vertical intercepts of the following equations.

a. $3x + 4y = 24$
b. $5x - 2y = 35$

Solution

a. To find the horizontal intercept, x-intercept, we will set $y = 0$ and solve for x.

$$3x + 4(0) = 24$$
$$3x = 24$$
$$x = 8$$

So the horizontal intercept is (8, 0).

To find the vertical intercept, y-intercept, we will set $x = 0$ and solve for y.

$$3(0) + 4y = 24$$
$$4y = 24$$
$$y = 6$$

So the vertical intercept is (0, 6).

b. To find the horizontal intercept we will set $y = 0$ and solve for x.

$$5x - 2(0) = 35$$
$$5x = 35$$
$$x = 7$$

So the horizontal intercept is (7, 0).

To find the vertical intercept we will set $x = 0$ and solve for y.

$$5(0) - 2y = 35$$
$$-2y = 35$$
$$y = -\frac{35}{2} = -17.5$$

So the vertical intercept is (0, −17.5).

EXAMPLE 3 PRACTICE PROBLEM

Find the horizontal and vertical intercepts of $8x + 2y = 40$.

GRAPHING LINES USING INTERCEPTS

Graphing lines that are in the general form can be done most easily using the intercepts. To graph a line, you only need to plot 2 points on the line, and the intercepts are two points that are easy to find.

EXAMPLE 4 GRAPHING LINES USING INTERCEPTS

For each equation, find the intercepts and graph the line.

a. $2x - 5y = 20$

b. $5x + 3y = 40$

Solution

a. First, we will find both intercepts.

$$2x - 5(0) = 20$$ To find the horizontal intercept, make $y = 0$ and solve for x.
$$2x = 20$$
$$x = 10$$
$$(10, 0)$$ Horizontal intercept.
$$2(0) - 5y = 20$$ To find the vertical intercept, make $x = 0$ and solve for y.
$$-5y = 20$$
$$y = -4$$
$$(0, -4)$$ Vertical intercept.

Now plot these points and draw the line.

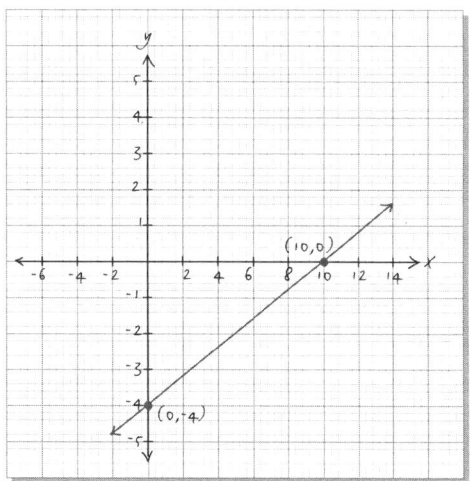

b.   First, we will find both intercepts.

$$5x + 3(0) = 40$$ To find the horizontal intercept, make
                        $y = 0$ and solve for x.
$$5x = 40$$

$$x = 8$$

$$(8, 0)$$ Horizontal intercept.

$$5(0) + 3y = 40$$ To find the vertical intercept, make
                        $x = 0$ and solve for y.
$$3y = 40$$

$$y = \frac{40}{3} \approx 13.3$$

$$\left(0, \frac{40}{3}\right)$$ Vertical intercept.

Now plot these points and draw the line.

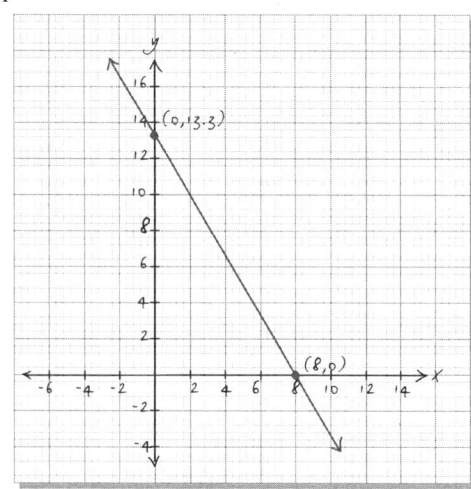

EXAMPLE   4   PRACTICE PROBLEM

Find the intercepts and graph the line $3x - 4y = 36$.

HORIZONTAL AND VERTICAL LINES

Concept Connection

Why is division by zero undefined?

If you have a group of 15 people can you divide them into zero groups? No.

$\frac{15}{0}$ is undefined.

Dividing by zero would be trying to put things into no piles. There has to be somewhere to put them. Undefined is the same as does not exist (DNE).

When you divide zero by something else you are dividing nothing into piles so each pile has nothing in it. Therefore dividing zero by any real number is always zero.

A graphing calculator will say one of two things when you use zero in division.

Divide by zero:

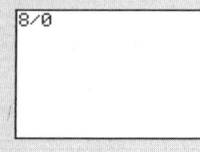

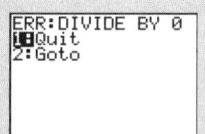

You get an error because it is undefined.

Divide into zero:

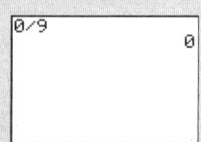

The answer is zero.

The general form of a line leads us to two unique types of lines, vertical and horizontal lines. When either A or B in the general form have the value zero, we get the equation of a horizontal or vertical line.

When A = 0, in the general form, the x term of the equation is eliminated.

$$Ax + By = C$$
$$0x + By = C$$
$$By = C$$

If you isolate the y, you get an equation of the form

$$y = \frac{C}{B} \quad \text{or} \quad y = k$$

where k is a constant. An equation of the form $y = k$ will result in a horizontal line. The following graph is the line $y = 3$.

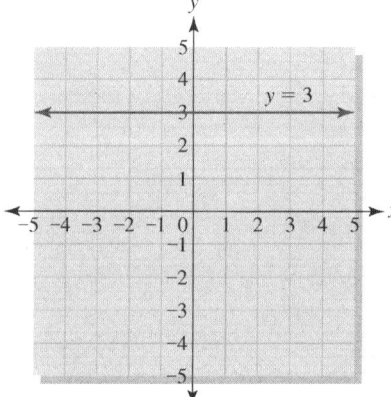

A horizontal line with the equation $y = 3$.

All of the points that lie on the horizontal line above will have a y value of 3. If we look at any two points on this line, there will be no vertical change. If we pick the points $(-4, 3)$ and $(2, 3)$ and calculate the slope, we get

$$(-4, 3) \quad \text{and} \quad (2, 3)$$
$$m = \frac{3 - 3}{2 - (-4)}$$
$$m = \frac{0}{6}$$
$$m = 0$$

Because there will never be any vertical change, the slope will always be zero. If you consider the general form for this line and build a table, you can see that the value of x will not affect the value of y.

x	$0x + y = 3$	(x,y)
-2	$0(-2) + y = 3$ $y = 3$	$(-2, 3)$
0	$0(0) + y = 3$ $y = 3$	$(0, 3)$
15	$0(15) + y = 3$ $y = 3$	$(15, 3)$

When $B = 0$, in the general form, the y term of the equation is eliminated.

$$Ax + By = C$$
$$Ax + 0y = C$$
$$Ax = C$$

If you isolate the x, you get an equation of the form

$$x = \frac{C}{A} \quad \text{or} \quad x = k$$

where k is a constant. An equation of the form $x = k$ will result in a vertical line. The following graph is the vertical line $x = 3$.

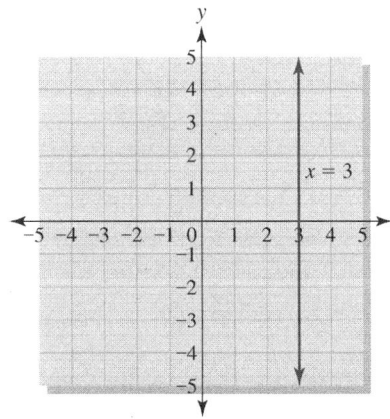

A vertical line with the
equation $x = 3$

All of the points that lie on the vertical line above have 3 as the x value. If we pick any two points on this line, the slope calculation will have no horizontal change. Using the points (3, 5) and (3, 1) gives us the slope calculation

$$(3, 5) \quad \text{and} \quad (3, 1)$$

$$m = \frac{5 - 1}{3 - 3}$$

$$m = \frac{4}{0} \quad \text{is undefined}$$

The slope of this line is undefined, since any two points on the line will result in the denominator being zero.

If you consider the general form for this line and build a table, you can see that the value of y will not affect the value of x.

y	$x + 0y = 3$	(x, y)
-7	$x + 0(-7) = 3$ $x = 3$	$(3, -7)$
4	$x + 0(4) = 3$ $x = 3$	$(3, 4)$
18	$x + 0(18) = 3$ $x = 3$	$(3, 18)$

DEFINITION

Horizontal LIne: A horizontal line has an equation of the form $y = k$ and a slope $m = 0$.

Vertical LIne: A vertical line has an equation of the form $x = k$ and a slope m undefined.

EXAMPLE 5 GRAPHING HORIZONTAL AND VERTICAL LINES

Sketch the graph of the following lines.

a. $x = 2$

b. $y = -4$

Solution

a. This equation is of the form $x = k$ so it will be a vertical line at $x = 2$.

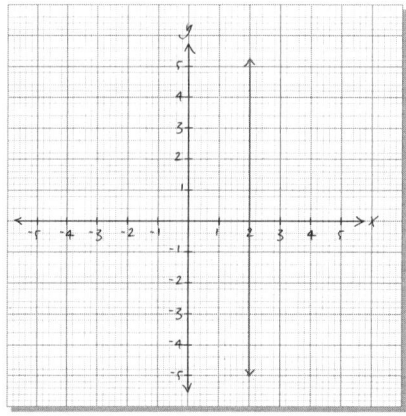

b. This equation is of the form $y = k$ so it will be a horizontal line at $y = -4$.

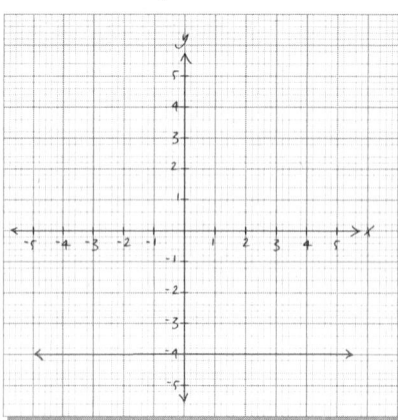

EXAMPLE 5 PRACTICE PROBLEM

Sketch the graph of the following lines.

a. $x = -3$

b. $y = 1.5$

1.4 Exercises

For Exercises 1 through 10, rewrite the equations in the general form of a line.

1. $y = 5x + 8$

2. $y = 7x + 20$

3. $y = -4x + 15$

4. $y = -3x + 9$

5. $y = \frac{2}{3}x - 8$

6. $y = \frac{3}{7}x - 4$

7. $y = \frac{1}{2}x + \frac{2}{5}$

8. $y = \frac{2}{9}x - \frac{7}{4}$

9. $y = -\frac{4}{5}x - \frac{1}{3}$

10. $y = -\frac{2}{7}x - \frac{7}{5}$

11. In section 1.1 exercises, we investigated the equation $B = 0.55m + 29.95$ which represented the cost in dollars for renting a 10-ft truck from Budget and driving it m miles.

    a. Find the vertical intercept and explain its meaning in the context of this situation.

    b. Find the horizontal intercept and explain its meaning in the context of this situation.

12. The cost in dollars to rent a backhoe for h hours can be represented by the equation $C = 85h + 55$.

    a. Find the vertical intercept and explain its meaning in the context of this situation.

    b. Find the horizontal intercept and explain its meaning in the context of this situation.

13. The percentage of girls a years older than 10 who are sexually active can be represented by the equation $P = 5a + 7$.

    a. Find the P-intercept and explain its meaning in the context of this situation.

    b. Find the a-intercept and explain its meaning in the context of this situation.

14. The percent P of companies that are still in business t years after the fifth year in operation can be represented by the equation $P = -3t + 50$.

    a. Find the P-intercept and explain its meaning in the context of this situation.

    b. Find the t-intercept and explain its meaning in the context of this situation.

15. A salesperson at a jewelry store earns 5% commission on all of her sales in addition to her $500 base monthly salary.

    a. Find an equation for the monthly salary that the salesperson earns if she makes s dollars in sales.

    b. Find the vertical intercept and explain its meaning in the context of this situation.

    c. Find the horizontal intercept and explain its meaning in the context of this situation.

16. A loan officer at a bank earns a $100 commission on each of her loans in addition to her $2000 base monthly salary.

    a. Find an equation for the monthly salary that the loan officer earns if she has l loans a month.

    b. Find the vertical intercept and explain its meaning in the context of this situation.

    c. Find the horizontal intercept and explain its meaning in the context of this situation.

17. The population of Maine in thousands t years since 2000 can be represented by the equation $P = 9.5t + 1277$.

Source: Model derived from data found in the Statistical Abstract 2007.

    a. Find the P-intercept and explain its meaning in the context of this situation.

    b. Find the t-intercept and explain its meaning in the context of this situation.

18. The population of Minnesota in thousands t years since 2000 can be represented by the equation $P = 41t + 4934$.

Source: Model derived from data found in the Statistical Abstract 2007.

   a. Find the P-intercept and explain its meaning in the context of this situation.

   b. Find the t-intercept and explain its meaning in the context of this situation.

19. The number of Florida residents who are enrolled in Medicare in thousands t years since 2000 can be represented by the equation $M = 44t + 2798$.

Source: Model derived from data found in the Statistical Abstract 2007.

   a. Find the M-intercept and explain its meaning in the context of this situation.

   b. Find the t-intercept and explain its meaning in the context of this situation.

20. The number of Vermont residents who are enrolled in Medicare in thousands t years since 2000 can be represented by the equation $M = 1.3t + 89$.

Source: Model derived from data found in the Statistical Abstract 2007.

   a. Find the M-intercept and explain its meaning in the context of this situation.

   b. Find the t-intercept and explain its meaning in the context of this situation.

For Exercises 21 through 30, find the intercepts and graph the line. Label the vertical and horizontal intercepts.

21. $2x + 4y = 8$

22. $4x + 2y = 20$

23. $3x - 5y = 15$

24. $5x - 2y = 20$

25. $4x + 6y = 30$

26. $10x + 8x = 16$

27. $-3x + 2y = 16$

28. $-4x + 3y = 18$

29. $-2x - 4y = -20$

30. $-7x - 5y = -70$

For exercises 31 through 56, graph the lines using any method. Label the vertical and horizontal intercepts.

31. $y = \frac{3}{7}x - 6$

32. $y = \frac{4}{5}x - 10$

33. $y - 5 = 3(x - 4)$

34. $y + 4 = 2(x + 3)$

35. $y - 4 = \frac{1}{2}(x + 1)$

36. $y + 7 = \frac{2}{3}(x - 9)$

37. $y = 7$

38. $y = 2.5$

39. $x = 9$

40. $x = 3.5$

41. $y = -4.5$

42. $y = -6.5$

43. $x = -8$

44. $x = -15$

45. $y + 3x = 6\left(\frac{1}{2}x - 2\right)$

46. $2y - 4x = -2(2x + 7)$

47. $5y + x = 2y + 3(y - 10)$

48. $y + x = 2x - 8 + y$

49. $y + 2 = 2(x + 4) - 2x$

50. $y - 5 = 3x + y - 14$

51. $y = \frac{2}{15}x - 6$

52. $y = \frac{3}{11}x - 6$

53. $10x - 25y = 100$

54. $2x - 100y = 4000$

55. $y = 0.001x - 20$

56. $y = -0.002x + 14$

Finding Equations of Lines

FINDING EQUATIONS OF LINES

Now that we have learned the basic characteristics of a line and know how to find the slope of a line, we can use these skills to find the equation of any line we may need. Unless we are given the slope of the line we are looking for, we will need two points to find the slope and thus determine the equation of the line that we want. In previous sections, we found several equations using the written descriptions of the situation. In each of those exercises, the slope was given to you. In other applications, you may not be given the slope but instead be given two points that you can use to calculate the slope.

Once you have the slope, you can use either the slope-intercept form

$$y = mx + b$$

or **point-slope formula**

$$y - y_1 = m(x - x_1)$$

to find the equation. The point-slope formula uses the slope m and one additional point (x_1, y_1) to find the equation of a line. After you substitute the values for m and the additional point, you can isolate y to put the equation into slope-intercept form.

STEPS TO FIND THE EQUATION OF A LINE USING THE POINT-SLOPE FORMULA $y - y_1 = m(x - x_1)$

1. Use any two points to calculate the slope.
2. Substitute in the slope and a point into the point-slope formula.
3. Write the equation in slope-intercept form.
4. Check your equation by plugging in the points to make sure they are solutions.

STEPS TO FIND THE EQUATION OF A LINE USING THE SLOPE-INTERCEPT FORM $y = mx + b$

1. Use any two points to calculate the slope.
2. Substitute in the slope and a point to find the value of b.
3. Write the equation in slope-intercept form.
4. Check your equation by plugging in the points to make sure they are solutions.

Say What?

WHAT IS DEPRECIATION?

Depreciation is a term used in accounting to describe how to calculate the way an asset like a car loses value over time.

Straight line depreciation means that the value decreases the same amount each year, and thus the value of the asset can be calculated using a linear equation.

EXAMPLE 1 FINDING AN EQUATION FROM AN APPLICATION

A business purchased a car in 2008 for $35,950. For tax purposes, the value of the car in 2011 was $20,550. If the business is using straight line depreciation, find the equation of the line that gives the value of the car based on the age of the car in years.

Solution

First we need to define the variables that we will use. We are discussing the value of the car for tax purposes and its age in years, so we can define the variables as follows:

a = The age of the car in years.

V = The value of the car, for tax purposes, in dollars.

Now the question has given us two examples of the value of the car: In 2008, the car was new, or 0 years old and worth $35,950, and in 2011, the car was 3 years old and worth $20,550. These two quantities represent the points

$$(0, 35950) \qquad (3, 20550)$$

Using these two points, we can calculate the slope of the equation.

$$m = \frac{y_2 - y_1}{x_2 - x_1}$$

$$m = \frac{20550 - 35950}{3 - 0}$$

$$m = \frac{-15400}{3}$$

$$m \approx -5133.33$$

The first point we were given is actually the vertical intercept, so we have the value of b. Therefore, our equation is

$$V = -5133.33a + 35950$$

We can check this equation by using both points in the equation and verifying that they are solutions.

$$V = -5133.33a + 35950$$
$$(0, 35950)$$
$$35950 = -5133.33(0) + 35950$$
$$35950 = 35950$$

$$(3, 20550)$$
$$20550 = -5133.33(3) + 35950$$
$$20550 \approx 20550.01 \qquad \text{Not exact because of rounding.}$$

Both points work, so we have the correct equation.

EXAMPLE [2] FINDING AN EQUATION FROM AN APPLICATION

Compassion International is a charity that helps children all over the world who live in poverty. In 2002, they had donations of about \$123.0 million, and in 2005, they had donations of about \$215.8 million. Assume that Compassion International's donations are growing at a constant rate, and find an equation to represent this situation.

Source: www.CharityNavigator.org

Solution

Since we are told the donations are growing at a constant rate, we know that we need to find a linear equation. We are given two examples of the donations in different years, so we can use them as our two points. First, we need to define the variables we want to use. Let

t = the years since 2000.

D = the donations in millions of dollars that Compassion International receives during a year.

With these definitions, we get the two points (2, 123.0) and (5, 215.8). Now we need to find a slope and vertical intercept to write the equation.

$$m = \frac{215.8 - 123.0}{5 - 2} = \frac{92.8}{3} \approx 30.93$$

Use the two points to find the slope. Round to one extra decimal place.

$$D = 30.93t + b$$

$$123.0 = 30.93(2) + b$$

$$123.0 = 61.86 + b$$

$$\underline{-61.86 \quad -61.86}$$

$$61.14 = b$$

Substitute the slope into m and use a given point to help find b.

$$D = 30.93t + 61.14$$

Substitute the slope and the vertical intercept into the final equation.

$$123.0 = 30.93(2) + 61.14$$

$$123.0 = 123.0$$

Check that both the given points satisfy the equation you found.

$$215.8 = 30.93(5) + 61.14$$

$$215.8 \approx 215.79$$

Therefore the equation $D = 30.93t + 61.14$ represents the donations D in millions of dollars received by Compassion International t years since 2000.

EXAMPLE (2) PRACTICE PROBLEM

The number of sports cards hobby stores has been declining steadily. In 1995, there were 4,500 stores, and in 2005, there were only 1,500. Find an equation for the line that gives the number of sports cards hobby stores given the number of years since 1995.

Source: Today's Local News July 7, 2007.

EXAMPLE **3** FINDING THE EQUATION OF A LINE

a. Use $y = mx + b$ to find the equation of the line that passes through the points $(2, 5)$ and $(4, 9)$.

b. Use the point-slope formula, $y - y_1 = m(x - x_1)$, to find the equation of the line that passes through the points $(3, 1)$ and $(7, -9)$.

Solution

a. First, we can find the slope using the two given points.

$$m = \frac{9 - 5}{4 - 2} = \frac{4}{2} = 2$$

Now, using the slope and one of the points, we can substitute them into the slope-intercept form of a line and find b.

$$m = 2$$
$$\text{point} = (2, 5)$$
$$y = 2x + b$$
$$5 = 2(2) + b$$
$$5 = 4 + b$$
$$1 = b$$

We can substitute the slope and the y-intercept into the slope-intercept form, and we will have the equation of the line.

$$y = 2x + 1$$

We can check our equation by substituting in the points we were given and confirming that they are solutions to the equation.

$$y = 2x + 1$$
$$5 = 2(2) + 1 \qquad \text{The point (2, 5) is a solution.}$$
$$5 = 5$$
$$9 = 2(4) + 1 \qquad \text{The point (4, 9) is a solution.}$$
$$9 = 9$$

b. First, we can find the slope using the two given points.

$$m = \frac{-9 - 1}{7 - 3} = \frac{-10}{4} = -2.5$$

Now, using the point-slope formula with the slope and one of the points we get:

$$m = -2.5$$
$$\text{point} = (3, \ 1)$$
$$y - 1 = -2.5(x - 3)$$
$$y - 1 = -2.5x + 7.5$$
$$\underline{+1 \qquad\qquad\quad +1}$$
$$y = -2.5x + 8.5$$

We can check our equation by substituting in the points we were given and confirming that they are solutions to the equation.

$$y = -2.5x + 8.5$$

$$1 = -2.5(3) + 8.5 \qquad \text{The point (3, 1) is a solution.}$$

$$1 = 1$$

$$-9 = -2.5(7) + 8.5 \qquad \text{The point (7, -9) is a solution.}$$

$$-9 = -9$$

EXAMPLE ③ PRACTICE PROBLEM

a. Use $y = mx + b$ to find the equation of the line that passes through the points (5, 4) and (20, 7).

b. Use the point-slope formula, $y - y_1 = m(x - x_1)$, to find the equation of the line that passes through the points (1, 4) and (5, 12).

EXAMPLE ④ FINDING THE EQUATION OF A LINE

a. Find the equation of the line that passes through the points in the table.

x	y
2	7
4	15
6	23
8	31

b. Find the equation of the line shown in the graph.

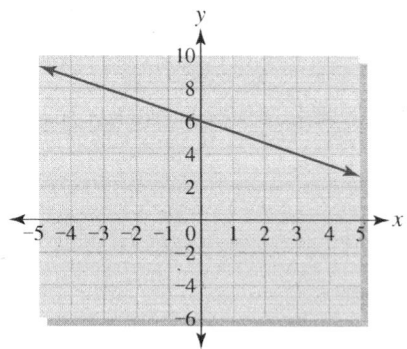

Solution

a. First we can find the slope of the line from the table and then use the slope and one of the points listed in the table in the point-slope formula to find our equation.

x	y	Slope
2	7	$\frac{15 - 7}{4 - 2} = \frac{8}{2} = 4$
4	15	$\frac{23 - 15}{6 - 4} = 4$
6	23	$\frac{31 - 23}{8 - 6} = 4$
8	31	

$$m = 4$$
$$\text{point} = (2, 7)$$
$$y - 7 = 4(x - 2)$$

Substitute the value of the slope and a point into the point-slope formula.

$$y - 7 = 4x - 8$$
$$y = 4x - 1$$

Write the equation in slope-intercept form.

We can check this equation by substituting in a couple of points from the table into the equation.

$$y = 4x - 1$$
$$15 = 4(4) - 1 \qquad \text{The point (4, 15) is a solution.}$$
$$15 = 15$$
$$23 = 4(6) - 1 \qquad \text{The point (6, 23) is a solution.}$$
$$23 = 23$$

b. Using the graph, we can estimate that the points $(-3, 8)$ and $(3, 4)$ are on the line.

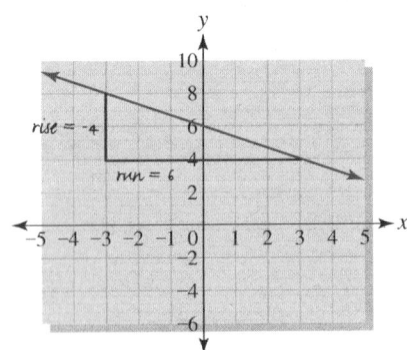

With a rise of -4 and a run of 6 we have a slope of
$$m = \frac{-4}{6} = -\frac{2}{3}$$

Using this slope and estimating the y-intercept to be $(0, 6)$, we get the equation .

$$y = -\frac{2}{3}x + 6$$

The graph is decreasing from left to right, so the slope should be negative, and the y-intercept seems to be $(0, 6)$, so we should have the correct equation for the line.

EXAMPLE ④ PRACTICE PROBLEM

a. Find the equation of the line that passes through the points in the table.

x	y
1	6
4	1
7	−4
10	−9

PARALLEL AND PERPENDICULAR LINES

When looking at two lines, we notice that a couple of special relationships occur. We say that two lines are parallel if they never cross. Two lines are said to be perpendicular if they cross at a right angle (90°).

Parallel lines must go in the same direction, and thus have the same slope, or they would eventually cross.

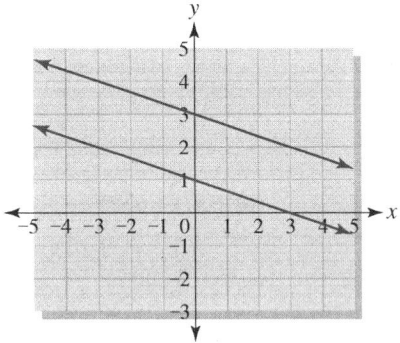

When graphing functions on the calculator, you need to be aware that the screen is a rectangle and will distort graphs by stretching them out along the horizontal axis. This will happen whenever the window is not set to the correct proportions that will account for the window's being wider than it is tall. The calculator has a ZOOM feature called ZSquare that will take any window and change it to the correct proportions to display a graph correctly.

The following two screens show the same pair of perpendicular lines (lines that cross at a 90 degree angle).

Standard window:

X min −10, X max 10
Y min −10, Y max 10

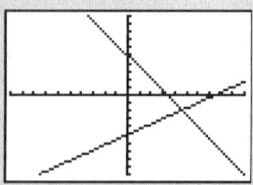

The standard window does not show these lines as perpendicular.

Window after ZSquare:

X min −15.161, X max 15.161
Y min −10, Y max 10

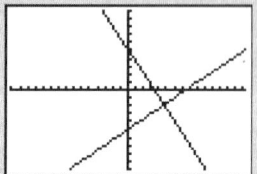

The square window does show these lines correctly as perpendicular.

Perpendicular lines form a right angle at the point where they intersect. For lines to meet at a right angle, their slopes must be opposite reciprocals of one another. For example, if the slope of one line is $m = \frac{2}{3}$ then the slope of a line perpendicular to it must be $m = -\frac{3}{2}$.

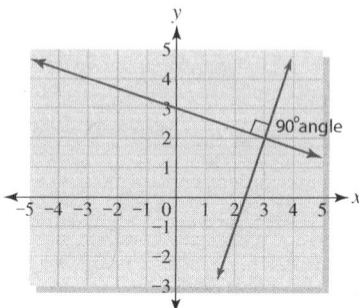

DEFINITION:

Parallel Lines: Two lines are parallel if their slopes are equal.
$$m_1 = m_2$$

Perpendicular Lines: Two lines are perpendicular if their slopes are negative reciprocals.
$$m_1 = -\frac{1}{m_2}$$

We can use the relationship between the slopes of parallel lines and between the slopes of perpendicular lines to help us find equations.

EXAMPLE **5** EQUATIONS OF PARALLEL AND PER-PENDICULAR LINES

a. Find the equation of the line that goes through the point (7, 10) and is parallel to the line $y = 2x + 5$.

b. Find the equation of the line that goes through the point (2, 6) and is perpendicular to the line $-3x - y = 8$.

c. Find the equation of the line that goes through the point (5, 8) and is parallel to the x-axis.

Solution

a. Since the line we want is parallel to the line $y = 2x + 5$, it will have the same slope $m = 2$. Using this slope and the given point, (7, 10), in the point-slope formula we get:
$$y - 10 = 2(x - 7)$$
$$y - 10 = 2x - 14$$
$$y = 2x - 4$$

b.   Since the line we want is perpendicular to the line $-3x - y = 8$, we need to know the slope of this line. To find the slope we will put the line into slope-intercept form.

$$-3x - y = 8$$
$$-y = 3x + 8$$
$$y = 3x - 8$$

This line has a slope of $m = 3$ and the line we want will have the opposite reciprocal slope $m = -\frac{1}{3}$. Using this slope and the given point, (2, 6), in the point-slope formula we get:

$$y - 6 = -\frac{1}{3}(x + 2)$$

$$y - 6 = -\frac{1}{3}x - \frac{2}{3}$$

$$y = -\frac{1}{3}x + \frac{16}{3}$$

c.   Since the line is parallel to the x-axis it will be a horizontal line and have a slope of zero. The line must go through the point (5, 8) so it will have the y-value 8. Therefore we get the line $y = 8$.

EXAMPLE   ⑤   PRACTICE PROBLEM

a.   Find the equation of the line that goes through the point (1, 5) and is parallel to the line $y = 4x + 1$.

b.   Find the equation of the line that goes through the point (4, −3) and is perpendicular to the line $y = -\frac{2}{5}x - 9$.

INTERPRETING THE CHARACTERISTICS OF A LINE, A REVIEW

EXAMPLE   ⑥   INTERPRETING SLOPE AND INTERCEPTS

Using the application and equation found in example 1, page 74, answer the following:

$$V = -5133.33a + 35950$$

a.   What is the slope of the equation? What does it represent in this situation?

b.   What is the vertical intercept of the equation? What does the slope represent in this situation?

c.   What is the horizontal intercept of the equation? What does it represent in this situation?

Solution

a. The slope is

$$m = -5133.33$$

$$m = \frac{-5133.33}{1}$$

$$m = \frac{\$(-5133.33)}{1 \text{ year of age}}$$

The slope means that the value of the car for tax purposes is decreasing $5133.33 per year of age.

b. The vertical intercept was given as (0, 35950) and means that when the car was new, zero years old, it was worth $35,950.

c. To find the horizontal intercept we will make $V = 0$ and solve for a.

$$0 = -5133.33a + 35950$$

$$\underline{-35950 \qquad\qquad\qquad\qquad -35950}$$

$$-35950 = -5133.33a$$

$$\frac{-35950}{-5133.33} = \frac{-5133.33a}{-5133.33}$$

$$7.003 \approx a$$

The horizontal intercept is (7.003, 0) and means that a car that is 7 years old is worth nothing for tax purposes.

EXAMPLE 7 INTERPRETING SLOPE AND INTERCEPTS

Using the application and equation found in example 2, page 75, answer the following:

a. What is the slope of the equation you found, and what does the slope mean in this context?

b. What is the vertical intercept of the equation you found, and what does that intercept mean in this context?

Solution

a. The slope of this equation is 30.93. The donations received by Compassion International are increasing by about $30.93 million per year.

b. The vertical intercept of this equation is (0, 61.14). In 2000, Compassion International had about $61.14 million in donations.

EXAMPLE 7 PRACTICE PROBLEM

Using the application and equation found in example 2 practice problem, page 75, answer the following:

a. What is the slope of the equation? What does the slope represent in this situation?

b. What is the vertical intercept of the equation? What does that intercept represent in this situation?

c. What is the horizontal intercept of the equation? What does that intercept represent in this situation?

EXAMPLE 8 PUTTING IT ALL TOGETHER IN AN APPLICATION

a. A local cellular phone company has a pay-as-you-talk plan that costs $15 per month and $0.23 per minute that you talk on the phone. Write an equation for the total monthly cost C of this plan if you talk for m minutes.

b. What is the slope of the equation you found in part a? What does it represent in this situation?

c. What is the vertical intercept of the equation you found in part a? What does it represent in this situation?

Solution

a. We are looking for a monthly cost, so we are considering only one month at a time. Since the plan costs $15 for the month, we will take this cost and add to it the cost for the minutes used. It costs $0.23 per minute so we will have a cost equation of

$$C = 0.23m + 15$$

where C is the total monthly cost in dollars when m minutes are used.

b. The slope of this equation is

$$m = 0.23 = \frac{0.23}{1} = \frac{\$0.23}{1\,\text{minute}}$$

This slope means that the total monthly cost will increase by $0.23 per minute used.

c. The equation is in slope-intercept form so the vertical intercept is (0, 15) and means that if you do not use any minutes in a month the plan will still cost you $15.

EXAMPLE 8 PRACTICE PROBLEM

In 2000 there were 709.5 thousand cases of Chlamydia infection, a sexually transmitted disease, in the United States. In 2003, there were 877.5 thousand cases of Chlamydia infection in the United States.
Source: Health, United States, 2005 by CDC.

a. Assuming the number of cases follows a linear trend, find an equation to represent this situation.

b. What is the slope of the equation you found, and what does it mean in this context?

c. What is the vertical intercept of the equation you found, and what does it mean in this context?

Skill Connection

CREATING TWO POINTS TO FIND THE EQUATION OF THE LINE

In Example 8 we used the wording to determine the slope of the equation. Another way to find this equation would be to create two points and use them to calculate the slope and find the equation.

In this situation the monthly cost would be $15 if you did not use the phone, so the point (0, 15) is on the line. If you use the phone for 1 minute, it will cost you an additional 23 cents, for a total cost of $15.23. Therefore, the point (1, 15.23) is also on the line.

Using these two points, you can calculate the slope and find the equation.

1.5 Exercises

For Exercises 1 through 14, find the equation of the line shown in the given graph.

1.

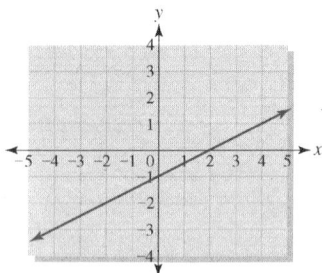

2.

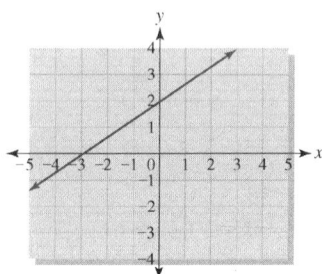

3.

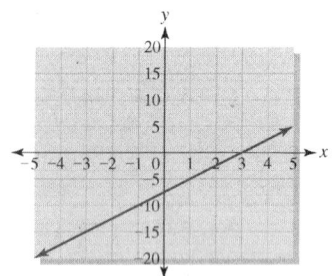

4.

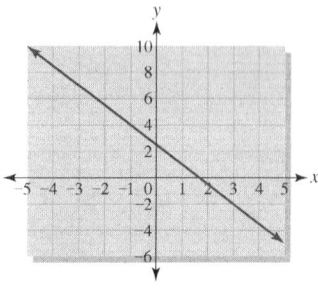

5.

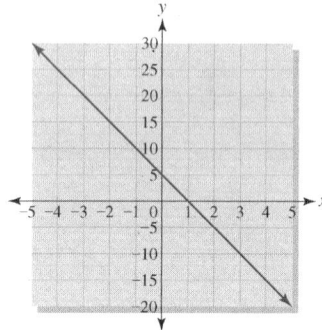

6.

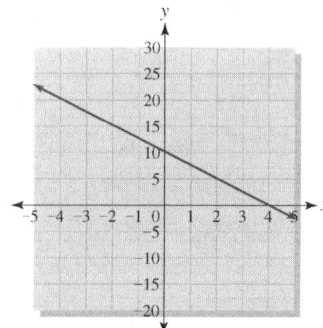

7.

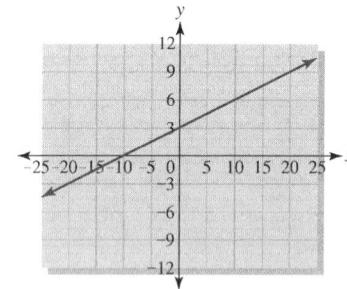

8.

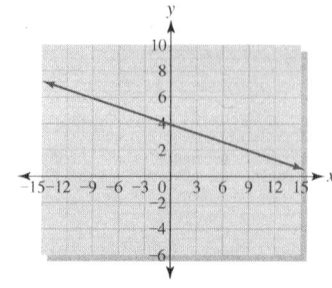

9.

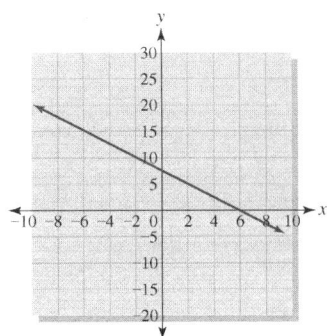

10.

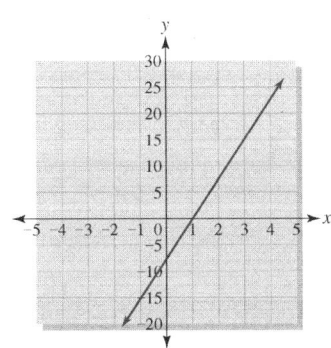

11.

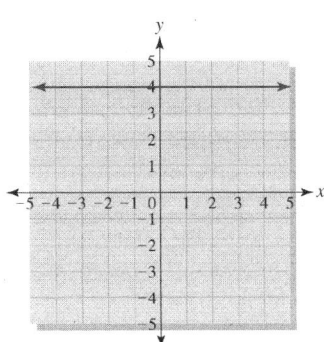

12.

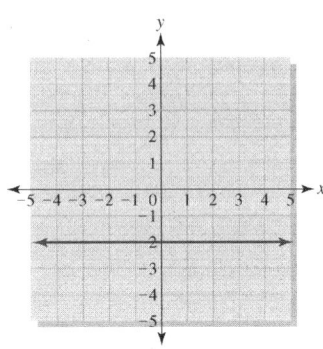

13.

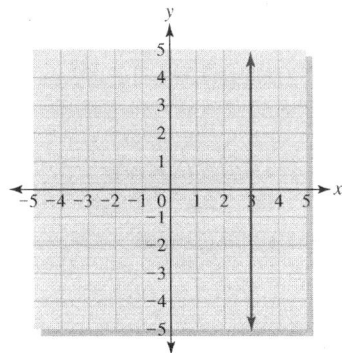

14.

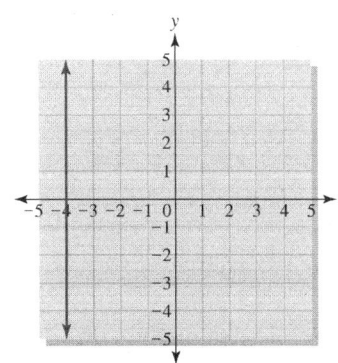

15. If 10 shirts cost $110 and 30 shirts cost $280 find a linear equation that gives the cost for *n* shirts.

16. If 15 sports jerseys cost $360 and 25 jerseys cost $500 find a linear equation that gives the cost for *j* jerseys.

17. The U.S. Census Bureau projects the populations of each state through 2030. They project the population of the state of Washington to be 6951 thousand in 2015 and 7996 thousand in 2025. Assuming the population growth is linear, find an equation that gives the population of Washington *t* years since 2000.

Source: Statistical Abstract 2007.

18. The U.S. Census Bureau projects the population of Nevada to be 2691 thousand in 2010 and 3863 thousand in 2025. Assuming the population growth is linear find an equation that gives the population of Washington *t* years since 2000.

Source: Statistical Abstract 2007.

19. According to Kristine Clark Ph.D., R.D. director of sports nutrition and assistant professor of nutritional sciences at Penn State University, a women's optimal weight is 100 pounds at 5 feet and 130 pounds at 5.5 feet. Use this information to find a linear model for a women's optimal weight in pounds based on the number of inches her height is above 5 feet.

Source: Natural Health March 2006.

20. According to the same Kristine Clark Ph.D., R.D., the optimal weight for a 5 foot man is 106 pounds and 142 pounds for a 5.5 foot man. Use this information to find a linear model for a man's optimal weight in pounds based on the number of inches above 5 feet.

Source: Natural Health March 2006.

21. In 2001, 408 teenagers underwent gastric bypass surgery. In 2003, 771 teenagers underwent gastric bypass surgery. Assuming the number of teens having gastric bypass surgery is growing at a constant rate, find an equation for the number of teenagers having gastric bypass surgery t years since 2000.

Source: Associated Press March 2007.

22. The number of hospitals in the United States has been declining for many years. In 2001 there were 5,801, but in 2004, there were only 5,759. Assuming this trend is linear, find an equation for the number of hospitals in the U.S. t years since 2000.

Source: Statistical Abstract 2007.

For Exercises 23 through 36, find the equation of the line using the given information. Write the equation in slope-intercept from.

23. Find the equation of the line that passes through the points (1, 3) and (4, 12).

24. Find the equation of the line that passes through the points (2, 4) and (7, 24).

25. Find the equation of the line that passes through the points (0, 7) and (8, 16).

26. Find the equation of the line that passes through the points (0, −8) and (2, 3).

27. Find the equation of the line that passes through the points (7, 6) and (21, −1).

28. Find the equation of the line that passes through the points (−5, −2) and (−3, −10).

29. Find the equation of the line that passes through the points (3, 5) and (7, 14).

30. Find the equation of the line that passes through the points (−2, 4) and (5, 18).

31. Find the equation of the line that passes through the points (−4, −5) and (−1, 7).

32. Find the equation of the line that passes through the points (−8, −3) and (−12, −15).

33. Find the equation of the line that passes through the points (7, −3) and (7, 9).

34. Find the equation of the line that passes through the points (4, 5) and (4, 12).

35. Find the equation of the line that passes through the points (2, 8) and (4, 8).

36. Find the equation of the line that passes through the points (−8, −15) and (10, −15).

For exercises 37 through 42 find the equation of the line passing through the points given in the table.

37.

x	y
0	7
2	11
4	15
6	19

38.

x	y
−4	−18
0	−4
4	10
8	24

39.

x	y
−2	29
0	15
2	1
4	−13

40.

x	y
-3	14.25
0	9
3	3.75
6	-1.5

41.

x	y
2	-18
5	-6
8	6
11	18

42.

x	y
4	7
9	-5.5
14	-18
19	-30.5

For Exercises 43 through 54 determine without graphing if the two given lines are parallel, perpendicular or neither. Check your answer by graphing both lines.

43. $y = 3x + 5$
$y = 3x - 7$

44. $y = 4x + 8$
$y = 4x - 3$

45. $2x + 3y = 15$
$y = \frac{3}{2}x + 4$

46. $y = 2x + 5$
$y = \frac{1}{2}x + 8$

47. $2x - 5y = 40$
$-4y = 10x + 10$

48. $y = 0.25x - 9$
$x = 4y + 3$

49. $4x - 3y = 20$
$12x - 9y = 30$

50. $y + 7 = 4(x + 3) - x$
$y = -\frac{1}{3}x + 2$

51. $3x + 2y = 8$
$-6y = 9x + 12$

52. $2x + 5y = 20$
$10x - 4y = 20$

53. $-5x + y = 7$
$2y = 10x - 9$

54. $0.5x + y = 1.75$
$4y = -2x + 7$

55. Find the equation of the line that goes through the point $(2, 8)$ and is parallel to the line $y = 4x - 13$.

56. Find the equation of the line that goes through the point $(4, 6)$ and is parallel to the line $y = -3x + 24$.

57. Find the equation of the line that goes through the point $(-6, 8)$ and is parallel to the line $y = \frac{2}{3}x + \frac{4}{5}$.

58. Find the equation of the line that goes through the point $(-4, -9)$ and is parallel to the line $y = \frac{1}{2}x - \frac{1}{4}$.

59. Find the equation of the line that goes through the point $(1, 7)$ and is perpendicular to the line $y = 2x - 1$.

60. Find the equation of the line that goes through the point $(6, 5)$ and is perpendicular to the line $y = 3x + 8$.

61. Find the equation of the line that goes through the point $(5, 1)$ and is perpendicular to the line $y = -4x + 5$.

62. Find the equation of the line that goes through the point $(-4, -3)$ and is perpendicular to the line $y = -6x - 11$.

63. Find the equation of the line that goes through the point (2, 3) and is perpendicular to the line $y = \frac{1}{5}x - 8$.

64. Find the equation of the line that goes through the point (−9, 3) and is perpendicular to the line $y = -\frac{3}{7}x - 14$.

65. The population of Washington state can be estimated by the equation $P = 75.8t + 5906.2$, where P represents the population of Washington in thousands of people t years since 2000.

Source: Model derived from data found in the US Statistical Abstract 2006.

  a. Find the slope of the equation and give its meaning in the context of this situation.

  b. Find the vertical intercept for the equation and give its meaning in the context.

  c. Find the horizontal intercept for the equation and give its meaning in the context.

66. The population of New York state can be estimated by the equation $P = 62.6t + 19005.4$, where P represents the population of New York in thousands of people t years since 2000.

Source: Model derived from data found in the US Statistical Abstract 2006.

  a. Find the slope of the equation and give its meaning in the context.

  b. Find the vertical intercept for the equation and give its meaning in the context.

  c. Find the horizontal intercept for the equation and give its meaning in the context.

67. The annual amount of public expenditure on medical research in the United States can be estimated by the equation $M = 4.02t + 25.8$, where M represents billions of dollars of public money spent on medical research t years since 2000.

Source: Model derived from data found in the US Statistical Abstract 2006.

  a. Find the slope of the equation and give its meaning in the context.

  b. Find the vertical intercept for the equation and give its meaning in the context.

  c. Find the horizontal intercept for the equation and give its meaning in the context.

68. The amount of wind energy produced in Alaska can be estimated by the equation $W = 0.225t + 0.45$, where W represents the megawatts of wind energy produced in Alaska t years since 2000.

Source: American Wind Energy Association. (AWEA.org)

  a. Find the slope of the equation and give its meaning in the context.

  b. Find the vertical intercept for the equation and give its meaning in the context.

  c. Find the horizontal intercept for the equation and give its meaning in the context.

69. The monthly profit for a small used car lot can be estimated by the equation $P = 500c - 6000$, where P represents the profit in dollars for this small used car lot when c cars are sold in a month.

  a. Find the slope of the equation and give its meaning in the context.

  b. Find the vertical intercept for the equation and give its meaning in the context.

  c. Find the horizontal intercept for the equation and give its meaning in the context.

70. The monthly profit for a coffee shop can be estimated by the equation $P = 0.75c - 3500$, where P represents the profit in dollars for this coffee shop when c customers visit in a month.

  a. Find the slope of the equation and give its meaning in the context.

  b. Find the vertical intercept for the equation and give its meaning in the context.

  c. Find the horizontal intercept for the equation and give its meaning in the context.

71. Dan gives surfing lessons over the summer and earns $30 for each 1 hour lesson given. His surfboards and supplies for the summer cost him $700.

  a. Write an equation for the profit in dollars that Dan makes from giving s surf lessons.

  b. Find the slope of the equation in part a and give its meaning in the context.

  c. Find the vertical intercept of the equation and give its meaning in the context.

d. Find the horizontal intercept of the equation and give its meaning in the context.

72. Janell tutors math students and earns $20 an hour for each session. She spends about $100 a month on transportation and other costs.

   a. Write an equation for the monthly profit in dollars that Janell makes from tutoring for h hours a month.

   b. Find the slope of the equation in part a and give its meaning in the context.

   c. Find the vertical intercept of the equation and give its meaning in the context.

   d. Find the horizontal intercept of the equation and give its meaning in the context.

73. The Parent Teacher Association (PTA) at Mission Meadows Elementary School is starting a recycling program to help raise money for a new running track to be installed on the campus. For each pound of aluminum cans recycled, they earn $1.24. The PTA started their fund raising with $2000 donated from the parents.

   a. Find an equation for the total amount the PTA has raised for the track depending on how many pounds of aluminum cans they recycle.

   b. What is the slope of the equation you found? Give its meaning in the context of the situation.

   c. Find the vertical intercept of the equation and give its meaning in the context of this situation.

   d. Find the horizontal intercept of the equation and give its meaning in the context of this situation.

74. Through the year 2050, the population of the United States is expected to grow by about 2.8 million people per year. In 2006, the population of the United States was estimated to be 298.4 million.

   Source: United States Census Bureau

   a. Find an equation to represent the population of the United States t years since 2000.

   b. Use the equation, you found in part a to estimate the population of the United States in 2020.

   c. When will the population of the United States reach 400 million?

   d. What is the vertical intercept of the equation you found and what does it mean in this context?

75. The percentage of Americans who have been diagnosed with diabetes has been growing steadily over the years. In 2002, 4.8% of Americans had been diagnosed with diabetes. In 2004, 5.1% of Americans had been diagnosed with diabetes.

   Source: CDC Diabetes Program

   a. Assuming that the percentage of Americans diagnosed with diabetes continues to grow at a constant rate, find an equation to represent this situation.

   b. Use the equation that you found in part a to estimate the percentage of Americans who will have been diagnosed with diabetes in 2010.

   c. What is the slope of the equation you found, and what does it mean in this context?

76. A toy manufacturer finds that, if they produce 1000 toy cars an hour, 1% of the cars are defective. If production is increased to 1500 toys an hour, 1.5% of the cars are defective.

   a. Assuming the percent of cars that are defective is linearly related to the number of cars produced an hour, find an equation for the percentage of cars that are defective if t toys are produced an hour.

   b. Use the equation you found in part a to find the percent of cars that are defective if 2500 cars are produced an hour.

   c. What is the slope of the equation you found, and what does it mean in the context of this situation?

77. Use the equation that you found in exercise 15 to answer the following questions.

   a. Use the equation to find how much 50 shirts will cost.

   b. What is the slope of the equation, and what does it mean in the context of this situation?

   c. Find the vertical intercept of the equation and give its meaning in the context of this situation.

   d. Find the horizontal intercept of the equation and give its meaning in the context of this situation.

78. Use the equation that you found in exercise 16 to answer the following questions.

   a. Use the equation to find how much 40 jerseys will cost.

b. What is the slope of the equation, and what does it mean in the context of this situation?

c. Find the vertical intercept of the equation and give its meaning in the context of this situation.

d. Find the horizontal intercept of the equation and give its meaning in the context of this situation.

79. Use the equation that you found in exercise 17 to answer the following questions.

a. What does the equation predict the population of Washington will be in 2030?

b. What is the slope of the equation, and what does it mean in the context of this situation?

c. Find the vertical intercept of the equation and give its meaning in the context of this situation.

d. Find the horizontal intercept of the equation and give its meaning in the context of this situation.

80. Use the equation that you found in exercise 18 to answer the following questions.

a. What does the equation predict the population of Nevada will be in 2030?

b. What is the slope of the equation, and what does it mean in the context of this situation?

c. Find the vertical intercept of the equation and give its meaning in the context of this situation.

d. Find the horizontal intercept of the equation and give its meaning in the context of this situation.

81. Use the equation that you found in exercise 19 to answer the following questions.

a. What does the equation give as the optimal weight of a woman who is 6 feet tall?

b. What is the slope of the equation, and what does it mean in the context of this situation?

c. Find the vertical intercept of the equation and give its meaning in the context of this situation.

d. Find the horizontal intercept of the equation and give its meaning in the context of this situation.

82. Use the equation that you found in exercise 20 to answer the following questions.

a. What does the equation give as the optimal weight of a man that is 6 feet tall?

b. What is the slope of the equation, and what does it mean in the context of this situation?

c. Find the vertical intercept of the equation and give its meaning in the context of this situation.

d. Find the horizontal intercept of the equation and give its meaning in the context of this situation.

83. Use the equation that you found in exercise 21 to answer the following questions.

a. What does the equation predict as the number of teenagers who had gastric bypass surgery in 2005?

b. What is the slope of the equation, and what does it mean in the context of this situation?

c. Find the vertical intercept of the equation and give its meaning in the context of this situation.

d. Find the horizontal intercept of the equation and give its meaning in the context of this situation.

84. Use the equation that you found in exercise 22 to answer the following questions.

a. What does the equation predict as the number of hospitals in the U.S. in 2010?

b. What is the slope of the equation, and what does it mean in the context of this situation?

c. Find the vertical intercept of the equation and give its meaning in the context of this situation.

d. Find the horizontal intercept of the equation and give its meaning in the context of this situation.

SECTION

1.6

Finding Linear Models

Creating scatterplots by hand can be very tedious, slow, and inaccurate. Using technology to help us create scatterplots and investigate models is a great help. In this section, we will discuss how to create scatterplots on a Texas Instruments graphing calculator. We will also find algebraic linear models using the data and these scatterplots. Algebraic models are equations that can represent the graphical model we learned about in Section 1.2. These models can be used more accurately and often more easily than a graphical model. In this text, the word *model* will refer to an algebraic model unless otherwise stated.

USING A CALCULATOR TO CREATE SCATTERPLOTS

After you have defined the variables and adjusted the data, you are ready to create a scatterplot. To create a scatterplot, or statplot, on the calculator you need to accomplish the following tasks.

i.   Clear all equations in Y= screen.

ii. Input the adjusted data into the calculator.

iii. Set a window (tell the calculator where to look in order to see the data).

iv. Set up the stat plot and graph it.

EXAMPLE   **1**   CREATING A SCATTERPLOT ON THE CALCULATOR

Create a scatterplot for the population data for Texas in various years given in the chart.

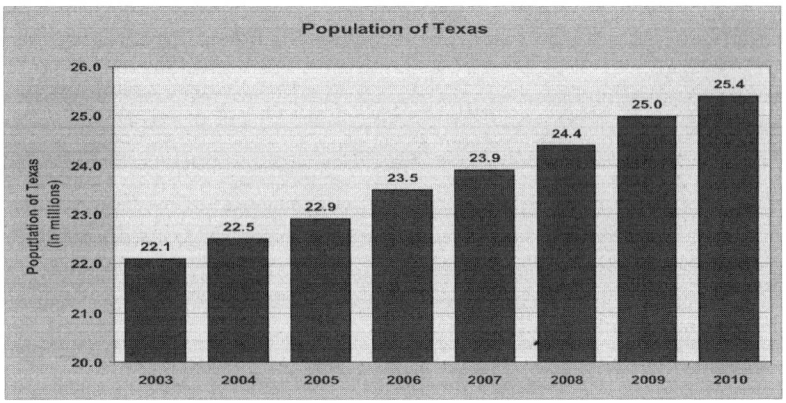

Source: Estimates and projections from the Texas State Data Center.

Solution

First, we will define the variables as follows.

t = Years since 2000.

P = The population of Texas (in millions).

The adjusted data are shown below.

t	Population of Texas (in millions)
3	22.1
4	22.5
5	22.9
6	23.5
7	23.9
8	24.4
9	25.0
10	25.4

Using Your TI Graphing Calculator

When putting sets of data into the calculator we will use the lists in the STAT menu. To get to the Lists:

[STAT], [1](Edit)

To clear old data from lists:

Using the arrow buttons, move the cursor up to the title of the list (such as L1) and press [CLEAR],[ENTER].

Important note:
Do not try to clear a list using the [DEL] button. This will erase the entire list. If you do erase a list or if all of your lists are not available, go to

[STAT], [5](SetUpEditor), [ENTER], [ENTER]

When you return to the stat list screen, all the original lists will be back.

To input these data into the calculator, you push the [STAT] button and then [1] (EDIT) for the edit screen. You should get the following two screens.

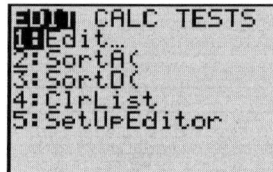 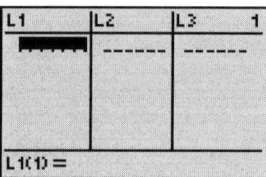

The second screen shows you the first three lists (L1, L2, and L3) in which we can input data. You will see a black cursor in the first list ready for your first input value. Now you just need to enter your first input value and then press [ENTER]. Repeat for all the input data and then use the right arrow button to move over to list 2 and do this again for the output data.

Now that the data have been entered, the window (the graphing area that the calculator is going to show you) needs to be set. Press the [WINDOW] button at the top of the calculator and enter Xmin, Xmax, Ymin, and Ymax values. The best way to choose these values is as follows.

- Xmin = A slightly smaller value than your smallest input.
- Xmax = A slightly larger value than your largest input.
- Ymin = A slightly smaller value than your smallest output.
- Ymax = A slightly larger value than your largest output.

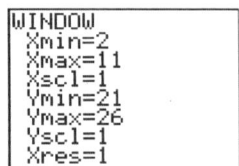

While there are many windows that will show you your data, you just need to be sure that the min and max are including your data values. Note that Xscl, Yscl, and Xres can all stay 1, since these settings will not affect the scatterplot. Now that the window is fixed, we need to set up the statplot and graph it. To set up the statplot, we need to press the [STAT PLOT] button, which is actually in yellow or blue above the [Y=] button. To activate a yellow or blue feature, you need to first press the yellow or blue [2nd] button on the left side of the calculator. If you press [2nd] and then [STAT PLOT], you get the following screen.

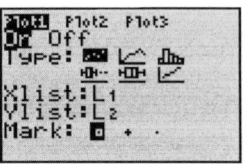

On the STAT PLOT screen, you will want to select the plot you want to work with. You can either highlight the plot you want and press [ENTER] or press the number of the plot. By pressing [1], you will get the following plot setup screen.

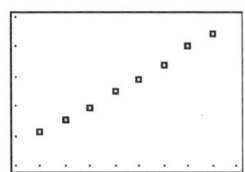

On the plot setup screen, you will want to use the arrow buttons to move around. First, turn the plot on by moving the cursor over the word **On** and pressing [ENTER]. The Type of plot should be the first one: a scatterplot. The Xlist and Ylist have to have the same names for the lists into which you put your input and output data, respectively. The standard list names are available above the numbers 1 thorough 6. To enter L1 use the [2nd] button and then [1]. The Mark that you use should be squares or plus signs. The dot is too hard to read. Once your screen is set like the one above, press the [GRAPH] button on the top right corner of the calculator, and you should get a scatterplot like the following.

This entire process will get easier with practice and will be a vital part of many problems throughout this textbook.

Using Your TI Graphing Calculator

Common Error Messages:

ERR:DIM MISMATCH

This error happens when the numbers of data in your lists are not the same. Go back to the stat edit screen and check your data. This can also happen if your statplot is set up to use the wrong lists.

ERR: WINDOW RANGE

This error happens when your window is not set correctly. Usually, your Xmin is larger than your Xmax or Ymin is larger than Ymax.

ERR: SYNTAX

This error occurs when you have entered something in the equation incorrectly. Use the goto option to see what the error is. This most often happens when a subtraction is used instead of the negative sign. The negative sign is to the left of the enter button.

EXAMPLE PRACTICE PROBLEM

Use a graphing calculator to create a scatterplot of the following set of data.

Year	Population of Arizona (in thousands)
2001	5,296
2002	5,439
2003	5,578
2004	5,740
2005	5,939
2006	6,239
2007	6,432
2008	6,623
2009	6,812
2010	7,000

Year	Number of Deaths
1994	3532
1995	3262
1996	2894
1997	2601
1998	2283
1999	2093

Source: Centers for Disease Control and Prevention.

Year	Population of Texas (in millions)
2003	22.1
2004	22.5
2005	22.9
2006	23.5
2007	23.9
2008	24.4
2009	25.0
2010	25.4

Source: Estimates and projections from the Texas State Data Center.

FINDING LINEAR MODELS

Because we are creating scatterplots with the calculator, we cannot draw by hand an "eyeball best fit" line on the calculator screen, so we need another method to create a model from our data. In Section 1.3, we learned about the slope-intercept form of a line and how to find the equation of a line. In perfect linear relationships, the data will have an equal amount of change between them. In real-life data, this change should still be somewhat equal but will typically have some variation in it. The number of homicides per year was decreasing by about 300 deaths per year. The population of Texas was increasing by about 0.4 million people per year. These amounts of change are somewhat consistent throughout the data. Although linearly related data will have a fairly consistent amount of change for each equal amount of change in inputs, you will want to look at the scatterplot to pick two points that will make a good "eyeball best fit" line. We will combine the steps to find an equation of a line from Section 1.5 with the scatterplots to find a linear model and check its reasonableness.

MODELING STEPS

1. Define the variables and adjust the data (if needed).
2. Create a scatterplot.
3. Select a model type.
4. **Linear Model:** Pick two points and find the equation of the line.
5. Write the equation of the model.
6. Check your model by graphing it with the scatterplot.

t	Population of Texas (in millions)
2003	22.1
2004	22.5
2005	22.9
2006	23.5
2007	23.9
2008	24.4
2009	25.0
2010	25.4

EXAMPLE 2 FINDING A LINEAR MODEL

Find a model for the population of Texas data in Example 1.

Solution

Step 1 Define the variables and adjust the data (if needed).

We will define the variables the same way as in example 1, so that we have:

t = Years since 2000.

P = The population of Texas (in millions).

The adjusted data are shown below.

t	Population of Texas (in millions)
3	22.1
4	22.5
5	22.9
6	23.5
7	23.9
8	24.4
9	25.0
10	25.4

Step 2 Create a scatterplot.

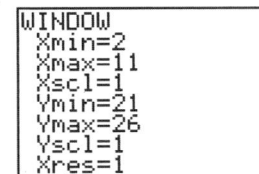 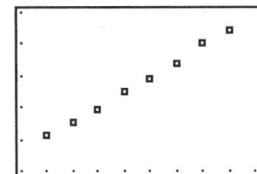

Step 3 Select a model type.

These data seem to be linearly related, so we will find a linear model.

Step 4 Linear Model: Pick two points and find the equation of the line.

Using a clear ruler, we can choose the second point and the last point because they seem to line up well with the rest of the data. First, we will calculate the slope.

$$(4, 22.5) \quad \text{and} \quad (10, 25.4)$$

$$m = \frac{25.4 - 22.5}{10 - 4}$$

$$m = \frac{2.9}{6}$$

$$m \approx 0.483$$

The data points given in the table have 1 decimal place so we should round to 2 or more decimal places. In this case, we are rounding to 2 additional decimal places.

This time we will substitute the slope and one of the points into the point-slope formula and find the equation of the line. We could also use $y = mx + b$ as we did in example 1.

$$P - 22.5 = 0.483(t - 4)$$ Substitute the slope and
$$P - 22.5 = 0.483t - 1.932$$ a point.

$$\underline{ + 22.5 \qquad\qquad + 22.5}$$ Isolate P to put the equation
$$P = 0.483t + 20.568$$ into slope-intercept form.

Step 5 Write the equation of the model.

The model is:

$$P = 0.483t + 20.568$$

Step 6 Check your model by graphing it with the scatterplot on the calculator. To check your model, you will want to enter the model into the Y= screen and graph it with the data. To do this, press the [Y=] button and enter the equation using the [X, T, θ, n] button for the input variable. After you enter the equation, press the [GRAPH] button to view the graph.

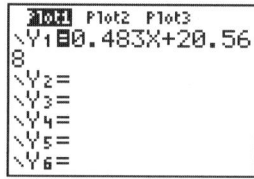 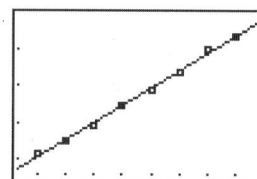

This equation gives us a line that comes close to all the points in the data and looks like a good fit, with the points that are not on the model well balanced on the top and bottom of the line. If the line did not fit well, we could adjust the slope or vertical intercept to try to get a better fit.

EXAMPLE   2   PRACTICE PROBLEM

Find a model for the population of Arizona data from example 1 practice problem.

Year	Population of Arizona (in thousands)
2001	5,296
2002	5,439
2003	5,578
2004	5,740
2005	5,939
2006	6,239
2007	6,432
2008	6,623
2009	6,812
2010	7,000

Once we have found a model, it can be used to extrapolate information about the situation, and the characteristics of the line can be interpreted just as we did for equations in previous sections. We can also use the equation to find the domain and range of the model instead of estimating values from the graph. As with the graphical models we found in Section 1.2, we will determine a domain based on the given data, being careful to avoid model breakdown. The range will then be calculated using the model by substituting the inputs that will give us the lowest and highest output values based on the given domain.

EXAMPLE 3 FINDING AND USING A LINEAR MODEL

The following chart gives the revenues in the Asia Pacific countries for Nike, Inc. for various years.

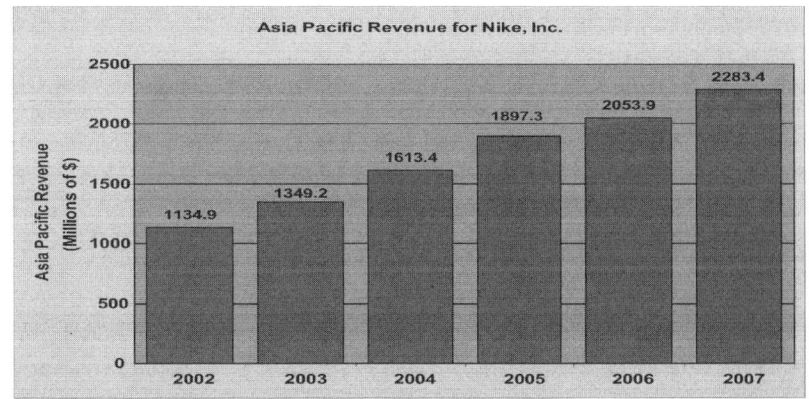

Source: Nike, Inc. Annual Reports 2004 and 2007.

a. Find a model for the data.

b. Using your model, estimate the Asia Pacific revenues for Nike in 2010.

c. What is the slope of your model, and what does it mean in this context?

d. Determine a reasonable domain and range for the model.

Solution

a.

Step 1 Define the variables and adjust the data (if needed).

t = Years since 2000.

R = The Asia Pacific revenues for Nike, Inc. (in millions of dollars).

Adjust the data

t	Asia Pacific Revenues (in millions of $)
2	1134.9
3	1349.2
4	1613.4
5	1897.3
6	2053.9
7	2283.4

Step 2 Create a scatterplot.

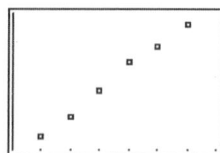

Step 3 Select a model type.

These data seem to be linearly related, so we will find a linear model.

Step 4 Linear Model: Pick two points and find the equation of the line.

Using a clear ruler, we can choose the first point and the next to last point because they seem to line up well with the rest of the data. First, we will calculate the slope.

$$(2, 1134.9) \quad \text{and} \quad (6, 2053.8)$$

$$m = \frac{2053.8 - 1134.9}{6 - 2}$$

$$m = \frac{918.9}{4}$$

$$m = 229.725$$

Substitute the slope into the slope-intercept form, and use one point to find the output value for the vertical intercept, b.

$$R = mt + b$$

$R = 229.725t + b$ **Substitute in the slope for m.**

$1134.9 = 229.725(2) + b$ **Substitute the point (2, 1134.9)**

$1134.9 = 459.45 + b$

$\underline{-459.45 \quad -459.45}$ **Subtract 459.45 from both sides.**

$675.45 = b$

Step 5 Write the equation of the model.

Using the slope and b we found in step 4, we write the equation.

$$R = 229.725t + 675.45$$

Step 6 Check your model by graphing it with the scatterplot on the calculator.

To check our model, we will put the equation into the Y= screen and graph it with the scatterplot to see how it fits the data.

 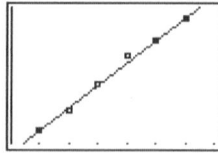

This equation gives us a line that comes close to all the points in the data and looks like a good fit, with the points that are not on the model well balanced on the top and bottom of the line. If the line did not fit well, we could adjust the slope or vertical intercept to try and get a better fit.

b. The year 2010 is represented by $t = 10$, so we substitute 10 for t and solve for R.

$$R = 229.725(10) + 675.45$$

$$R = 2972.7$$

In 2010, Nike, Inc. should have about $2972.7 million revenue from the Asia Pacific countries.

c.   The slope for this model is

$$m = \frac{229.725}{1} = \frac{\$229.725\,\text{million}}{1\,\text{year}}$$

The Asia Pacific revenues for Nike, Inc. are increasing by approximately $229.725 million per year.

d.   The input values from the data go from 2 to 7, so if we expand beyond the data a few years in both directions, we have one possible domain of [0, 10]. Now the range can be found by looking for the lowest and highest points on the model within that domain. On the graph, the lowest point within the domain occurs when $t = 0$. Since we now have an equation, we can calculate the output for this point.

$$R = 229.725(0) + 675.45$$
$$R = 675.45$$

The highest point within this domain occurs at $t = 10$. Using the equation, we get

$$R = 229.725(10) + 675.45$$
$$R = 2972.7$$

If we use these output values, the range is [675.45, 2972.7].

Many mistakes that occur in the modeling process can be avoided by checking your work along the way. When calculating the slope in step 4, be sure to think about the slope you found and consider whether it agrees with the trend that you see in your scatterplot. If your data are decreasing from left to right, the slope should be negative, and if the data are increasing from left to right, the slope should be positive.

EXAMPLE   3   PRACTICE PROBLEM

The following chart gives the total prize money given out at professional rodeo events in the United States.

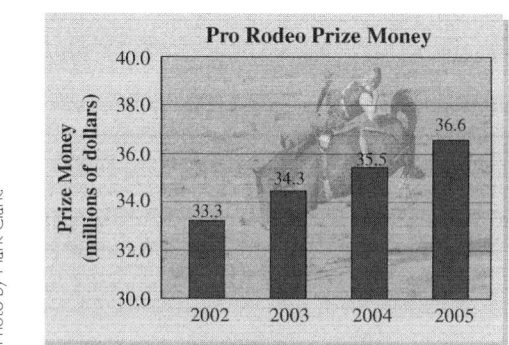

Source: Statistical Abstract 2007.

a.   Find a model for the data.

b.   Using your model, estimate the total prize money given out in 2009.

c.   What is the slope of your model, and what does it mean in this context?

d.   Determine a reasonable domain and range for this model.

1.6 Exercises

For Exercises 1 through 8, choose the window that will best display the given data in a scatterplot.

1.

x	y
2	5
4	12
10	45
12	58

a. Xmin = 2, Xmax = 12
   Ymin = 5, Ymax = 58

b. Xmin = 1, Xmax = 13
   Ymin = 0, Ymax = 60

c. Xmin = -10, Xmax = 20
   Ymin = -10, Ymax = 80

2.

x	y
8	10
16	20
20	25
35	40

a. Xmin = 0, Xmax = 60
   Ymin = 0, Ymax = 80

b. Xmin = 8, Xmax = 35
   Ymin = 10, Ymax = 40

c. Xmin = 5, Xmax = 38
   Ymin = 5, Ymax = 45

3.

x	y
-20	1.2
-17	1.5
-12	1.7
-9	1.8

a. Xmin = -8, Xmax = -21
   Ymin = 0, Ymax = 5

b. Xmin = -30, Xmax = 10
   Ymin = 1, Ymax = 5

c. Xmin = -22, Xmax = -8
   Ymin = 1, Ymax = 2

4.

x	y
-42	-3.25
-36	-3.27
-22	-3.28
-12	-3.29

a. Xmin = -45, Xmax = -10
   Ymin = -3.3, Ymax = -3.24

b. Xmin = -10, Xmax = -45
   Ymin = -2, Ymax = -3

c. Xmin = -45, Xmax = -10
   Ymin = -4, Ymax = -3

5.

x	y
100	50
115	200
130	600
135	1000

a. Xmin = 50, Xmax = 200
   Ymin = 0, Ymax = 1050

b. Xmin = 90, Xmax = 145
   Ymin = 40, Ymax = 1010

c. Xmin = 95, Xmax = 140
   Ymin = 0, Ymax = 1050

6.

x	y
0	2002
1500	2010
2800	2015
8700	2018

a. Xmin = 0, Xmax = 8700
   Ymin = 1000, Ymax = 3000

b. Xmin = -20, Xmax = 8720
   Ymin = 1950, Ymax = 2050

c. Xmin = -500, Xmax = 9500
   Ymin = 2000, Ymax = 2020

7.

x	y
2.25	120
2.26	130
2.27	140
2.28	150

a. Xmin = 2.245, Xmax = 2.285
   Ymin = 115, Ymax = 155

b. Xmin = 2.2, Xmax = 2.33
   Ymin = 100, Ymax = 175

c. Xmin = 2, Xmax = 3
   Ymin = 90, Ymax = 200

8.

x	y
-1.1	-2000
-0.9	-3000
-0.5	-4000
-0.2	-5000

a. Xmin = 0, Xmax = -2
   Ymin = -1000, Ymax = -6000

b. Xmin = -1.5, Xmax = 0
   Ymin = -5500, Ymax = -1500

c. Xmin = -1.2, Xmax = -0.1
   Ymin = -5400, Ymax = -1600

9. The worlds production of beef and pork is given in the following chart.

Drawing Adapted from Mark R. Vogel. "Where's The Beef?" Found at http//www .seekingsources.com/cuts_of_beef.htm. Reprinted with Permission from Mark R. Vogel.

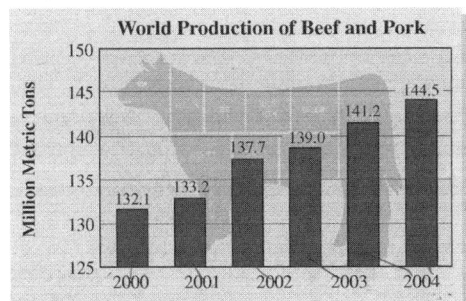

a. Find a model for these data.

b. According to your model, what was the world production of beef and pork in 2006?

c. When will the world production of beef and pork reach 175 million metric tons?

d. Give a reasonable domain and range for this model.

e. What is the slope of your model? Explain its meaning in this context.

10. The gross profit for Exxon Mobil Corporation for various years is given here.

Year	Gross Profit (in Billions of $)
2002	37
2003	47
2004	61
2005	75
2006	84

Source: http://finance.google.com

a. Find a model for these data.

b. Give a reasonable domain and range for this model.

c. According to your model, when will Exxon Mobil Corp.'s gross profit be 100 billion dollars?

d. What is the slope of your model? Explain its meaning in this context.

e. According to your model, what was Exxon Mobil Corp.'s gross profit in 2008?

11. The number of cases reported for the sexually transmitted disease Chlamydia for several years is given in the table.

Year	Reported Chlamydia Cases (in thousands)
2000	702
2001	783
2002	835
2003	877
2004	929

Source: U.S. Center for Disease Control and Prevention

a. Find a model for these data.

b. What is the slope of this model? Explain its meaning in this context.

c. Estimate the number of Chlamydia cases that were reported in 2006.

12. According to the North American Transportation Statistics Database the total number of vehicle-miles driven on the rode is given below.

Year	Vehicle Road Miles (in billions)
2000	2747 27.5
2001	5798 58
2002	2856 28.6
2003	2891 29
2004	2963 29.6
2005	2989 29.9

Source: North American Transportation Statistics Database.

a. Find a model for these data.

b. Give a reasonable domain and range for this model.

c. What were the total vehicle road miles be in 2008?

d. What is the slope of your model? Explain its meaning in this context.

13. The amount of chicken consumption by Americans is given in the following chart.

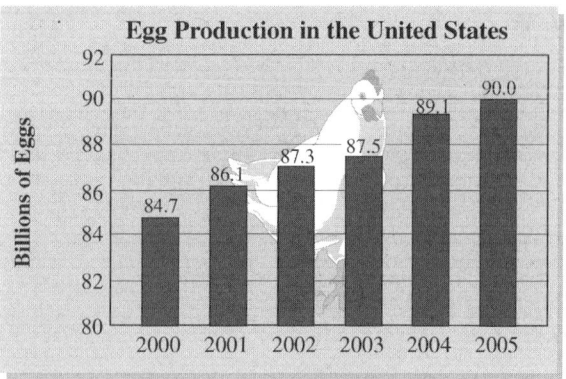

Source: Statistical Abstract 2007.

  a. Find a model for these data.

  b. According to your model, how many eggs were produced in 2008?

  c. Give a reasonable domain and range for this model.

  d. What is the slope of your model? Explain its meaning in this context.

14. The total revenue for Costco Wholesale Corporation is given here.

Year	Total Revenue (in millions of $)
2002	37,863
2003	42,546
2004	48,110
2005	52,952
2006	60,151
2007	64,400

Source: http://finance.google.com.

  a. Find a model for these data.

  b. What is the slope of this model? Explain its meaning in this context.

  c. Estimate the total revenue for Costco Wholesale Corporation in 2010.

While the Internet was growing in popularity in the late 1990's, many countries saw an increase in the number of Internet users. The number of European Internet users,

in millions, is given in the following chart. Use these data for Exercises 15 through 18.

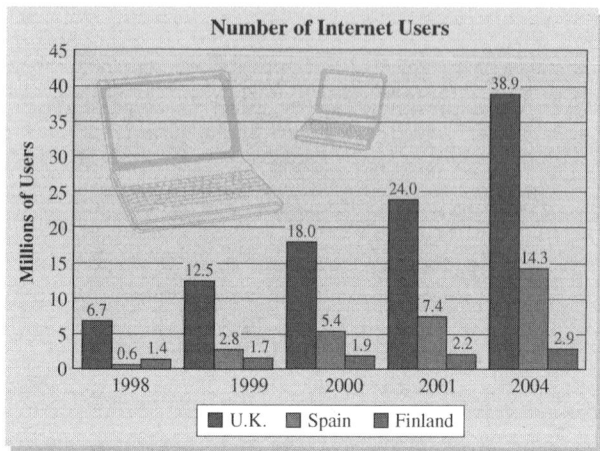

Source: NetStatistica.com and Internet worldstats.com.

15. Find models for the number of Internet users in the United Kingdom, Spain, and Finland.

16. Give a reasonable domain for the models you found in Exercise 15.

17. According to your models found in Exercise 15, which country has the fastest-growing Internet user population? Explain.

18. What is the horizontal intercept for Spain's Internet users model found in Exercise 15? Explain its meaning in this context.

19. The revenues for FedEx for several years are given in the table.

Year	Revenue (in millions of $)
2000	18,257
2001	19,629
2002	20,607
2003	22,487
2004	24,710
2005	29,363
2006	32,294
2007	35,214

Source: FedEx.com.

a. Find a model for these data.

b. Give a reasonable domain and range for this model.

c. What is the slope of your model? Explain its meaning in this context.

d. Estimate the revenue for FedEx in 2010.

20. The gross profit for United Parcel Service, Inc. has been increasing steadily for the past several years. The gross profits for UPS are given below.

Years Since 2000	Gross Profit (in millions of $)
2	9,702
3	10,401
4	11,254
5	12,807
6	13,820

Source: http://finance.google.com

a. Find a model for these data.

b. Give a reasonable domain and range for this model.

c. What is the slope of your model? Explain its meaning in this context?

d. Estimate the gross profit for UPS in 2012.

e. What is the vertical intercept for this model and what does it mean in this context?

21. The number of gallons of milk consumed in the United States is given in the following table.

Year	Milk Consumed (in millions of gallons)
2000	9297
2002	9244
2003	9192
2004	9141
2005	9119

Source: USDA/Economic research service.

a. Find a model for these data.

b. Give a reasonable domain and range for this model.

c. What is the vertical intercept for this model? Explain its meaning in this context.

d. In what year did Americans drink 9000 million gallons of milk?

e. What is the horizontal intercept for your model? Explain its meaning in this context.

22. The total operating expenses for Anheuser-Busch Companies, Inc. is given below.

Year	Total Operating Expenses (in millions of $)
2002	10,587
2003	10,947
2004	11,761
2005	12,549
2006	12,998

Source: http://finance.google.com

a. Find a model for these data.

b. Give a reasonable domain and range for this model.

c. Estimate the total operating expenses for Anheuser-Busch in 2009.

d. What is the slope of your model? Explain its meaning in this context.

23. The amount spent by individuals on health care expenses in the United States is given here.

Year	National Health Expenditures by Individuals (in billions of $)
2006	251
2007	266
2008	281*
2009	299*
2010	317*
2011	335*

*Projected
Source: Centers for Medicare and Medicaid Services.

a. Find a model for these data.

b. Give a reasonable domain and range for this model.

c. What is the slope of your model? Explain its meaning in this context.

d. Estimate the amount spent by individual Americans on health care in 2015.

e. When will the amount spent be $400 billion?

24. Quiksilver, Inc., the makers of the Quiksilver and Roxy clothing lines, has had steadily increasing profits over the past several years. The profits for Quiksilver, Inc. are given in the chart.

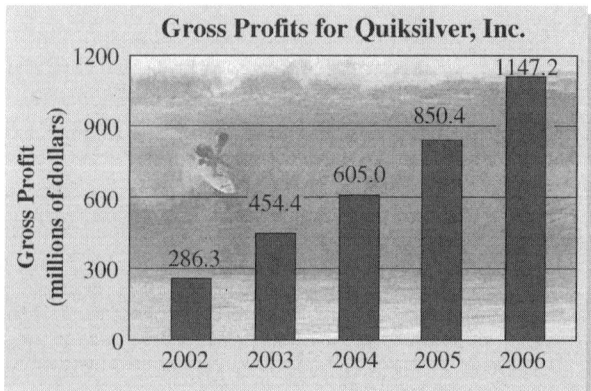

Gross Profits for Quiksilver, Inc.

Source: CBS.Marketwatch.com.

a. Find a model for these data.

b. Use your model to estimate the profit for Quiksilver, Inc. in 2009.

c. Give a reasonable domain and range for this model.

d. What is the slope for your model? Explain what this means in this context.

25. The population of the United States is given below.

Year	Population of the United States (in millions)
2000	282
2001	285
2002	288
2003	291
2004	294
2005	297
2006	299

Source: U.S. Census Bureau.

a. Find a model for these data.

b. Using your model estimate the population of the United States in 2010.

c. Give a reasonable domain and range for this model.

d. What is the vertical intercept for your model and what does it mean in this context?

e. Estimate when the U.S. population will reach 315 million.

f. Give a reasonable domain and range for this model.

26. The death rate (per 100,000 people) for motor vehicle–related injuries for males in the United States for various years is given below.

Year	Death Rate (per 100,000 people)
1999	20.87
2000	21.33
2001	21.47
2002	21.95

Source: Centers for Disease Control and Prevention.

a. Find a model for these data.

b. According to your model, when will the death rate be approximately 20 deaths per 100,000 people.

c. What is a reasonable domain and range for your model?

d. What does your model predict the death rate to be in 2005?

e. What is the vertical intercept for your model and what does it mean in this context?

Introduction to Functions

RELATIONS AND FUNCTIONS

The way that things in life are related to one another is important to understand. In mathematics, we are also concerned with how different sets are related. These relationships can be as simple as the relationship between a person and his/her height, age, or weight. We could look at the day of the week and the number of work absences at a certain company. We might want to know the names of the brothers and sisters of each student in a class. Another relationship that we might consider is the number of credits each student is taking this semester. All of these represent what we in mathematics would call a **relation.** A relation is any connection between the elements of a set of input(s) (domain) and the elements of a set of output(s) (range) and is typically represented by a set of ordered pairs or using an equation. The equation $y = 5x + 7$ represents a relation that relates the x-values with a corresponding y-value using arithmetic operations. Using this equation we can an infinite number of ordered pairs that represent the inputs and outputs of this relation.

DEFINITION

Relation: A set of ordered pairs.

Relation A: $(1, 5), (3, 7), (9, 4), (-2, 4), (3, -1)$

In a context, a relation may represent a correspondence between the elements of one set of quantities (domain) and the elements of another set of quantities (range).

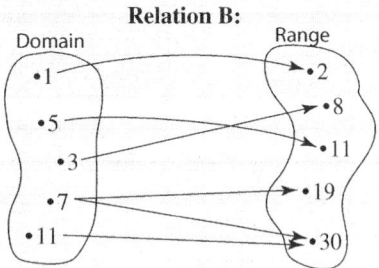

Relation B:

A special type of relation is one in which each input is paired with only one output. That is, when you put in one value, you only get out exactly one value. This type of relation is called a **function** in mathematics. In determining if a relation is a function it is important to consider whether each and every input has exactly one output associated with it. In the following example, we will determine which relations are functions and which are not.

DEFINITION

Function: A relation in which each input is related to only one output. For each input value in the domain, you must have one and only one output value in the range.

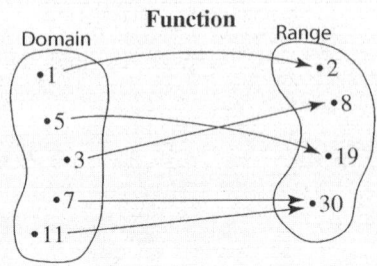

EXAMPLE 1 DETERMINE IF SETS AND WORD DESCRIPTIONS ARE FUNCTIONS

Determine whether the following descriptions of relations are functions or not. Explain your reasoning.

a. The set $S = \{(1, 3), (5, 7), (7, 9), (15, 17)\}$

b. The set $B = \{(2, 8), (2, 7), (3, 16), (4, 11)\}$

c.

Day	1	2	5	7
Height of Plant *(in cm)*	3	5	12	17

d.

Age of Student	7	8	7	6	6	9	10
Grade Level	2nd	3rd	3rd	1st	1st	4th	4th

e. The relationship between Monique's age, in days, and her height

f. The advertised prices of Sony 32-inch TVs in this Sunday's newspaper

Solution

In each part, we need to consider whether for each input value there is exactly one output.

a. The set S is a function, since each input has exactly one output value.

b. The set B is not a function, since the input 2 has two different output values.

c. This table is a function because each day has one plant height associated with it.

d. In this table, the age of the student could be associated with more than one grade level. The 7-year-olds in this table go to either second or third grade, so this relation is not a function.

e. If you consider just one age, Monique will have only one height, so this is a function.

f. Sony 50-inch TVs would be advertised for several different prices, so this is a relation but not a function.

EXAMPLE ① PRACTICE PROBLEM

Determine whether the following descriptions of relations are functions or not. Explain your reasoning.

a.

Units Produced	100	150	200	250	300	350	400
Total Cost (*in $*)	789	1565	2037	2589	3604	4568	5598

b.

First Name of Student	John	Mary	Mark	Fred	Juan	Karla	John
No. of Units	10	12	9	15	16	21	4

c. The amount of profit that a company makes each month of a year

d. The population of California each year

As you can see from Example 1, functions can be represented by sets of data or words. If you consider the data sets from the last two sections, you will see that they all represent functions, since each input value had exactly one output value. Any set of data can be considered a function if it holds to the requirement that each input has only one output associated with it. Functions can also be represented as formulas or by using graphs. Most of the equations that we will work with in this class will be functions.

When you are given an equation to consider, it may seem harder to determine whether the equation represents a function. When given an equation, look for anything that is out of the ordinary, such as a $\pm$ symbol that might result in two answers for any one input. Most equations and sets of data that we will consider in this class will be functions, but it is best to have a way of recognizing when an equation is not a function. One way to get an idea if an equation represents a function is to pick an arbitrary input and substitute it into the equation to see if one or more outputs come out. If the equation gives only one output it may be a function. You will need to consider if there are any values that you could use as inputs that would result in more than one output. If we consider the linear equation $y = 2x + 5$, any x value we substitute into this equation will give us only one output. For example $x = 3$ results in

$$y = 2(3) + 5$$
$$y = 11$$

The input $x = 3$ results in the output $y = 11$. If we picked any other x-value, it would also result in only one output. Since each input results in only one output, the equation represents a function. An equation such as $x = y^2$ would not satisfy the condition that each input have only one output.

$$4 = y^2$$

$$4 = (-2)^2 \qquad 4 = (2)^2$$

$$y = -2 \qquad y = 2$$

In this example, we can see the input $x = 4$ results in two outputs $y = 2$ and $y = -2$. Therefore, this equation is not a function.

Another way to determine whether an equation is a function or not is to look at its graph. Consider the following graph

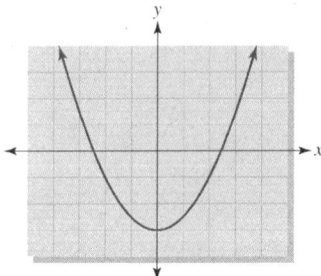

Every point that lies on this curve represents an input(x) and an output(y). Using this graph, we will determine whether each input(x) has exactly one output(y) associated with it. To determine if an input(x) has only one output(y), we pick an x-value on the horizontal axis, draw an imaginary vertical line at that location, and see how many times the vertical line crosses the curve.

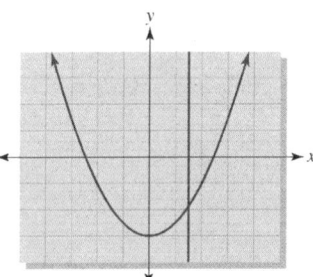

In this example, we chose the input $x = 1.5$ and drew a vertical line, which crosses the graph in only one place. If any vertical line that you draw crosses the graph only once, then you have the graph of a function. In this graph, you can see that all of the vertical lines cross the graph only once. This is the graph of a function.

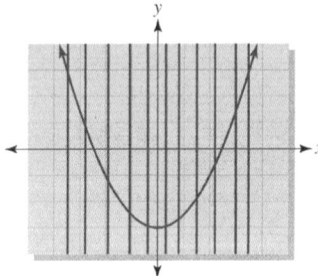

This process of testing vertical lines through a graph is called **the vertical line test.** This test can be used with most equations as long as you can graph them or put them in a graphing calculator and see its graph. When using the vertical line test on a calculator, you should be sure to get a viewing window that shows the overall characteristics of the graph, or you might incorrectly decide that the graph passes the vertical line test. Although the graphing calculator is a great tool, it can show us only what we ask it to. We can mislead ourselves if we are not thorough and careful.

DEFINITION

The vertical line test for a function: If any vertical line intersects a graph in at most one point, the graph represents a function.

An example of the vertical line test proving a curve is not a function is the graph

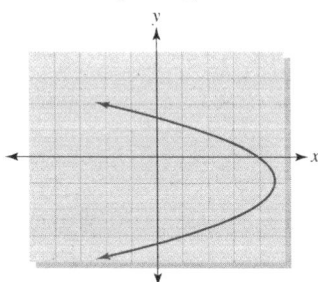

Drawing an arbitrary vertical line through the graph shows that it intersects the curve more than once. Therefore, this is not the graph of a function.

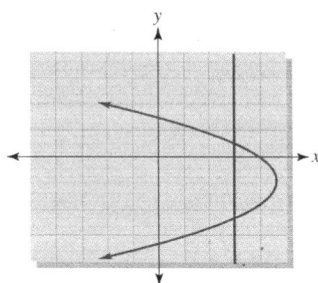

EXAMPLE 2 DETERMINE IF EQUATIONS AND GRAPHS ARE FUNCTIONS

Consider the following equations and graphs and determine whether they are functions or not.

a. $P = 2.57t + 65$

b. $W = 2g^2 + 5g - 9$

c. $y = 2x \pm (6x - 9)$

d.

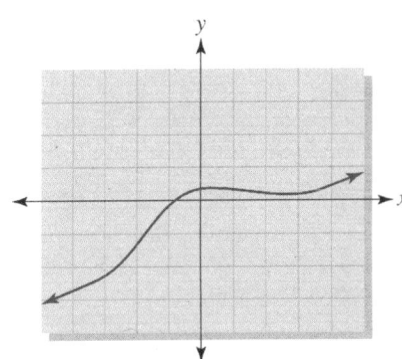

e.

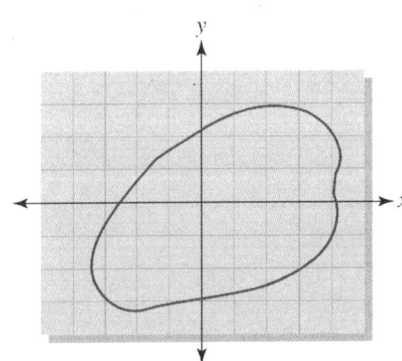

Solution

a. This is a linear equation, and for each input t we will get a single output P. Thus this equation represents a function.

b. This equation is not linear, yet it still has only one output, W, associated with each input value, g. Therefore, it also is a function.

c. This equation has a $\pm$ symbol, which means that almost all inputs will result in more than one output. For example, if $x = 5$, we get

$$y = 2(5) \pm (6(5) - 9) \qquad \text{Substitute } x = 5.$$
$$y = 10 \pm (30 - 9)$$
$$y = 10 \pm 21$$

$$y = 10 + 21 \qquad y = 10 - 21 \qquad \text{The } \pm \text{ symbol means}$$
$$y = 31 \qquad \qquad y = -11 \qquad \text{you need two equations.}$$
$$\text{We get two results.}$$

This equation gives two outputs for an input so this equation does not represent a function.

d.   This graph passes the vertical line test so it does represent a function.

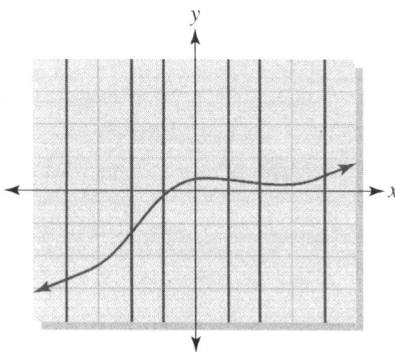

e.   This graph does not pass the vertical line test, so it does not represent a function.

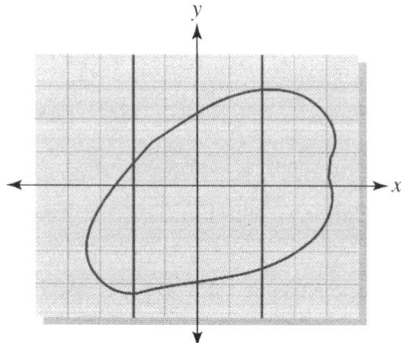

EXAMPLE   ②   PRACTICE PROBLEM

Determine whether or not the following tables, graphs, and equations represent functions. Explain your reasoning.

a.

Name of Student	Mary	Mark	Karla	Fred	Mark
Gender	Female	Male	Female	Male	Male

b.

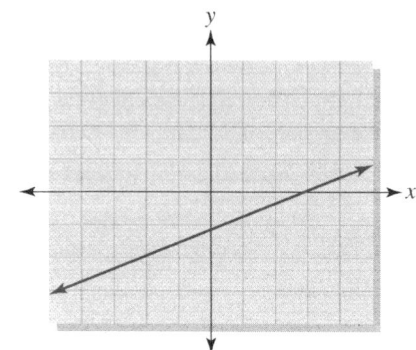

c.

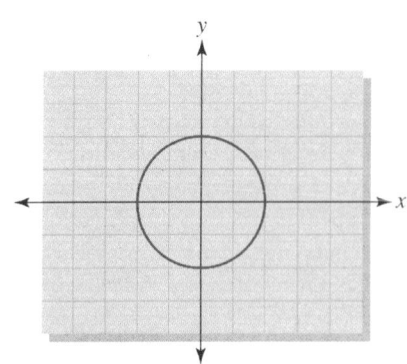

d. $C = 3.59u + 1359.56$

e. $H = 17.125 \pm \sqrt{3.5m}$ This can be entered in the equation as two equations: one using the plus the other using the minus.

FUNCTION NOTATION

Function notation was developed as a shorthand way of providing a great deal of information in a very compact form. If variables are defined properly with units and clear definitions, then function notation can be used to communicate what you want to do with the function and what input and/or output values you are considering.

Let's define the following variables:

P = Population of Hawaii (in millions)

t = Years since 2000

Then the population of Hawaii at time t can be represented by the following function.

$$P(t) = 0.013t + 1.210$$

$P(t)$ is read "P of t" and represents a function named P that depends on the variable t. In real-world applications, the variable in the same position as P represents the output variable and the variable in the same position as t represents the input variable. You can see this if you consider the P, outside the parentheses and thus the output, and the t, inside the parentheses and thus the input. There is not much of a difference between this function notation and the equation $P = 0.013t + 1.210$, but when you want to use this equation to represent the population as a function of time, it becomes much clearer to communicate using the function notation. If you are given the variable definitions above, you can make the following statements or ask the following questions in a couple of ways.

a.   In words:

    i.   Use the given equation to determine what the population of Hawaii was in 2005.

    ii.   Use the given equation to determine when the population of Hawaii will be 4 million.

b.   Using function notation these same statements can be written as:

    i.   Find $P(5)$.   **ii.**   Find $P(t) = 4$.

Using the function notation allows you to communicate what the input variable or output variable is equal to without words. $P(5)$ is asking you to substitute 5 for the input variable t and determine the value of the function $P(t)$. $P(t) = 4$ tells you that the output variable P is equal to 4 and directs you to determine the value of the input variable t that gives you a population of 4 million people.

Function notation can be a simple way to communicate information in a short way, but you must be careful when interpreting the information. Be sure to know how the variables are defined and use these definitions as a basis for interpreting any results.

EXAMPLE   3   INTERPRETING FUNCTION NOTATION IN AN APPLICATION

Given the following definitions, write sentences interpreting the following mathematical statements.

$G(t)$ = The number of guests at a local beach resort during year t

$P(b)$ = The profit, in millions of dollars, from the sale of b bombers

a. $G(2010) = 2005$

b. $P(10) = 7$

Solution

a.   In this notation, it is important to consider the location of each number. The 2010 is inside the parentheses, so it must be the value of t. Therefore, 2010 must be the year, and because 2005 is what $G(t)$ is equal to, this must represent the number of guests at the beach resort. The final interpretation might say, "In 2010, there were 2005 guests at this local beach resort."

b.   10 must be the number of bombers, since it is in the parentheses, and 7 must be the profit in millions. The profit from the sale of 10 bombers is 7 million dollars.

EXAMPLE   3   PRACTICE PROBLEM

Given the following definitions, write sentences interpreting the following mathematical statements.

$C(m)$ = The cost, in hundreds of dollars, for producing m, miracle mops.

$P(t)$ = The population of Michigan, in millions, t, years since 2000.

a. $C(2500) = 189$

b. $P(10) = 10.4$

If you are careful in determining which number represents which variable, you can interpret the meaning of the results by referring back to the definitions of each variable. Be sure to pay close attention to the units involved in each problem. If the profit from making bombers was measured in dollars and not millions of dollars, the bombers probably would not be made.

WRITING MODELS IN FUNCTION NOTATION

Function notation will change the modeling process given in Section 1.6 only in that you should now write your model in function notation and use that notation in showing your work. This will change step 5 of the modeling process to say "Write the equation of the model using function notation."

MODELING STEPS

1. Define the variables and adjust the data (if needed).
2. Create a scatterplot.
3. Select a model type.
4. **Linear Model:** Pick two points and find the equation of the line.
5. Write the equation of the model using function notation.
6. Check your model by graphing it with the scatterplot.

EXAMPLE   4   MODELS USING FUNCTION NOTATION

The following chart gives the population of Colorado, in millions, for various years.

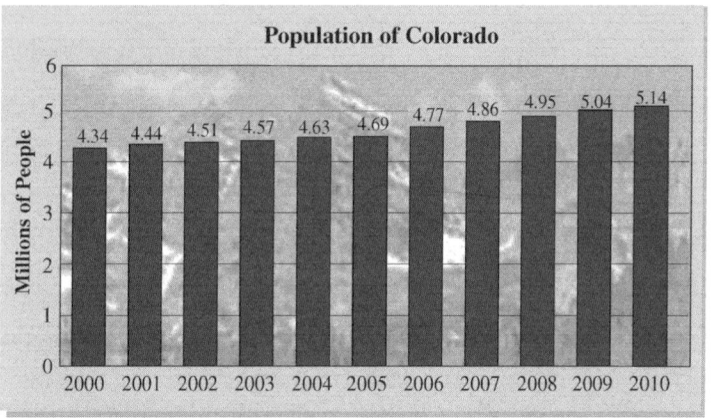

Source: Colorado Department of Local Affairs.

Let P be the population of Colorado, in millions, t years since 2000.

a.   Find a model for these data. Write your model in function notation.

b.   Determine a reasonable domain and range for your model.

c.   Find $P(15)$ and interpret its meaning in this context.

d.   Find when $P(t) = 5.35$ and interpret its meaning in this context.

Solution

a.   **Step 1** Define the variables and adjust the data.

$t =$ Years since 2000.

$P =$ Population of Colorado in millions.

Steps 2 Create a scatterplot.

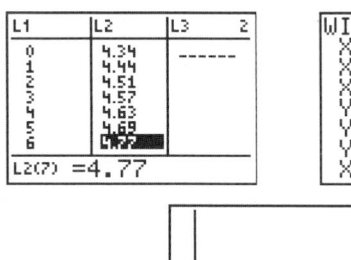 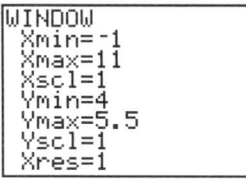

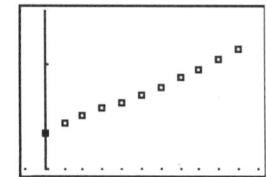

Steps 3 and 4 These data appear to be linear, so pick two points and find the equation of the line.

We could pick the first and ninth points since a line through these points would appear to go through the data well.

$$(0, 4.34) \quad \text{and} \quad (8, 4.95)$$

$$m = \frac{4.95 - 4.34}{8 - 0}$$

$$m \approx 0.076$$

We have calculated the slope, and the first point (0, 4.34) is a vertical intercept so we can write the equation using these values.

$$P = 0.076t + 4.34$$

Step 5 Write the equation of the model using function notation.

$$P(t) = 0.076t + 4.34$$

Step 6 Check the model by graphing it with the scatterplot.

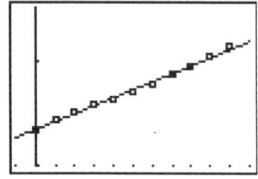

This model fits the data pretty well with all the points that it misses well balanced above and below the line. If you choose different points you will usually get a slightly different model. Remember we are looking for a good "Eye-ball" best fit line.

b. From the data given and the reasonable fit that we obtained with our model, we should be able to expand our domain beyond the data, giving us a possible domain of $[-2, 12]$. Looking at the graph, we can see that the lowest output value is going to be on the left side of the domain and the highest output value is going to be on the right side of the domain. This tells us that we need to evaluate the function at $t = -2$ and $t = 12$ to get the lowest and highest range values, respectively.

$$P(-2) = 0.076(-2) + 4.34$$
$$P(-2) = 4.188$$

$$P(12) = 0.076(12) + 4.34$$
$$P(12) = 5.252$$

Since $P(-2) = 4.188$, the lowest point, and $P(12) = 5.252$, the highest point, the range corresponding to the domain is [4.188, 5.252].

c. Substituting 15 for the input variable results in
$$P(15) = 0.076(15) + 4.34$$
$$P(15) = 5.48$$

The 15 inside the parentheses means 15 years since 2000, so the year is 2015. The result 5.48 is the population of Colorado in millions, so, together, they indicate that in 2015 the population of Colorado will be about 5.48 million.

d. Setting the function equal to 5.35 results in

$$5.35 = 0.076t + 4.34$$

| -4.34 | -4.34 | Subtract 4.34 from both sides of the equation. |

$$1.01 = 0.076t$$

$$\frac{1.01}{0.076} = \frac{0.076t}{0.076}$$

Divide both sides of the equation by 0.076.

$$13.289 \approx t$$

Check the solution.

$$P(13.289) = 0.076(13.289) + 4.34$$
$$P(13.289) \approx 5.35$$

This means that the population of Colorado will reach approximately 5.35 million in about 2013.

The most common function names in mathematics are f, g, and h. The function names $f(x)$, $g(x)$, and $h(x)$ will often replace the output variable y. These function names help to distinguish between different equations. If we have the following equations in slope-intercept form, we cannot easily distinguish between them.

$$y = 2x + 7 \qquad y = -4.7x - 8.6 \qquad y = 3x - 6$$

But if we write these using function notation, we can more easily distinguish them when we want to evaluate one of the equations.

$$f(x) = 2x + 7 \qquad g(x) = -4.7x - 8.6 \qquad h(x) = 3x - 6$$

Now, if we want to evaluate $f(x)$ at $x = 4$, we can write $f(4)$ and the reader would know which function to use and what to substitute in for x.

$$f(4) = 2(4) + 7$$
$$f(4) = 15$$

EXAMPLE **5** USING FUNCTION NOTATION

Let

$$f(x) = -3x + 5 \qquad g(x) = 2.5x - 9.7 \qquad h(x) = 4x^2 - 19$$

Find the following.

a. $f(7)$

b. $h(3)$

c. $g(x) = 12.3$

d. $f(x) = 11$

Solution

a. 7 is the input value. Substitute 7 for x and solve.

$$f(7) = -3(7) + 5$$
$$f(7) = -21 + 5$$
$$f(7) = -16$$

b. Using the function h, substitute 3 for the input variable x.

$$h(3) = 4(3)^2 - 19$$
$$h(3) = 4(9) - 19$$
$$h(3) = 36 - 19$$
$$h(3) = 17$$

c. We are given that the value of the function is 12.3, so set $g(x)$ equal to 12.3 and solve for the input variable.

$$12.3 = 2.5x - 9.7$$
$$\underline{+9.7 \qquad\qquad +9.7}$$
$$22 = 2.5x$$
$$\frac{22}{2.5} = \frac{2.5x}{2.5}$$
$$8.8 = x$$

$$g(8.8) = 2.5(8.8) - 9.7 \qquad \textbf{Check the solution.}$$
$$g(8.8) = 12.3$$

d. We are given that the value of the function is 11, so set $f(x)$ equal to 11 and solve for the input variable.

$$11 = -3x + 5$$
$$\underline{-5 \qquad\qquad -5}$$
$$6 = -3x$$
$$\frac{6}{-3} = \frac{-3x}{-3}$$
$$-2 = x$$

$$f(-2) = -3(-2) + 5 \qquad \textbf{Check the solution.}$$
$$f(-2) = 11$$

EXAMPLE PRACTICE PROBLEM

Let

$$f(x) = -4.25x + 5.75 \qquad g(x) = 4x + 8$$

Find the following.

a. $f(3)$

b. $g(x) = 20$

EXAMPLE **6** USING FUNCTION NOTATION

Use the graph to estimate the following.

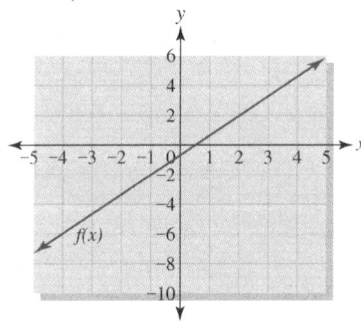

a. $f(-4)$

b. x such that $f(x) = -2$

Solution

a. -4 is inside the parentheses so it is an input or x value and we need to find the output or y value. According to the graph when $x = -4$ the line has a y value of -6. Therefore $f(-4) = -6$.

b. The line has an output of -2 when x is -1.

DOMAIN AND RANGE OF FUNCTIONS

When choosing a domain and range for a model in the context of an application, you must consider model breakdown something to be avoided. When you are considering the domain and range of a function that is not in the context of an application, model breakdown will not need to be considered. This makes the domain and range much less restricted and allows for as broad a domain and range as possible. The only restrictions to the domain of a function will be any real number that results in the function being undefined. Because all nonvertical linear equations are defined for all real numbers, their domain will be all real numbers. Therefore, because a linear graph will continue to go up and down forever, the range of any non-horizontal linear function will also be all real numbers. All real numbers can also be expressed by using interval notation: $(-\infty, \infty)$.

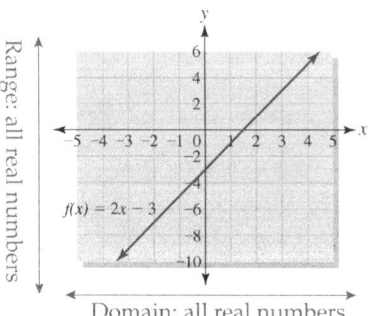

DEFINITIONS

Domain of a function: The set of all real numbers that make the function defined. Avoid division by zero and negatives under a square root.
Range of a function: The set of all possible output values resulting from values of the domain.

Domain and range of linear functions that are not vertical or horizontal:

Domain: all real numbers or $(-\infty, \infty)$

Range: all real numbers or $(-\infty, \infty)$

The horizontal line $y = k$ will have a domain of all real numbers but will have a range of only $\{k\}$. This is because no value of x will affect the output of the function, which is always k. A vertical line $x = k$ has a domain and range that are just the reverse. The domain will be $\{k\}$, since that is the only value x can have, but the range will be all real numbers.

EXAMPLE   7   DOMAIN AND RANGE OF FUNCTIONS

Determine the domain and range of the following functions.

a. $f(x) = 5x + 2$ **b.** $g(x) = -0.24x + 9$ **c.** $h(x) = 10$

Solution

a. Because this is a linear function, all real number inputs will result in real number outputs. Therefore, the domain is $(-\infty, \infty)$, and its range is also $(-\infty, \infty)$.

b. Because this is also a linear function, its domain is $(-\infty, \infty)$, and its range is $(-\infty, \infty)$.

c. This function represents a horizontal line, so its domain is still $(-\infty, \infty)$, but the range is [10], since the only output value for this function is 10.

EXAMPLE   7   PRACTICE PROBLEM

Determine the domain and range of the following functions.

a. $f(x) = -2x + 7$ **b.** $g(x) = -8$ **c.** $h(x) = 6x - 2$

Say What?

TO INFINITY AND BEYOND!

This well known saying has helped many to know the word infinity, In mathematics, infinity is a quantity that is unlimited.

We use the infinity symbol ∞ or negative infinity symbol $-\infty$ in interval notation to indicate that the interval is unlimited.

1.7   Exercises

For each of the relations in Exercises 1 through 10, specify the input and output variables and their definition and units. Determine whether or not each relation is a function.

1. $G(a)$ = The grade level of students when they are a years old.

2. $S(a)$ = The salary, in dollars, of a person who is a years old.

3. $H(a)$ = The heights, in inches, of children attending Mission Meadows Elementary School who are a years old.

4. $P(w)$ = The postage, in dollars, it takes to mail a first-class package weighing w ounces.

5. $I(t)$ = The interest earned, in dollars, on an investment after t years.

6. $P(t)$ = The price, in dollars, of Nike shoes during the year t.

7. $S(y)$ = The song at the top of the pop charts during the year y.

8. $B(m)$ = The number of students in this class who have a birthday during the mth month of the year.

9. $T(t)$ = The amount of taxes, in dollars, you paid in year t.

10. $A(m)$ = The number of tourists visiting Arizona, in thousands, during the mth month of 2012.

Determine whether the tables in Exercises 11 through 16 represent functions or not. Assume that the input is in the left column.

11.

Month	Cost (in dollars)
Jan.	5689.35
Feb.	7856.12
May	2689.15
June	1005.36

12.

Year	Number of SBA Loans to Minority-Owned Small Businesses
2000	11,999
2002	14,304
2003	20,183
2004	25,413
2005	29,717

Source: U.S. Small Business Administration.

13.

Age	Death Rate for HIV (per 100,000)
1–4 years	0.2
5–14 years	0.2
15–24 years	0.5
25–34 years	7.2
35–44 years	13.9
45–54 years	10.9
55–64 years	4.9
65–74 years	2.2
75–84 years	0.6

Source: Centers for Disease Control and Prevention

14.

Time (in hours)	Cost (in dollars per hour)
1 – 4	12.00
5 – 9	15.00
10 – 14	18.00
15 – 19	21.00

15. *Yes*

Day of the Week	Amount Spent on Lunch (*in $*)
Monday	4.78
Tuesday	5.95
Wednesday	0
Thursday	4.99
Friday	15.26
Monday	5.75
Tuesday	6.33
Wednesday	0
Thursday	4.25
Friday	20.36

16.

Person's Height	5'10"	6'2"	5'5"	5'7"	5'10"	6'1"
Person's Weight (*in kg*)	86.4	92	70	82	91	90

17. List the domain and range of the relation given in Exercise 11.

18. List the domain and range of the relation given in Exercise 12.

19. List the domain and range of the relation given in Exercise 14.

20. List the domain and range of the relation given in Exercise 16.

21. Use the relation given in Exercise 13 to determine the death rate from HIV for 20 year olds.

22. Use the relation given in Exercise 13 to determine the death rate from HIV for 30 year olds.

23. Use the relation given in Exercise 12 to find the number of small business loans made to minority-owned small businesses in 2001.

24. Use the relation given in Exercise 12 to find the number of small business loans made to minority-owned small businesses in 2005.

25. Use the relation in Exercise 15 to determine the amount spent on lunch on Tuesday.

26. Use the relation in Exercise 15 to determine the amount spent on lunch on Friday.

Determine whether the equations in Exercises 27 through 32 represent functions.

27. $W = -9.2g + 7.5$

28. $Q = 2v^2 - 6.5$

29. $8 = x^2 + y^2$

30. $y = \dfrac{23}{5}x - \dfrac{2}{5}$

31. $K = 5c^3 + 6c^2 - 9$

32. $Z = \pm(6d + 9)$

Determine whether the graphs in Exercises 33 through 36 represent functions.

33.

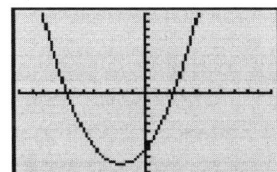

34.

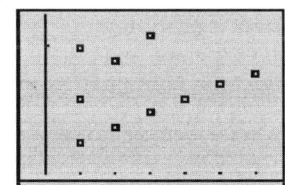

35.

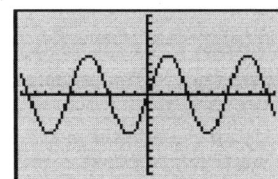

36.

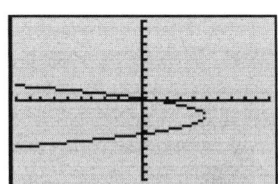

37. $W(d)$ is the weight in kilograms of a person d days after starting a diet. Write sentences interpreting the following mathematical statements.

   a. $W(0) = 86.5$

   b. $W(10) = 82$

   c. $W = 75$ when $d = 30$

   d. $W(100) = 88$

38. $F(t)$ is the value of fresh fruits exported (in millions of dollars) in year t. Write sentences interpreting the following mathematical statements.

Source: Statistical Abstract 2001.

   a. $F(1995) = 1973$
   b. $F = 1971$ when $t = 1996$
   c. $F(2000) = 2077$

39. $P(s)$ is the population, in millions, of state s in 2006. Write the following statements in function notation.

Source: U.S. Census Bureau.

   a. The population of Ohio was 11.48 million in 2006.
   b. The population of Texas was 23,507,783 in 2006.
   c. The population of Wyoming was 515,004 in 2006.

40. $F(t)$ is the amount of citrus fruits exported from the United States in millions of pounds in year t. Write the following statements in function notation.

Source: U.S. Census Bureau.

   a. In 2001 2,392 million pounds of citrus fruits were exported from the United States.
   b. In 2003 2,245,000,000 pounds of citrus fruits were exported from the United States.
   c. In 2005 1.937 billion pounds of citrus fruits were exported from the United States.

41. The bar graph below gives the average monthly Social Security benefit for retired workers in dollars for various years.

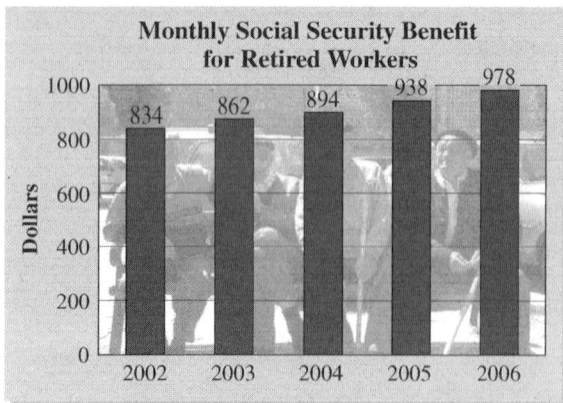

Source: Social Security Administration.

   a. Let $B(t)$ be the average monthly Social Security benefit in dollars t years since 2000. Find a model for the data given. Write your model in function notation.
   b. Find $B(9)$. (Remember to explain its meaning in this context.)
   c. Find t such that $B(t) = 1100$. Interpret your result.
   d. Give a reasonable domain and range for this model.
   e. What is the B intercept for this model and what does it mean in this context?

42. The amount Americans spend on athletic and sports footwear is given in the table.

Year	Amount Spent on Athletic & Sport Footwear (in billions of dollars)
2001	13.8
2002	14.1
2003	14.4
2004	14.8
2005	15.0

Source: Statistical Abstract 2007.

   a. Let $F(t)$ be the amount spent by Americans on athletic and sport footwear t years since 2000. Find a model for these data. Write your model in function notation.
   b. Give a reasonable domain and range for this model.
   c. Find the F intercept for this model and explain its meaning in this context.
   d. In 2000 Americans spent $13.5 billion on athletic and sport footwear. How does this compare to your result from part c?
   e. Find $F(12)$. Interpret your result.
   f. Find t such that $F(t) = 20$. Interpret your result.

43. The number of people enrolled in Medicare given in the table.

Year	Medicare Enrollees (in millions)
2002	40.5
2003	41.2
2004	41.9
2005	42.5

Source: Statistical Abstract 2007.

a. Let $M(t)$ be the number of Medicare enrollees t years since 2000. Find a model for these data. Write your model in function notation.

b. Give a reasonable domain and range for this model.

c. Find the M intercept for this model and explain its meaning in this context.

d. Find $M(10)$. Interpret your result.

e. Find t such that $M(t) = 50$. Interpret your result.

44. The bar graph below gives the number of farms in the United States for various years.

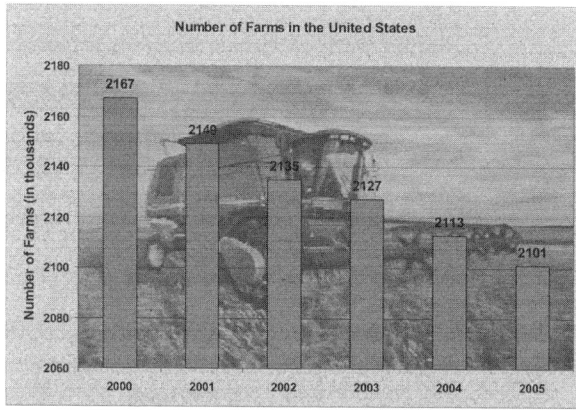

Source: Statistical Abstract 2007.

a. Let $F(t)$ be the number of farms in the United States in thousands t years since 2000. Find a model for the data given. Write your model in function notation.

b. Find $F(11)$. (Remember to explain its meaning in this context.)

c. Find t such that $F(t) = 1700$. Interpret your result.

d. Give a reasonable domain and range for this model.

e. What is the F intercept for this model and what does it mean in this context?

45. $f(x) = 2x - 7$

a. Find $f(5)$.

b. Find $f(-10)$.

c. Find x such that $f(x) = -1$.

d. Give the domain and range for $f(x)$.

46. $g(x) = 5x + 12$

a. Find $g(3)$.

b. Find $g(-7)$.

c. Find x such that $g(x) = 47$.

d. Give the domain and range for $g(x)$.

47. $h(x) = \dfrac{2}{3}x + \dfrac{1}{3}$

a. Find $h(15)$.

b. Find $h(-9)$.

c. Find x such that $h(x) = 4$.

d. Give the domain and range for $h(x)$.

48. $f(x) = \dfrac{1}{5}x + \dfrac{3}{5}$

a. Find $f(20)$.

b. Find $f(12)$.

c. Find x such that $f(x) = 10$.

d. Give the domain and range for $f(x)$.

49. $g(x) = -18$

a. Find $g(2)$.

b. Find $g(-11)$.

c. Give the domain and range for $g(x)$.

50. $h(x) = 12.4$

a. Find $h(5)$.

b. Find $h(-123)$.

c. Give the domain and range for $h(x)$.

51. $f(x) = 3.2x - 4.8$

a. Find $f(2)$.

b. Find $f(-14)$.

c. Find x such that $f(x) = -10$.

d. Give the domain and range for $f(x)$.

52. $g(x) = -4.3x - 5$

    a. Find $g(15)$.

    b. Find $g(-20)$.

    c. Find x such that $g(x) = -45.6$.

    d. Give the domain and range for $g(x)$.

53. $h(x) = 14x + 500$

    a. Find $h(105)$.

    b. Find x such that $h(x) = -140$.

    c. Give the domain and range for $h(x)$.

54. $f(x) = 25x - 740$

    a. Find $f(30)$.

    b. Find $f(-19)$.

    c. Find x such that $f(x) = -240$.

    d. Give the domain and range for $f(x)$.

For Exercises 55 and 60, use the graph of the function to answer the questions.

55.

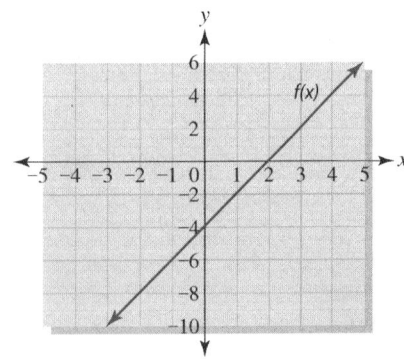

    a. Estimate $f(3)$.

    b. Estimate $f(-2)$.

    c. Estimate x such that $f(x) = -6$.

    d. Give the domain and range for $f(x)$.

    e. Estimate the vertical and horizontal intercepts for $f(x)$.

56.

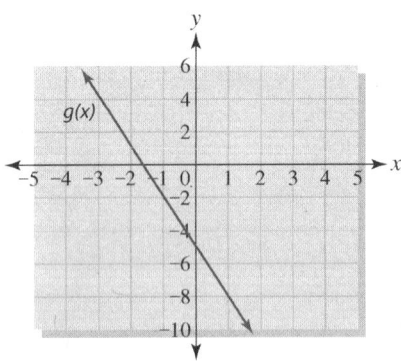

    a. Estimate $g(1)$.

    b. Estimate $g(-3)$.

    c. Estimate x such that $g(x) = 1$.

    d. Give the domain and range for $g(x)$.

    e. Estimate the vertical and horizontal intercepts for $g(x)$.

57.

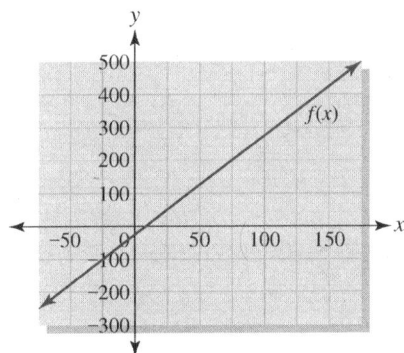

    a. Estimate $f(25)$.

    b. Estimate $f(100)$.

    c. Estimate x such that $f(x) = 200$.

    d. Give the domain and range for $f(x)$.

    e. Estimate the vertical and horizontal intercepts for $f(x)$.

58.

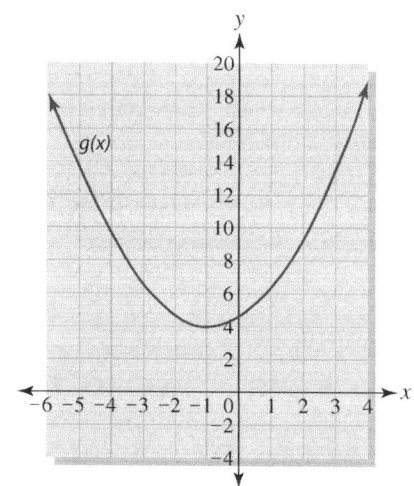

a. Estimate $h(5)$.
b. Estimate $h(20)$.
c. Estimate x such that $h(x) = 20$.
d. Give the domain and range for $h(x)$.
e. Estimate the vertical and horizontal intercept for $h(x)$.

59.

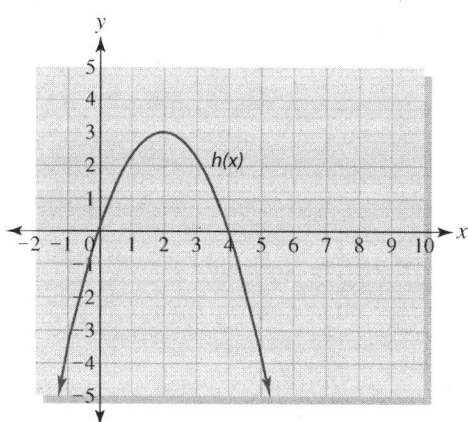

a. Estimate $h(2)$.
b. Estimate $h(5)$.
c. Estimate x such that $h(x) = -1$.
d. Give the domain and range for $h(x)$.
e. Estimate the vertical and horizontal intercepts for $h(x)$.

60.

a. Estimate $g(3)$.
b. Estimate $g(-4)$.
c. Estimate x such that $g(x) = 6$.
d. Give the domain and range for $g(x)$.
e. Estimate the vertical and horizontal intercept for $g(x)$.

Chapter 1 Summary

Section 1.1 Solving Linear Equations

- In **solving** equations, your goal is to isolate the variable on one side of the equation.
- Be sure to write **complete sentence** answers to any problems that are presented in a context.
- **Define all variables** you use to create an equation.
- Include **units** for all variable definitions and answers.

EXAMPLE

The number of minutes an hour a local radio station can play music depends on the number of commercials that are played during that hour. Math Rocks 101.7 FM follows the equation below when considering how much time they have to play music.

$$M = 60 - 0.5C$$

where M is the number of minutes of music the station can play in an hour if they play C thirty-second commercials that hour.

a. Find the number of minutes of music the station can play in an hour when they play 12 thirty-second commercials.

b. If the radio station claims to have at least 53 minutes of music each hour, what is the most number of thirty-second commercials the station can play an hour?

Solution

a. Since there are 12 thirty-second commercials being played, $C = 12$, and we get

$$M = 60 - 0.5(12)$$
$$M = 54$$

With 12 thirty-second commercials an hour the station can play 54 minutes of music.

b. If the station needs to play 53 minutes of music, $M = 53$, so we will get

$$53 = 60 - 0.5C \qquad \text{Subtract 60 from both sides.}$$
$$-7 = -0.5C \qquad \text{Divide both sides by 0.5.}$$
$$14 = C$$

The station can play 14 thirty-second commercials an hour if they want to play 53 minutes of music.

EXAMPLE

Hip Hop Math Records is a new record label that is paying "Math Dude", a new music artist, $10,000 to record a CD and $1.50 for every CD sold.

a. Write an equation for the amount of money "Math Dude" will make from this deal depending on how many CDs are sold.

b. Use your equation to find how much money "Math Dude" will make if he sells 100,000 CDs.

c. Use your equation to find out how many CDs "Math Dude" must sell to make himself a million dollars.

Solution

a. Let M be the money Hip Hop Math Records will pay "Math Dude" in dollars.

Let C be the number of CDs sold by "Math Dude."

$$M = 10000 + 1.5C$$

b. If he sells 100,000 CDs, $C = 100,000$, so we get

$$M = 10000 + 1.5(100000)$$
$$M = 160000$$

If "Math Dude" sells 100,000 CDs, Hip Hop Math Records will pay him $160,000.

c. If "Math Dude" is going to make a million dollars, we must have $M = 1,000,000$, so we get

$$1000000 = 10000 + 1.5C$$
$$990000 = 1.5C$$
$$660000 = C$$

For "Math Dude" to make a million dollars, he must sell 660,000 CDs.

Section 1.2 Using Data to Create scatterplots

- To create a **scatterplot** from data, you need to determine the independent and dependent variables.

- The **independent variable** will be placed on the horizontal axis. This variable is often called the *input variable*.

- The **dependent variable** will be placed on the vertical axis. This variable is often called the *output variable*.

- An "eyeball best fit" line can be used as a **graphical model.**

- The **vertical intercept** is where the graph crosses the vertical axis. It will be a point of the form $(0, b)$.

- The **horizontal intercept** is where the graph crosses the horizontal axis. It will be a point of the form $(a, 0)$.

- The **domain** will include all values of the input variable that will result in reasonable output values.

- The **range** will include all values of the model that come from the inputs in the domain.

EXAMPLE 3

The gross profit for PETsMART, Inc. is reported in the following table.

Year	2000	2001	2001	2003
Gross Profit (in millions of $)	529.0	673.5	788.0	909.9

Source: CBSMarketwatch.com.

a. Create a scatterplot of the data and draw an "eyeball best fit" line through the data.

b. Use your graphical model to estimate the gross profit for PETsMART, Inc. in 2004.

c. What is the vertical intercept of your model and what does it mean in this context?

d. What are a reasonable domain and range for your model?

Solution

a. Let t be time in years since 2000.
Let P be the gross profit of PETsMART, Inc. in millions of dollars.

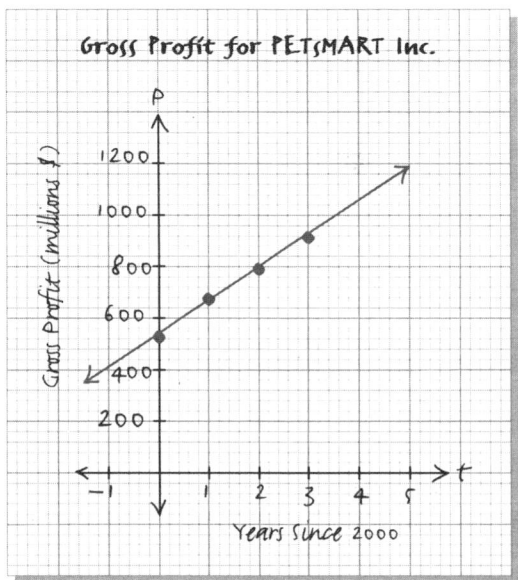

b. According to the graph in 2004 PETsMART, Inc. had a gross profit of about $1050 million dollars, or $1.05 billion.

c. The vertical intercept for this model is about $(0, 550)$. Therefore in 2000 PETsMART, Inc. had a gross profit of about $550 million.

d. Domain: $[-1, 4]$, range: $[400, 1050]$.

Section 1.3 Introduction to Graphing

- The **slope** of a line is the steepness of the line. Slope can be remembered as

$$m = \frac{\text{rise}}{\text{run}} = \frac{\text{change in } y}{\text{change in } x} = \frac{y_2 - y_1}{x_2 - x_1}$$

- The slope of a line is a constant and represents the amount that the output variable changes per a change in the input variable.

- The **slope-intercept form** of a line is $y = mx + b$, where m represents the slope and $(0, b)$ is the vertical intercept.

- You can graph an equation by plotting points.

- You can graph a linear equation using the slope and y-intercept.

EXAMPLE 4

Use the graph to find the following.

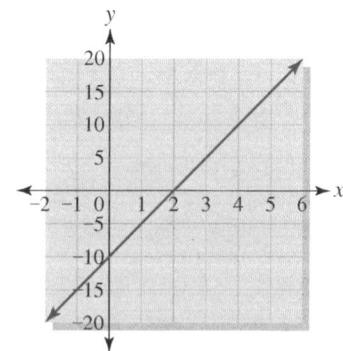

a. Find the slope of the line.

b. Estimate the vertical intercept.

c. Estimate the horizontal intercept.

Solution

a. The points $(1, -5)$ and $(4, 10)$ are on the line, so we get

$$\frac{10 - (-5)}{4 - 1} = \frac{15}{3} = 5$$

The slope is 5.

b. The vertical intercept is about $(0, -10)$.

c. The horizontal intercept is about $(2, 0)$

EXAMPLE 5

Graph the line $y = \frac{2}{3}x - 6$ using the slope and y-intercept.

Solution

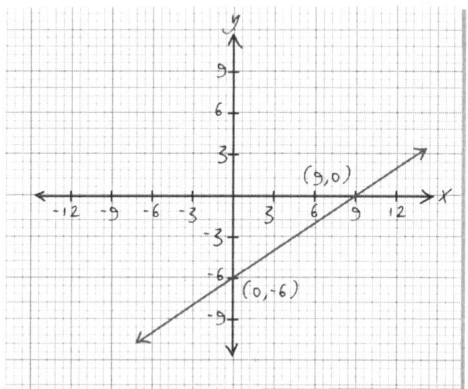

EXAMPLE 6

Using the equation you found in Example 2 part a above, explain the meaning of the slope in the context of the situation.

Solution

The slope is $m = 1.5$ and means that the amount that Math Dude will make from recording and selling the CD's will increase by $1.50 per CD sold.

Section 1.4 The General Form of Lines

- The **General Form of a Line** is $Ax + By = C$ where A, B, and C are integers and A is non-negative.

- To find an intercept make the "other" variable zero and solve.

- When interpreting an intercept be sure to interpret both parts of the point.

- A **horizontal line** has an equation of the form $y = k$.

- A **vertical line** has an equation of the form $x = k$.

- The slope of a horizontal line is zero. The slope of a vertical line is undefined.

EXAMPLE 7

The cost C in dollars to produce L strands of Christmas lights can be modeled by $C = 3.5L + 10000$.

a. Find the L-intercept and explain its meaning in the context of this situation.

b. Find the C-intercept and explain its meaning in the context of this situation.

Solution

a. To find the L-intercept we make $C = 0$ and solve.

$$0 = 3.5L + 10000$$
$$-10000 = 3.5L$$
$$-2857.14 \approx L$$

Therefore the L-intercept is $(-2857.14, 0)$. If we produce negative 2857.14 strands of lights the cost will be zero. This is model breakdown since we cannot produce a negative number of strands of lights.

b. To find the C-intercept we make $L = 0$ and we get $(0, 10000)$. This intercept means that if the company produces no strands of lights they still have costs of $10,000.

EXAMPLE 8

Find the intercepts and graph the line $2x + y = 4$.

Solution

To find the x-intercept we make $y = 0$ and solve.

$$2x + 0 = 4$$
$$2x = 4$$
$$x = 2$$

Therefore the x-intercept is $(2, 0)$.
To find the y-intercept we make $x = 0$ and solve. This leaves us with $y = 4$ so the y-intercept is $(0, 4)$.
Using these points we get the graph.

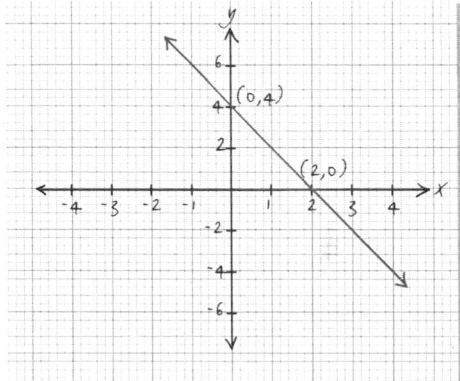

Section 1.5 Finding Equations of Lines

- The slope-intercept form $y = mx + b$ or the point-slope formula $y - y_1 = m(x - x_1)$ can be used to find the equation of a line.

- Use the following steps to find the equation of a line using the point-slope formula.

  1. Use any two points to calculate the slope.
  2. Substitute the slope and a point into the point-slope formula.
  3. Write the equation in slope-intercept form.
  4. Check your equation by plugging in the points to be sure they are solutions.

- Use the following steps to find the equation of a line using the slope-intercept form.

  1. Use any two points to calculate the slope.
  2. Use the slope and a point to find the value of b.
  3. Write the equation in slope-intercept form.
  4. Check your equation by plugging in the points to be sure they are solutions.

- Parallel lines have the same slopes and never intersect.

- Perpendicular lines have opposite reciprocal slopes and intersect at a right angle.

EXAMPLE 9

A car that is being depreciated is worth $35,000 new and worth $20,000 after 3 years. If linear depreciation is being used find an equation for the value of the car depending on its age.

Solution

First we will define the variables.
$a =$ age of the car in years.
$V =$ The value of the car in dollars.
The problem gives us two values for the car which result in the points $(0, 35000)$ and $(3, 20000)$. Using these two points we calculate the slope and get.

$$m = \frac{35000 - 20000}{0 - 3} = -5000$$

Using the point (0, 35000) as the vertical intercept, we get the equation $V = -5000a + 350000$.

EXAMPLE

Determine if the following set of lines is parallel, perpendicular or neither.

a. $y = 2x + 7$
    $x + 2y = 20$
b. $y = 0.25x + 5$
    $2x - 8y = 24$

Solution

a. First we will put the second equation into slope-intercept form so we can compare the slopes of the two equations.
$$x + 2y = 20$$
$$2y = -x + 20$$
$$y = -\frac{1}{2}x + 10$$

We can see from the slope-intercept form that the slope of the first equation is 2 and the slope of the second equation is $-\frac{1}{2}$. These two slopes are opposite reciprocals so these lines are perpendicular.

b. First we will put the second equation into slope-intercept form.
$$2x - 8y = 24$$
$$-8y = -2x + 24$$
$$y = 0.25x - 3$$

We can see that both lines have a slope of 0.25 so the lines are parallel.

Section 1.6 Finding Linear Models

* To find a **linear model,** you should follow these steps.

    1. Define the variables and adjust the data.
    2. Create a scatterplot.
    3. Select a model type.
    4. **Linear Model:** Pick two points and find the equation of the line.
    5. Write the equation of the model.
    6. Check your model by graphing it with the scatterplot.

EXAMPLE

The gross profit for PETsMART, Inc. is reported in the following table.

a. Find a model for the data.
b. Use your model to estimate the gross profit for PETsMART, Inc. in 2005.
c. What is the horizontal intercept of your model and what does it mean in this context?
d. What are a reasonable domain and range for your model?

Solution

a. 1. Define the variables and adjust the data.
    t = time in years since 2000.

    P = the gross profit of PETsMART, Inc. in millions of dollars.

Year	2000	2001	2002	2003
Gross Profit (in millions of $)	529.0	673.5	788.0	909.9

Source: CBSMarketwatch.com.

2. and 3. Create a scatterplot and select a model type.

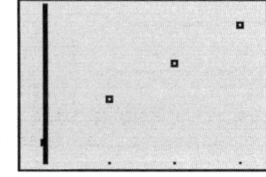

4. and 5. Pick two points and find the equation of the line, and write the equation of the model.

The first and third points seem to fit nice so we can use them to find the slope.

$$m = \frac{788 - 529}{2 - 0} = 129.5$$

The point (0, 529) is the vertical intercept. Therefore we get the equation $P = 129.5t + 529$.

6. Check your model by graphing it with the scatterplot.

This model seems to fit well.

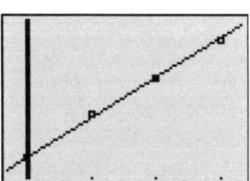

b. 2005 is represented by $t = 5$, so we get
$$P = 129.5(5) + 529 = 1176.5.$$
Therefore in 2005 PETsMART, Inc. should have a gross profit of about $1176.5 million dollars.

c. To find the horizontal intercept, we make $P = 0$ and get

$$0 = 129.5t + 529$$
$$-529 = 129.5t$$
$$-4.085 = t$$

Therefore in about 1996 PETsMART, Inc. made no gross profit. This is model breakdown.

d. Domain: $[-1, 4]$, range: $[399.5, 1047]$.

Section 1.7 Introduction to Functions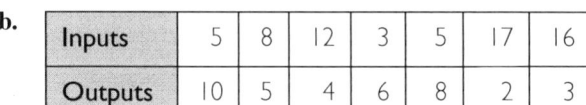

- A **relation** is any relationship between a set of inputs and outputs.
- A **function** is a relation in which each input has exactly one output associated with it.
- The **vertical line test** can be used to determine if a graph represents a function.
- Models should be written using function notation.
- The **domain** and **range** of a linear function that is not in a context and that is not vertical or horizontal are both **all real numbers.**

b.

Inputs	5	8	12	3	5	17	16
Outputs	10	5	4	6	8	2	3

c.

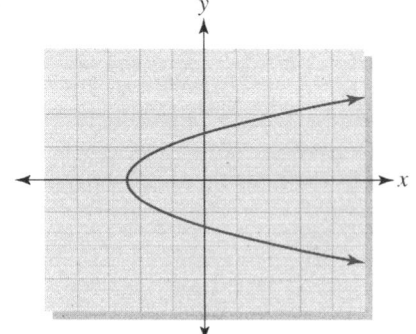

EXAMPLE 12

Determine whether the following represent functions. Explain why or why not.

a. $y = 3x + 9$

d.

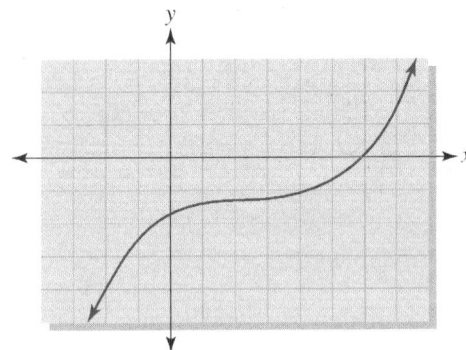

Solution

a. This is a function, since each input will result in only one output.

b. This table does not represent a function because the input value of 5 is associated with the outputs 10 and 8.

c. This graph is not a function because it fails the vertical line test. A vertical line will pass through the graph more than once.

d. This graph does represent a function, since it passes the vertical line test.

EXAMPLE

Let $D(w)$ be the number of deaths in Arkansas due to the West Nile virus during week w of the year.

a. Interpret the notation $D(7) = 6$.

b. Given that there were three deaths during the forty-fifth week of the year, write this in function notation.

Solution

a. During the seventh week of the year there were six deaths due to the West Nile virus in Arkansas.

b. $D(45) = 3$.

Chapter 1 Review Exercises

At the end of each exercise, you'll find the section number in brackets where the material is covered, in case you need help solving the problem.
For Exercises 1 through 4 solve the equation.

1. $2x + 5 = 7(x - 8)$ [1.1]

2. $\frac{1}{3}x - \frac{2}{3} = 5$ [1.1]

3. $0.4t + 2.6 = 0.8(t - 8.2)$ [1.1]

4. $-2(x + 3.5) + 4(2x - 1) = 7x - 5(2x + 3)$ [1.1]

5. The height of the grass used in the rough at a local golf course is given by

$$h = 0.75d + 4.5$$

where h is the height of the grass in inches d days after the grass was cut and fertilized.

   a. What is the height of the grass 1 week after it is cut and fertilized?

   b. For a coming tournament, the officials want the grass in the rough to be 12 inches high. How many days before the tournament should the golf course cut and fertilize this grass? [1.1]

6. The number of candles produced by the Holy Light Candle Company can be modeled by

$$C(t) = 0.56t + 4.3$$

where $C(t)$ is the number of candles, in millions, produced t years since 2010.

   a. Find the number of candles produced by Holy Light Candle Company in 2015.

   b. Find $C(3)$ and interpret its meaning in this context.

   c. Find when $C(t) = 10$ and interpret its meaning in this context. [1.7]

7.

Bobcat® is a registered trademark of the Bobcat Company—a unit of Ingersoll-Rand.

Pauley's Rental Company

Bobcat A300

$40.00 per hour

$15.00 per rental for insurance

   a. Use the ad to find an equation for the cost to rent a Bobcat tractor from Pauley's Rental Company.

   b. How much will it cost to rent the Bobcat for 2 hours?

   c. How much will it cost to rent the Bobcat for 3 days if a day is measured as 8 hours each? [1.1]

8. Satellite phone service is available in some airplanes for a $10 connection fee and a $10 per minute charge.

   a. Find an equation for the cost of a satellite phone call.

   b. Use your equation to find the cost of a 3 minute satellite phone call.

   c. If your credit card only has $300 left of available credit, how long of a satellite phone call can you make? [1.1]

9. A small manufacturing company makes custom putters for a chain of sporting goods stores. The total cost to produce different numbers of putters is given in the table.

Number of Putters	Total Cost (in dollars)
5	600.00
10	725.00
20	950.00
25	1100.00
50	1700.00

a. Find a model for these data.

b. Find the vertical intercept of your model. Explain its meaning in this context.

c. Find the total cost for 100 putters.

d. Give a reasonable domain and range for this model.

e. What is the slope of this model? Explain its meaning in this context. [1.6]

10. The gross profit for Costco Wholesale Corporation is given in the following chart.

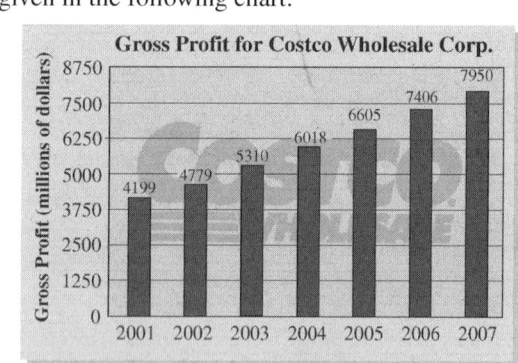

Source: http://finance.google.com.

a. Find a model for these data. Write your model using function notation.

b. Estimate when the gross profit for Costco will be 9 billion dollars.

c. What was the gross profit for Costco in 2009?

d. Give a reasonable domain and range for this model.

e. What is the slope of this model? Explain its meaning in this context. [1.7]

11. According to the U.S. Department of Labor Bureau of Labor Statistics, the percent of full-time workers in private industry who filed an injury case is given below.

Years Since 2000	Percent
3	4.7
4	4.5
5	4.4
6	4.2

Source: U.S. Bureau of Labor Statistics.

a. Find a model for these data. Write your model using function notation.

b. What is the slope for this model? Explain its meaning in this context.

c. What percent of full-time workers in private industry will have filed a injury case in 2003?

d. What is the vertical intercept for this model? What does it mean in this context?

e. What is the horizontal intercept for this model? What does it mean in this context? [1.7]

12. The number of kindergarten through twelfth grade students in Texas public schools is given below.

Year	Students
2000	4,002,227
2001	4,071,433
2002	4,160,968
2003	4,255,821
2004	4,328,028
2005	4,400,644
2006	4,521,043

Source: Texas Education Agency.

a. Create a scatterplot of these data by hand. [1.2]

b. Find a model for these data.

c. What is the slope for this model? Explain its meaning in this context.

d. How many kindergarten through twelfth grade students can Texas expect to have in its public schools in 2011? [1.6]

Costco® Wholesale Is a Registered Trademark of Costco Wholesale Corporation and Is Used with Permission.

For Exercises 13 through 16, solve each equation for the indicated variable.

13. Charles's Law: $V = bT$ for T.   [1.1]

14. Force: $F = ma$ for m.   [1.1]

15. $ax + by = c$ for x.   [1.1]

16. $2x - ay = b$ for y.   [1.1]

For Exercises 17 through 24, sketch the graphs of the equations by hand using any method you wish. Label the horizontal and vertical intercepts.

17. $y = 3x - 4$ [1.3]

18. $y = 2x + 3$ [1.3]

19. $2x + 3y = 24$ [1.4]

20. $5x - 6x = 42$ [1.4]

21. $y = x^2 + 5$ [1.3]

22. $y = x^2 - 6$ [1.3]

23. $y = \dfrac{1}{2}x + 7$ [1.3]

24. $y = -\dfrac{2}{7}x + 4$ [1.3]

25. Give the domain and range for the function $f(x) = 2x - 8$.   [1.7]

26. Give the domain and range for the function $h(x) = -\dfrac{2}{3}x + 14$ [1.7]

27. Give the domain and range for the function $g(x) = 12$.   [1.7]

28. Give the domain and range for the function $f(x) = -8$.   [1.7]

29. Find the equation of the line passing through the points (2, 7) and (7, 27).   [1.5]

30. Find the equation of the line passing through the points (4, 9) and (−3, 23).   [1.5]

31. Find the equation of the line that passes through the point (4, 10) and is parallel to the line $y = 3x - 12$.   [1.5]

32. Find the equation of the line that passes through the point (2, 16) and is parallel to the line $y = -0.5x + 4$.   [1.5]

33. Find the equation of the line that passes through the point (3, 7) and is perpendicular to the line $y = -4x + 7$.   [1.5]

34. Find the equation of the line that passes through the point (−2, 9) and is perpendicular to the line $y = \dfrac{2}{5}x + 7$.   [1.5]

35. Use the graph to find the following.

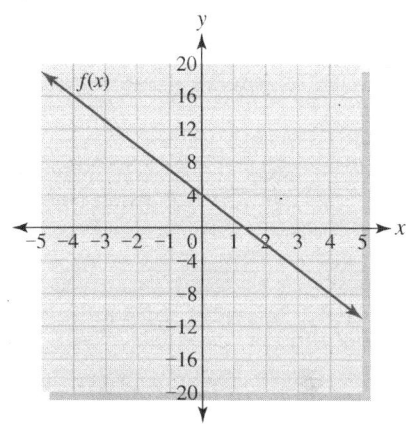

  a.   Estimate the y-intercept.

  b.   Estimate the x-intercept.

  c.   Find the slope of the line.   [1.3]

  d.   Find $f(4)$.

  e.   Find x such that $f(x) = 16$.

  f.   Find an equation for $f(x)$.   [1.7]

36. Use the graph to find the following.

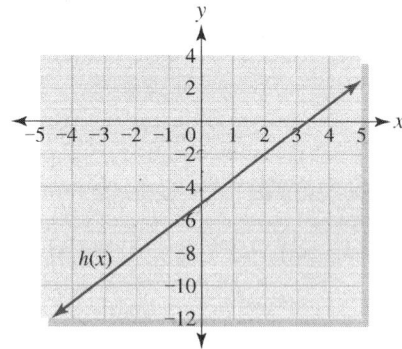

  a.   Estimate the vertical intercept.

  b.   Estimate the horizontal intercept.

  c.   Find the slope of the line.   [1.3]

  d.   Find $h(-2)$.

  e.   Find x such that $h(x) = -2$.

  f.   Find an equation for $h(x)$.   [1.7]

37. Use the table to find the following.

x	y
−3	4
0	6
3	8
6	10
9	12

  a. The slope of the line passing through the points

  b. The y-intercept

  c. The equation for the line passing through the points   [1.5]

38. Use the table to find the following.

x	y
−8	6.6
−2	3.9
0	3
3	1.65
7	−0.12

  a. The slope of the line passing through the points

  b. The y-intercept

  c. The equation for the line passing through the points   [1.5]

39. Let $E(h)$ be the cost in dollars for electricity to use holiday lights h hours a day. Explain the meaning of $E(6) = 13$ in the context of this situation.   [1.7]

40. The population of Michigan in thousands can be modeled by $P(t) = 30t + 9979$ where t is years since 2000.

  a. Find $P(10)$ and explain its meaning in the context of this situation.

  b. Find when the population of Michigan will reach 11,000 thousand.   [1.7]

Chapter 1 Test

1. The number of work-related injury cases in the U.S. private industry is given below.

Year	Number of Injury Cases (in thousands)
2003	4095
2004	4008
2005	3972
2006	3857

Source: U.S. Bureau of Labor Statistics.

a. Find a model for these data.

b. How many work-related injury cases were there in private industry in 2009?

c. What is the slope of your model? Explain its meaning in this context.

d. Give a reasonable domain and range for this model.

e. When was the number of work-related injury cases in private industry 5 million?

2. Sketch by hand the graph of the line
$$y = 4x - 2$$
Label the vertical and horizontal intercepts.

3. Sketch by hand the graph of the line
$$2x - 4y = 10$$
Label the vertical and horizontal intercepts.

4. Solve $ax - by = c$ for x.

5. Find the equation for the line passing through the points $(-4, 8)$ and $(6, 10)$.

6. Find the equation for the line passing through the point $(5, 8)$ and parallel to the line $y = 2x - 7$.

7. The number of paintings that John Clark sells each month can be modeled by
$$P = 2m + 30$$
where m is the number of months since he started selling paintings.

a. In month 6, how many paintings will John sell?

b. In what month will he sell 50 paintings?

8. The total revenue for Apple Inc. for the past several years is given below.

Years Since 2000	Total Revenue (in millions of $)
4	8279
5	13931
6	19315
7	24006

Source: http://finance.google.com

a. Find a model for these data. Write the model using function notation.

b. What is the vertical intercept for your model? Explain its meaning in this context.

c. Estimate the total revenue for Apple Inc. in 2010.

d. When might the total revenue be 30 billion dollars?

e. Give a reasonable domain and range for this model.

f. What is the slope of this model? Explain its meaning in this context.

9. Use the table to find the following.

x	y
−5	16
0	12
5	8
10	4
15	0

   a. The slope of the line passing through these points

   b. The y-intercept

   c. The x-intercept

   d. The equation for the line passing through these points

10. Use the graph to find the following.

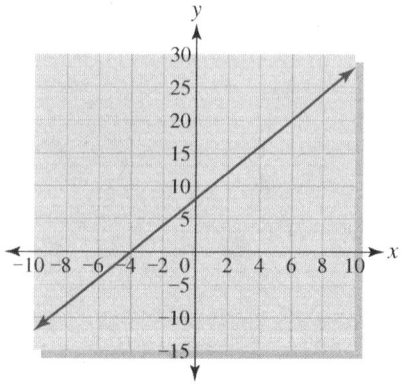

   a. The slope of the line

   b. The vertical intercept

   c. The horizontal intercept

   d. An equation for the line

11. Give the domain and range for the function $f(x) = -4x + 5$.

12. Solve the equation $W = ht^2$ for h.

13. Solve $1.5(x + 3) = 4x + 2.5(4x - 7)$

14. If $P(t)$ is the population of New York in millions t years since 2000. Explain the meaning of $P(10) = 19.4$ in the context of this situation.

15. Let $H(p) = 0.20p + 5$ be the hours you should study or a test on chapter 1 if you want a grade of $p\%$ on this test.

   a. Find $H(70)$ and explain its meaning in the context of this situation.

   b. Find how many hours you need to study to get a 100% on the exam.

Chapter 1 Projects

Group Experiment

Three or more people

What you will need:

- A tape measure
- A timer (stopwatch)

or, instead of a tape measure and timer,

- A Texas Instruments CBR unit or a Texas Instruments CBL unit with a Vernier motion detector probe.

WALKING A "STRAIGHT" LINE

It will be your job to create data that will best be modeled by a linear function. Using the tape measure and timer or the CBR/CBL unit, have someone walk while the others record the time and the distance from a fixed object (i.e., a wall or desk). Collect at least five points. Be sure to check your data so that you are satisfied that the data will be a linear function before you move on.

Write up

a. Describe how you collected the data. What was the most important thing to do to get a linear model?

b. Create a scatterplot of the data on the calculator or computer and print it out or neatly by hand on graph paper.

c. Find a model to fit the data.

d. What is the slope of your model and what does it represent in this experiment?

e. What is the vertical intercept of your model and what does it represent in this experiment?

f. What are a reasonable domain and range for your model? (Remember to consider any restrictions on the experiment.)

NEW FLOORING COSTS

Research Project

One or more people

What you will need:

- Prices from a flooring store for either wood flooring or the tile of your choice.
- Installation costs per square foot for the flooring of your choice.

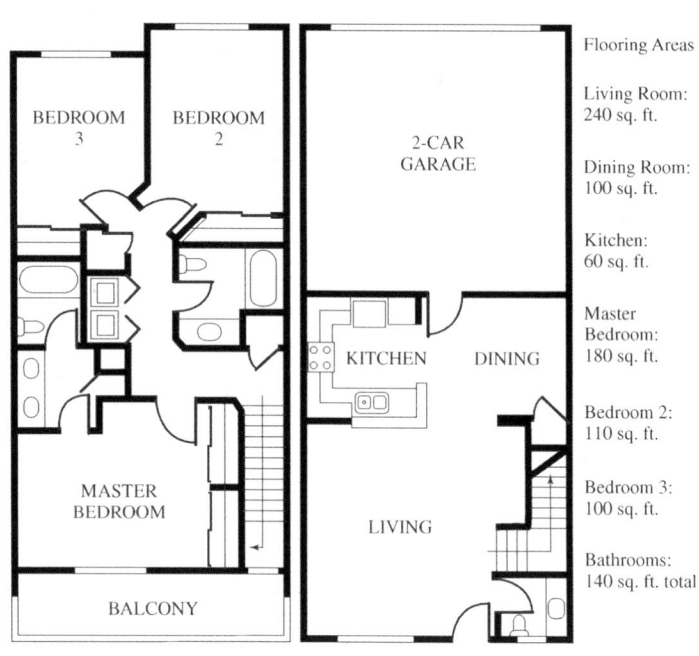

Flooring Areas

Living Room: 240 sq. ft.

Dining Room: 100 sq. ft.

Kitchen: 60 sq. ft.

Master Bedroom: 180 sq. ft.

Bedroom 2: 110 sq. ft.

Bedroom 3: 100 sq. ft.

Bathrooms: 140 sq. ft. total

You are given the task of estimating the cost of flooring for a new house that your family is buying. The house comes with very basic flooring options installed, and you need to decide in how many rooms you want to replace this basic flooring with upgraded wood floors or tile. You should consider both the cost for materials and cost for installation.

Write up

a. Describe the flooring you chose and its cost per square foot for materials and installation. Are there any fixed costs for delivery or other items that you should consider in your calculations?

b. Let f be the number of square feet of flooring to be purchased and installed. Find a function for the cost of materials.

c. Find a function for the cost of installing the flooring.

d. Estimate the cost for the materials if you install them in the living room, dining room, and kitchen.

e. Estimate the cost for installation only for the living room, dining room, and kitchen.

f. Use the two functions from parts b and c to create a new function that gives you the total cost for materials and installation of the flooring.

g. Estimate the total cost to put your flooring choice in all of the bedrooms.

h. What are a reasonable domain and range for your total cost function?

i. If your budget for flooring is $6000, choose what rooms to install the flooring in and give reasons for your choices. Include how much of your budget you used.

Research Project

One or more people

What you will need:

- Find data for a real-world situation that can be modeled with a linear function.

- You might want to use the Internet or library. Statistical abstracts and some journals and scientific articles are good resources for real-world data.

- You might want to do a search for statistical abstracts or the Centers for Disease Control and Prevention websites. These sites will have many possible data sets.

FIND YOUR OWN LINE

In this project, you are given the task of finding data for a real-world situation for which you can find a linear model. You may use the problems in this chapter to get ideas of things to investigate, but you should use data not discussed in this textbook.

Write up

a. Describe the data that you found and where you found the data set. Cite any sources you used.

b. Create a scatterplot of the data on the calculator or computer and print it out or neatly by hand on graph paper. Your scatterplot should be linear. If not, find another data set.

c. Find a model to fit the data.

d. What is the slope of your model and what does it represent in the context?

e. What is the vertical intercept of your model and what does it represent in the context?

f. What is the horizontal intercept of your model and what does it represent in the context?

g. What are a reasonable domain and range for your model?

h. Use your model to estimate an output value for an input value that will fit your model but that you did not collect in your original data.

Systems of Linear Equations and Inequalities

- Recognize a system of equations.
- Identify the solutions of a system of equations using the graph.
- Identify the solutions of a system of equations using numerical methods.
- Explain the meaning of a solution to a system of equations in a context.
- Recognize dependent and inconsistent systems of equations.
- Solve a system of linear equations using the substitution method.
- Solve a system of linear equations using the elimination method.
- Solve linear inequalities and explain the solution in a context.
- Set up a system of linear equations as an inequality when appropriate.
- Solve absolute value equations and inequalities.
- Graph a linear inequality with two variables.
- Graph a system of linear inequalities.

© BananaStock/Alamy

The number of cellular telephone subscribers has increased dramatically over the past twenty years. CTIA-The Wireless Association, an industry group, found that in 1985 there were an estimated 203,600 cell phone subscribers in the United States. That number grew to 233,041,000 in 2006. Comparing different cell phone plans is a valuable skill because choosing the best plan can save you money. In this chapter, we will discuss how you can analyze comparative data using systems of equations and inequalities. One of the chapter projects on page 291 will ask you to use graphing and problem solving to determine which cell phone service plan is the best value for you.

www   Mentor   iLrn

Introduction to Systems of Linear Equations

DEFINITION OF SYSTEMS

In many areas of life, we compare two or more quantities in an attempt to make a good decision about which is best for our situation. This type of decision-making process can often be simplified by using what in mathematics is called a **system of equations.** A system of equations is a set of equations that requires a solution that will work for all of the equations in the set. This is often the case when we are trying to determine when two options are going to be equal.

DEFINITION

System of Linear Equations: A set of two or more linear equations.
  Each equation in the system can contain one or more variables.
Solution of a System of Two Linear Equations: An ordered pair (or a set of ordered pairs) that is a solution to every equation in the system. On the graph, this is seen as the intersection point of the lines.

GRAPHICAL AND NUMERICAL SOLUTIONS

If we consider the system of equations

$$y = 2x + 5$$
$$y = 4x - 25$$

We can build tables of ordered pairs that satisfy each equation.

x	$y = 2x + 5$
0	5
5	15
10	25
15	35
20	45

x	$y = 4x - 25$
0	-25
5	-5
10	15
15	35
20	55

By looking at the two tables, you can see that the point (15, 35) is a solution to both equations and thus the solution to the system of equations. By graphing these two equations, we can see the solution to the system is the point where the two lines intersect.

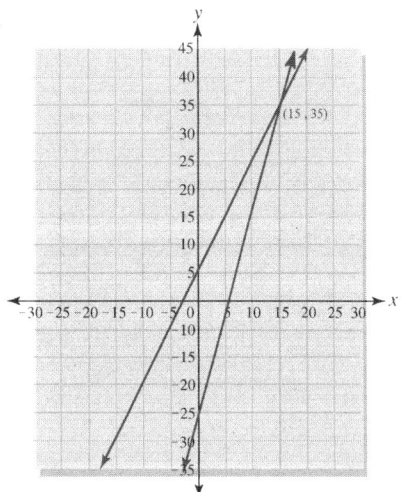

EXAMPLE 1 COMPARING TWO OPTIONS BY SOLVING A SYSTEM

Emily is moving to a new apartment and wants to rent a 10-foot truck. She called both U-Haul and Bob's Truck Rental to compare prices.

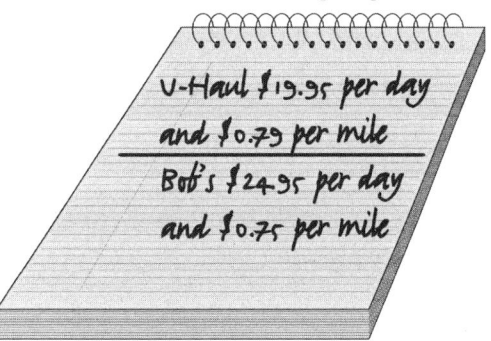

Use Emily's notes to answer the following.

a. Find equations for these two companies' total rental quotes for one day.

b. Graph the two equations on the same calculator window.

c. Find the distance traveled that will result in the two companies having the same cost.

Solution

a. First we define the variables:

$U(m)$ = The total cost in dollars to rent a 10-foot truck for a day from U-Haul

$B(m)$ = The total cost in dollars to rent a 10-foot truck for a day from Bob's

m = The number of miles traveled in the rented truck

We write the two equations as follows:

$$U(m) = 19.95 + 0.79m$$
$$B(m) = 24.95 + 0.75m$$

b. Put the function for Bob's into Y1 and the function for U-Haul into Y2 in the Y=screen, and remember to turn the stat plots off. To set an appropriate viewing window, let's consider the situation. We are renting the truck and cannot drive less than zero miles and will probably not drive more than 200 miles in one day, so set X-min and X-max to 0 and 200, respectively. We know we will be charged at least $19.95 for the rental, and if we drive the full 200 miles, the cost from each company reaches a maximum of a little less than $180, so let's set the Y-min and Y-max to 15 and 180, respectively.

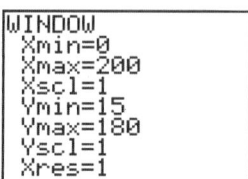 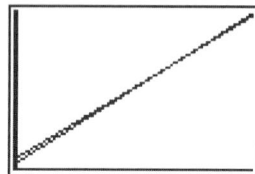

c. Using the trace button on the calculator, we estimate the solution to be about 130 miles at a cost of $122.29.

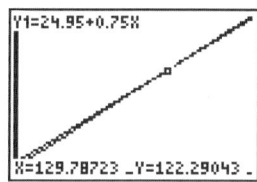

Therefore U-Haul and Bob's will both rent a 10-foot truck for a day at a cost of about $122.29 if you drive the truck 130 miles. To get a more precise solution to this system, we would need to zoom in on the graph so that the intersection becomes clearer or solve the system algebraically. The solution to this system tells us that U-Haul and Bob's charge the same amount if you rent a 10-foot truck for a day and drive it about 130 miles. Since U-Haul charges a lower day rate, they will be cheaper than Bob's for trips shorter than 130 miles. Bob's will result in a cheaper total cost for trips longer than 130 miles.

EXAMPLE ① PRACTICE PROBLEM

The number of American male smokers per thousand who are expected to die in the next 10 years from a heart attack or lung cancer can be modeled by the following two functions:

$$H(a) = 4.4a - 9$$
$$L(a) = 5.54a - 19.5$$

where $H(a)$ represents the number of American male smokers who are expected to die in the next 10 years from a heart attack per every 1000 male smokers aged a years over 40. $L(a)$ represents the number of American male smokers who are expected to die in the next 10 years from lung cancer per every 1000 male smokers aged a years over 40.

Source: Journal of the National Cancer Institute.

a. Graph both equations on the same window.

b. Find the age at which American male smokers have the same risk of dying from a heart attack as they do from lung cancer.

Solving systems of equations by graphing can be done by hand or using the graphing calculator. In the previous example, we solved the system by graphing the equations on the calculator. In the next example, we will graph the system by hand and find the solution.

EXAMPLE   **2** SOLVING SYSTEMS OF EQUATIONS BY GRAPHING

Solve the following system by graphing the equations by hand.

$$y = \frac{2}{3}x - 7$$

$$y = -x + 3$$

Solution

First we graph both equations on the same set of axes. The first line has a slope $m = \frac{2}{3}$ and a y-intercept of $(0, -7)$. The second graph has a slope $m = -1$ and a y-intercept of $(0, 3)$. Using these slopes and intercepts we get the graph.

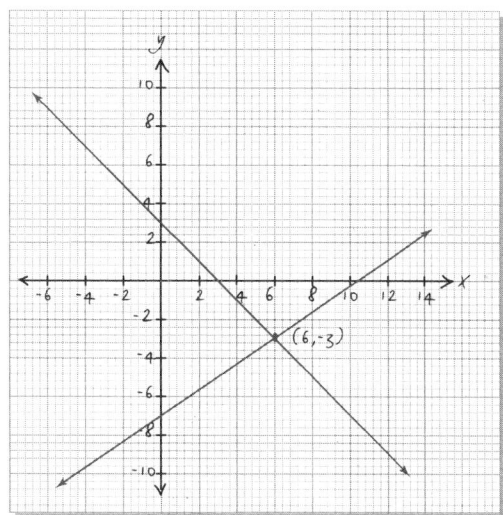

From the graph, we can see that the lines intersect at the point $(6, -3)$. Therefore, the solution to the system is $(6, -3)$. This solution should be checked by substituting the x and y values into both equations to make sure they are solutions to both equations.

First equation	Second Equation
$y = \frac{2}{3}x - 7$	$y = -x + 3$
$-3 = \frac{2}{3}(6) - 7$	$-3 = -(6) + 3$
$-3 = 4 - 7$	$-3 = -6 + 3$
$-3 = -3$	$-3 = -3$

Using Your TI Graphing Calculator

Using two stat plots:

To find a model for a second set of data you can either erase the first set of data or put the second set of data into the stat lists with the first set and turn on a second stat plot.

In Example 3 the input data for both sets of data are the same years so you only need to add the third list with the chicken consumption data.

After entering the data you can now turn on a Stat Plot 2 and make sure that Xlist is set to L1 and the Ylist is set to L3, where we put the chicken consumption data.

Now set your window to see both sets of data and create your model.

Sometimes it may be best to plot one set of data at a time when deciding on what points to pick for the modeling process.

EXAMPLE ② PRACTICE PROBLEM

Solve the following system by graphing the equations by hand.

$$y = \frac{1}{2}x - 2$$

$$y = -\frac{1}{4}x + 7$$

EXAMPLE 3 USING MODELS IN A SYSTEM OF EQUATIONS

The amounts of beef and chicken consumed by Americans are given in the table.

Year	Beef Consumption (in millions of pounds)	Chicken Consumption (in millions of pounds)
1996	25,861	21,845
1998	26,305	23,254
2000	27,338	25,606
2002	27,878	27,467
2004	28,534	28,678

Source: U.S. Department of Agriculture National Agricultural Statistics Service.

a. Find a model for the amount of beef consumed by Americans.

b. Find a model for the amount of chicken consumed by Americans.

c. Estimate the year in which the amount of beef consumed will be the same as the amount of chicken consumed.

Solution

a. Using the steps to find a linear model from Chapter 1, we define the variables.

B = The amount of beef consumed by Americans, in millions of pounds

t = years since 1990

Using the first and next to last points given, we get the model

$$B(t) = 336.17t + 23844$$

b. Now, for a model to find the amount of chicken consumed, we define the variables.

C = The amount of chicken consumed by Americans, in millions of pounds

t = years since 1990

Using the first and last points, we get the model

$$C(t) = 854.125t + 16720.25$$

c. Graph the two models on one window to see the intersection and trace the estimated answer.

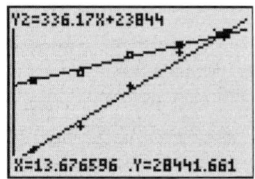

From this graph we can estimate the solution to be (13.68, 28,441.7). Thus in 2004 the amounts of chicken and beef consumed by Americans were approximately 28,442 million pounds.

Using the graph and trace on your calculator is one way to check the solutions to many of the equations we will solve in this book. Another way for you to find solutions to systems of equations is to use the table feature of your calculator. You can also use the table feature to check a solution to a system of equations. First enter both equations in the $Y =$ screen and then go to the table. Enter the input values you want to check, and as long as the y values that are returned are the same, you have found a solution. In Example 2, we graphed and traced to find the approximate solution of (13.68, 28,441.7). Using these two functions and values close to 13.68 in the table, we get

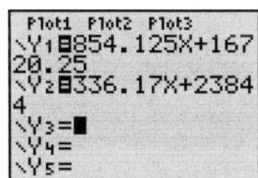 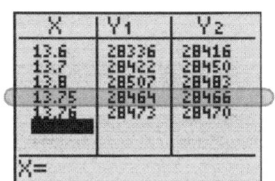

From this table, we can see that a more accurate solution would be about (13.75, 28,465). Because we did not find an exact solution, we averaged the y values to get an approximate solution. In this particular application, we round the answers to a whole number for the year, so the table does confirm that our solution is close.

EXAMPLE 4 SOLVING SYSTEMS OF EQUATIONS NUMERICALLY

Use the table on the calculator to numerically find the solution to the following systems of equations.

a. $y = 2x + 7$

$y = 5x - 3.5$

b. $y = -9x + 177$

$y = 13x - 65$

Solution

a. Using the table, we can guess a value for x to start and then guess again and again until we get the y values to be closer and eventually the same.

X	Y1	Y2
0	7	-3.5
2	11	6.5
4	15	16.5

X=

The first guesses were $x = 0$, $x = 2$, and $x = 4$. With the first two guesses, Y1 was greater than Y2. When $x = 4$,the relationship reversed. Now Y1 is less than Y2. This means that the place where they are equal must be between $x = 2$ and $x = 4$.

X	Y1	Y2
0	7	-3.5
2	11	6.5
4	15	16.5
3	13	11.5

X=

A guess of $x = 3$ resulted in Y1 being greater than Y2 again, so we need to guess again. The answer seems to fall somewhere between 3 and 4.

X	Y1	Y2
0	7	-3.5
2	11	6.5
4	15	16.5
3	13	11.5
3.5	14	14

X=

A guess of $x = 3.5$ results in Y1 equal to Y2, so we have found the solution to the system of equations.

From the last table we can see that the solution is (3.5, 14).

b. Using the same guessing technique, we get the table

X	Y1	Y2
0	177	-65
2	159	-39
4	141	-13
8	105	39
10	87	65
11	78	78
12	69	91

X=8

From this table, we can see that the solution is (11, 78).

EXAMPLE PRACTICE PROBLEM

Use the table on the calculator to numerically find the solution to the systems of equations.

a. $y = 2x - 8$

$y = -4x + 28$

b. $y = -\dfrac{3}{5}x + 9$

$y = 0.6(7 - x) + 2.8$

Remember that the table is a good way to get a solution numerically. However, this method can be difficult if you have no idea where the solution is. The table is a great way to check solutions to equations, and we will use it often in the remaining chapters of this book.

TYPES OF SYSTEMS

 **CONCEPT INVESTIGATION I** WHAT KIND OF SYSTEMS ARE THERE?

To start, turn off all your stat plots, set your window to a standard window using [zoom] [6] (ZStandard), and clear all equations from the Y= screen.

a. Graph the following systems one at a time, and answer the questions.

i. $y = 2x - 5$ How many solutions does this system have?
$y = -5x + 9$

ii. $w = \frac{2}{3}p - 2$ How many solutions does this system have?

$w = \frac{1}{9}p + 3$

iii. $3c - 2d = -13$ How many solutions does this system have?
$5c + d = -13$

b. Graph the following systems one at a time, and answer the questions.

i. $-x + 5y = -15$ How many solutions does this system have?

$y = \frac{3}{15}x + 2$ How are the equations in this system related?

ii. $P = 2.75t - 4.25$ How many solutions does this system have?
$P = 2.75t + 3$ How are the equations in this system related?

c. Graph the following systems one at a time, and answer the questions.

i. $6x + 2y = 14$ How many solutions does this system have?
$21x + 7y = 49$ How are the equations in this system related?

ii. $-3.5x - 2.75y = 4.5$ How many solutions does this system have?
$9.8x + 7.7y = -12.6$ How are the equations in this system related?

This concept investigation shows that systems of two linear equations can have three types of answers.

In mathematics we give the following names to these three types of systems.

- **Consistent system:** A system with at least one solution. The lines intersect in one place.

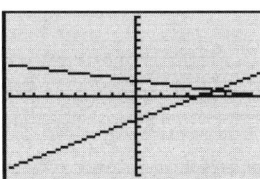

• **Inconsistent system:** A system with no solutions. The lines are parallel and, therefore, do not intersect. Use caution when looking at the calculator, because some lines might seem parallel but really are not. Recall that parallel lines have the same slope but different *y*-intercepts, so if you compare the slopes of the equations, you can determine if the system is inconsistent.

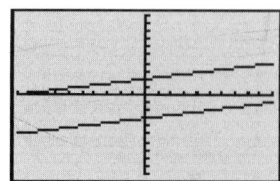

• **Dependent system:** A system with infinitely many solutions. The lines are the same and, therefore, intersect in infinitely many places, and each point on the line is a solution for the system. Use caution when looking at the calculator, because some lines might seem the same but really are not. For the lines to be the same, they must have the same slope and *y*-intercept. These systems are also considered consistent.

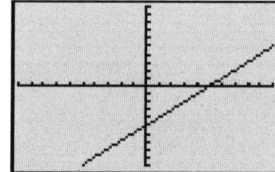

Getting an idea for the visual representation of these solutions will help in the next section when we solve systems algebraically.

EXAMPLE   **5**   DESCRIBING TYPES OF SYSTEMS OF EQUATIONS

For each of the following systems of equations, determine what type of system it is and give the solution.

a. $y = 3x + 5$

   $y = 3x - 2$

b. $2x + 5y = 20$

   $y = 4x - 18$

c.
   $y = \frac{1}{3}x + 4$

   $x - 3y = -12$

Solution

a.   By looking at these two equations, we can see that the slopes are both $m = 3$ but that they do not have the same *y*-intercept. Therefore, these lines are parallel, and the system is inconsistent and has no solutions. If we look at a table of points for these lines, the outputs will always be 7 units apart.

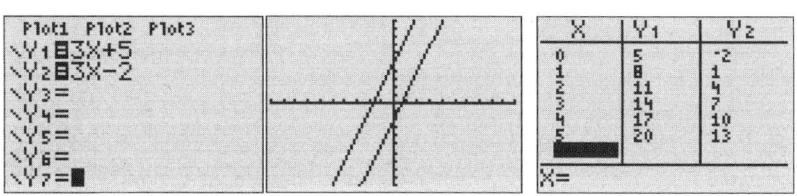

b. To compare these equations, we will put the first equation into slope-intercept form.

$$2x + 5y = 20$$
$$\underline{-2x \qquad\quad -2x}$$
$$5y = -2x + 20$$
$$\frac{5y}{5} = \frac{-2x + 20}{5}$$
$$y = -\frac{2}{5}x + 4$$

Now, if we compare the equations

$$y = -\frac{2}{5}x + 4$$
$$y = 4x - 18$$

we can see that the slopes are different, so the system is consistent, using the table, we can find the solutions numerically.

From the table, we can see that the solution is the point $(5, 2)$.

c. To compare these equations, we will put the second equation into slope-intercept form.

$$x - 3y = -12$$
$$\underline{-x \qquad\quad -x}$$
$$-3y = -x - 12$$
$$\frac{-3y}{-3} = \frac{-x - 12}{-3}$$
$$y = \frac{1}{3}x + 4$$

Now, if we compare the equations

$$y = \frac{1}{3}x + 4$$

$$y = \frac{1}{3}x + 4$$

we can see that the equations have the same slope and y-intercept, so they are the same line. This is a dependent system and has an infinite number of solutions.

Any solution of the equation $y = \frac{1}{3}x + 4$ is a solution to the system.

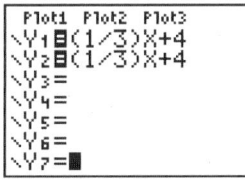 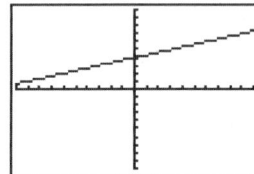

EXAMPLE ⑤ PRACTICE PROBLEM

For each of the following systems of equations, determine what type of system it is and give the solution.

a. $5x + y = 3$

$y = -5x - 7$

b. $y = \frac{1}{4}x + 7$

$y = 0.125(2x + 40) + 2$

2.1   Exercises

For Exercises 1 through 6 use the graphs to find the solution to the system of equations.

1.

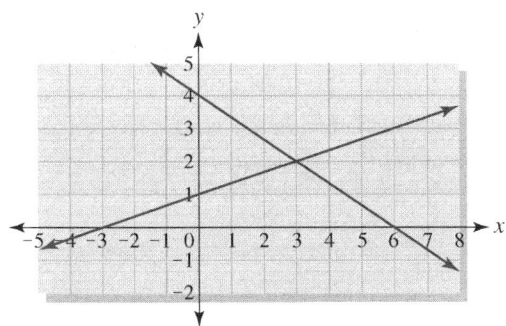

2.

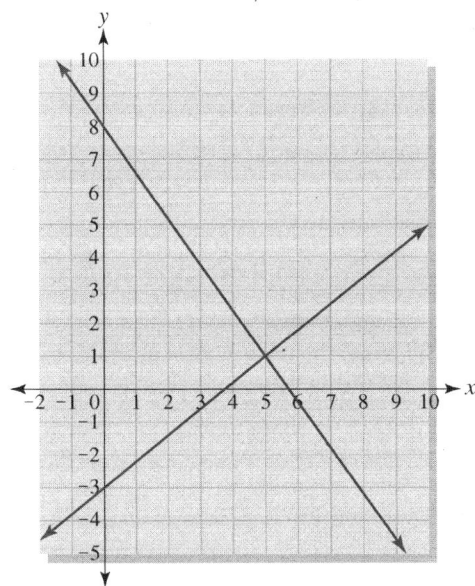

3.

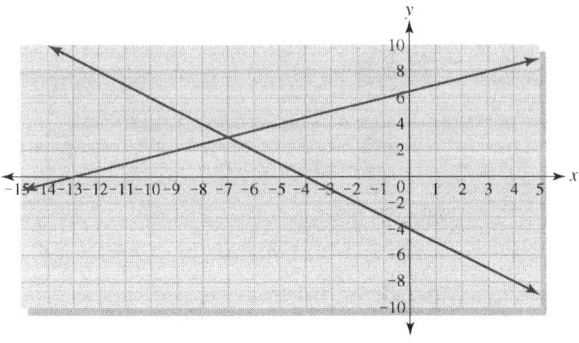

4.

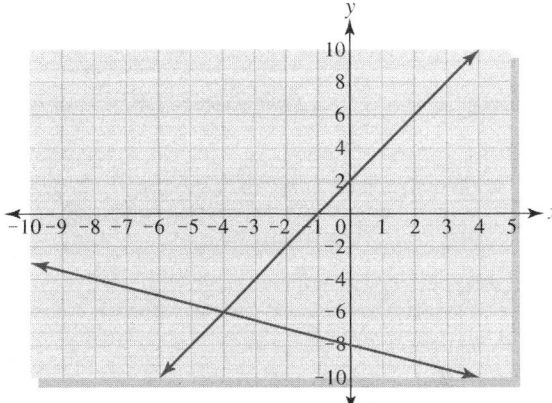

5.

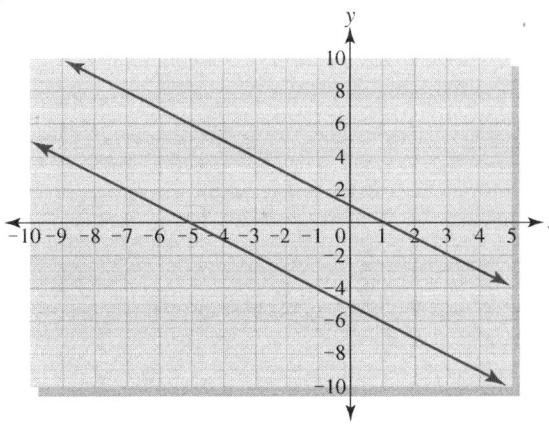

6.

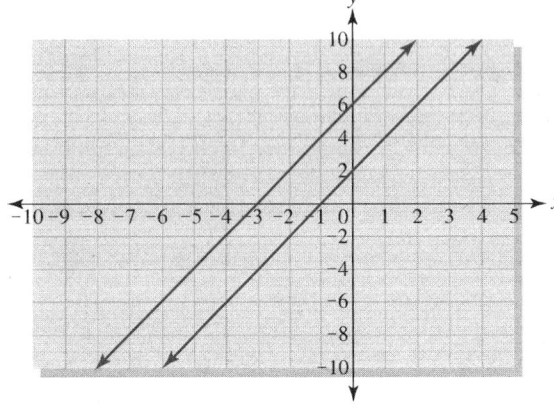

7. Use the graph to answer the questions

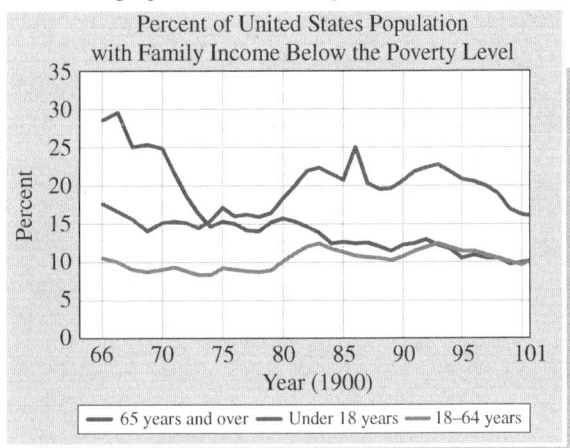

Source: http://www.cdc.gov/nchs/hus.htm.

a. Estimate when and at what percent the percentage of people under 18 years was the same as the percentage of people 65 years and over with family income below the poverty level.

b. Estimate when and at what percent the percentage of people 65 years and over with family income below the poverty level was first equal to the percentage of people 18–64 years old.

8. Use the graph to answer the questions.

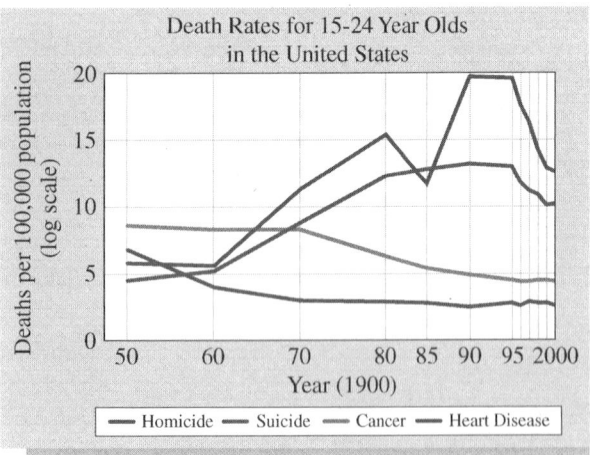

Source: http://www.cdc.gov/nchs/hus.htm.

a. Estimate the year in which the number of deaths from cancer was the same as the number from

suicide. Approximate the number of deaths per 100,000 population from each of these causes in that year.

b. Estimate the year in which the number of deaths from heart disease was the same as the number from homicide. Approximate the number of deaths per 100,000 population from each of these causes in that year.

9. The amount of manufacturing output per hour in dollars by the United States and Norway is given in the table.

Year	United States ($)	Norway ($)
2000	147.7	105.9
2003	175.5	121.6
2004	187.8	128.8
2005	194.0	132.4

Source: Statistical Abstract 2008.

a. Find models for the data given.

b. Graph both models on the same calculator screen.

c. Estimate when the amount of manufacturing output per hour by the United States will be equal to the amount of manufacturing output per hour by Norway.

10. The number of women enrolled in U.S. colleges has been increasing steadily since the 1970s. The numbers of women and men enrolled in U.S. colleges are given in the following bar charts.

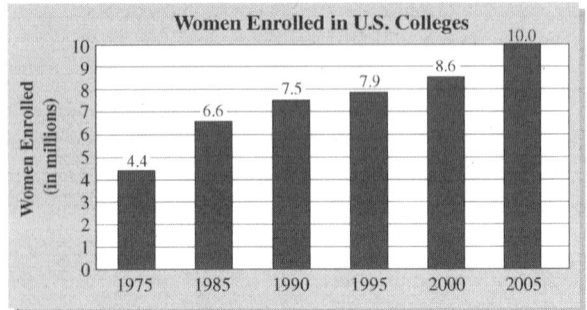

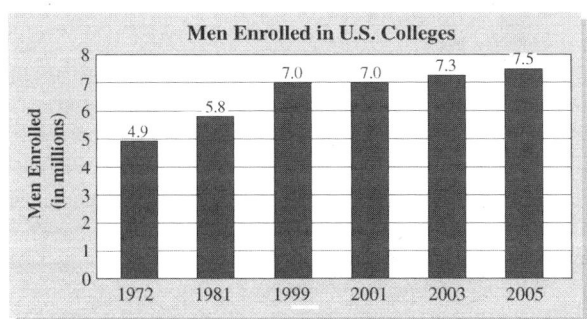

Source: Statistical Abstract 2001 and 2008.

a. Find models for the numbers of women and men enrolled in U.S. colleges.

b. Graph both models on the same calculator screen.

c. Estimate when the number of women enrolled was the same as the number of men enrolled.

11. The percentages of white male and female Americans 25 years old or older who are college graduates are given in the table.

Year	Males (%)	Females (%)
1980	21.3	12.3
1985	24	16.3
1990	25.3	19
1995	27.2	21
2000	28.5	23.9
2005	29.4	26.8

Source: Statistical Abstract 2001 and 2008.

a. Find models for the data given.

b. Graph both models on the same calculator screen.

c. Estimate when the percentage of white American males with college degrees will be the same as the percentage of white American women with college degrees.

12. The number of bachelor's degrees earned by blacks and Hispanics in the United States is given in the bar chart.

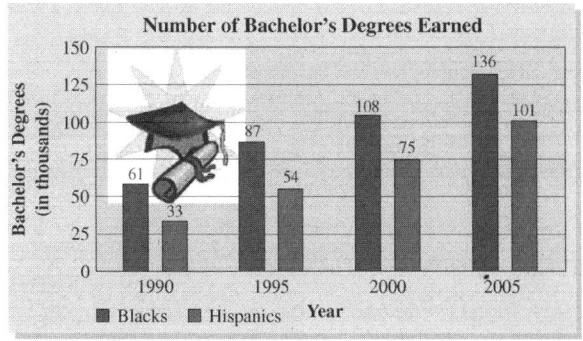

Source: Statistical Abstract 2008.

a. Find models for the bachelor's degrees earned data given.

b. Graph both models on the same calculator screen.

c. Determine in what year blacks and Hispanics will earn the same number of bachelor's degrees.

13. Hope's Pottery makes clay vases to sell at a local gallery. Hope has determined the following models for her monthly revenue and costs associated with making and selling these vases.

© Royalty-Free/CORBIS

$$R(v) = 155v$$
$$C(v) = 5000 + 65v$$

where $R(v)$ represents the monthly revenue in dollars for selling v vases and $C(v)$ represents the monthly costs in dollars for producing and selling v vases.

a. Graph both functions on the same calculator screen.

b. Determine the break-even point for Hope's Pottery. (The break-even point is when the revenue equals the costs.)

14. Jim's Carburetors rebuilds carburetors for local auto repair shops. Jim has determined the following models for his monthly revenue and costs associated with rebuilding these carburetors.

$$R(c) = 65c$$
$$C(c) = 2300 + 17c$$

where $R(c)$ represents the monthly revenue in dollars for rebuilding c carburetors and $C(c)$ represents the monthly costs in dollars for rebuilding c carburetors.

a. Graph both functions on the same calculator screen.

b. Determine the break-even point for Jim's Carburetors.

15. *La Opinion,* a Spanish-language newspaper in Los Angeles County, California, is one of the area's fastest-growing daily newspapers. The *Long Beach Press Telegram* is another large daily newspaper in Los Angeles County. The daily circulation for both of these newspapers is given by the following functions:

$$O(t) = 7982t + 28{,}489$$
$$L(t) = 726t + 97{,}395$$

where $O(t)$ represents the daily circulation of La Opinion t years since 1990 and $L(t)$ represents the daily circulation of the Long Beach Press Telegram t years since 1990.

Source: Los Angeles Almanac 2001.

a. Use graphing to estimate when the two newspapers had the same daily circulation.

b. Compare the slopes of these two functions and describe what they represent in this context.

16. The number of doctoral degrees given to black and Asian or Pacific Islanders in the United States can be modeled by the following functions:

$$B(t) = 126t + 1084.5$$
$$A(t) = 95.78t + 1592.9$$

where $B(t)$ represents the number of doctoral degrees earned by blacks t years since 1990 and $A(t)$ represents the number of doctoral degrees

earned by Asian and Pacific Islanders t years since 1990.

Source: Statistical Abstract 2008.

a. Use graphing to estimate when blacks and Asian and Pacific Islanders will earn the same number of doctoral degrees.

b. Compare the slopes of these two functions and describe what they represent in this context.

17. A local BMW dealer has two salary options for its sales force. The sales force can choose either of the following two salary options:

$$O_1(s) = 250 + 0.03s$$
$$O_2(s) = 0.03s + 200$$

where $O_1(s)$ represents weekly salary option 1 in dollars when s dollars in sales are made per week and $O_2(s)$ represents weekly salary option 2 in dollars when s dollars in sales are made per week. Find what sales level will give a salesperson the same salary with either option.

18. Harry's Flooring has two salary options for its salespeople. Each salesperson can choose which option to base his or her salary on:

$$O_1(s) = 500 + 0.07s$$
$$O_2(s) = 0.07s + 700$$

where $O_1(s)$ represents monthly salary option 1 in dollars when s dollars in sales are made per month and $O_2(s)$ represents monthly salary option 2 in dollars when s dollars in sales are made per month. Find what sales level would result in both salary options having the same monthly salary.

For Exercises 19 through 26, graph each system by hand and find the solution(s) to the system. Label each system as consistent, inconsistent, or dependent.

19. $y = 2x + 4$

$y = -\dfrac{1}{3}x + 11$

20. $y = x - 2$

$y = \dfrac{1}{2}x + 2$

21. $y = \dfrac{2}{3}x - 4$

$y = -\dfrac{1}{3}x - 10$

22. $y = \dfrac{1}{5}x + 4$

$y = -\dfrac{2}{5}x + 1$

23. $y = 0.5x + 2$
$\quad x - 2y = 8$

24. $y = x - 5$
$\quad 3x - 3y = -24$

25. $n = \dfrac{2}{3}m + 7$
$\quad 6m - 9n = -63$

26. $p = 0.25t - 4$
$\quad 2t + 8p = -32$

For Exercises 27 through 36, graph each system by hand or using a graphing calculator and find the solution(s) to the system. Label each system as consistent, inconsistent, or dependent.

27. $x + y = -6$
$\quad -2x + y = 3$

28. $w = 8d + 54$
$\quad w = \dfrac{3}{2}d + \dfrac{17}{2}$

29. $p = 2.5t + 6$
$\quad p = \dfrac{5}{2}t - 6$

30. $3x - 4y = 8$
$\quad 0.75x - y = -2$

31. $y = \dfrac{1}{3}x + 5$
$\quad 2x - 6y = -30$

32. $H = 3c + 6.5$
$\quad -6c + 2H = 13$

33. $R = 2.75t + 6.35$
$\quad R = -1.5t + 12.45$

34. $D = 3.5c + 6.5$
$\quad D = 2.5c - 4.5$

35. $4x + 5y = -7$
$\quad 3x - 7y = 27$

36. $2t - w = 4$
$\quad w = -\dfrac{1}{2}t - 2$

For Exercises 37 through 44, solve the systems numerically using the given tables. Label each system as consistent, inconsistent, or dependent.

37.

X	Y₁	Y₂
0	5	19
1	8	15
2	11	11
3	14	7
4	17	3
5	20	-1

X=

38.

X	Y₁	Y₂
0	5	-23
-1	2	-19
-2	-1	-15
-3	-4	-11
-4	-7	-7
-5	-10	-3
-6	-13	1

X= -6

39.

X	Y₁	Y₂
0	3	-7
5	8	-2
10	13	3
15	18	8
20	23	13
25	28	18

X=

40.

X	Y₁	Y₂
0	15	4
1	13	2
2	11	0
3	9	-2
4	7	-4
5	5	-6

X=

41.

X	Y₁	Y₂
0	1	1
1	3	3
2	5	5
3	7	7
4	9	9
5	11	11

X=0

42.

X	Y₁	Y₂
0	-4	-4
2	-5.333	-5.333
4	-6.667	-6.667
6	-8	-8
8	-9.333	-9.333
10	-10.67	-10.67

X=

43.

X	Y₁	Y₂
0	5	-5
1	8	-2
2	11	1
3	14	4
4	17	7
5	20	10
6	23	13

X=6

44.

X	Y₁	Y₂
0	9	8
1	11	10
2	13	12
3	15	14
4	17	16
5	19	18
6	21	20

X=0

For Exercises 45 through 52, solve the systems numerically. Label each system as consistent, inconsistent, or dependent.

45. $p = 3t + 8$
$\quad p = 2t + 17$

46. $y = 5x - 20$
$\quad y = 2x + 1$

47. $2.5x - 0.5y = -23.5$
$\quad 7.5x - 1.5y = -23.5$

48. $R = 0.8t + 6$
$\quad R = \dfrac{4}{5}t + 40.5$

49. $C = 1.4n + 74$
$\quad 7n - 5C = -370$

50. $8t + 4p = 23$
$\quad 2p = -4t + 11.5$

51. $y = 2x - 20$
$\quad y = -4x + 70$

52. $y = 5x - 20$
$\quad y = -2x + 8$

Solving Systems of Equations Using the Substitution Method

In Section 2.1, we learned how to solve systems of two linear equations by graphing and numerically using tables. These methods can be done well on a graphing calculator. One drawback to solving some systems by graphing or numerically is that you have to have an idea of where the intersection is going to be in order to find it. Another concern when solving systems graphically or numerically is that we may not be able to find the exact answers. To avoid these problems, we can use algebraic methods that will find the exact solution whenever possible.

SUBSTITUTION METHOD

There are two basic algebraic ways to solve systems of equations: the **substitution method** and the **elimination method.** Because most of the models we find in this text are in $y = mx + b$ form, we will learn the substitution method first. The substitution method means to substitute the expression of a variable from one equation into that same variable in the other equation.

> ### SUBSTITUTION METHOD FOR SOLVING SYSTEMS OF EQUATIONS
>
> * Best used when a variable is already isolated in at least one equation.
> * Substitute the expression representing the isolated variable from one equation in place of that variable in the other equation.
> * Remember to find the values for both variables.
> * Check the solution in both equations.

EXAMPLE 1 SOLVING SYSTEMS USING THE SUBSTITUTION METHOD

Solve the system using the substitution method.
$$w = 5k + 7$$
$$w = -3k + 23$$

Solution

Because w is already isolated on the left side of at least one of the equations, we can substitute the expression from the first equation into the w for the second equation:
$$5k + 7 = -3k + 23$$

Now that we have substituted we can solve for k.

$$5k + 7 = -3k + 23$$
$$\underline{+3k \qquad\qquad +3k}$$
$$8k + 7 = 23$$
$$\underline{\qquad\quad -7 \quad -7}$$
$$8k = 16$$
$$\frac{8k}{8} = \frac{16}{8}$$
$$k = 2$$

We know that $k = 2$, so we can substitute 2 for k in either equation and find w.

$$w = 5k + 7$$
$$w = 5(2) + 7$$
$$w = 17$$

The solution to the system is the point $(2, 17)$. We should always check the solution in both equations to be sure that it is valid.

First Equation	Second Equation
$w = 5k + 7$	$w = -3k + 23$
$17 = 5(2) + 7$	$17 = -3(2) + 23$
$17 = 17$	$17 = 17$

EXAMPLE $\boxed{2}$ USING THE SUBSTITUTION METHOD TO SOLVE SYSTEMS WITH MODELS

Using the models from Section 2.1 for beef and chicken consumption by Americans, find the year when beef and chicken consumption was or will be the same.

$$B(t) = 336.17t + 23844$$
$$C(t) = 854.125t + 16720.25$$

where $B(t)$ is the amount of beef consumed by Americans in millions of pounds t years since 1990 and $C(t)$ is the amount of chicken consumed by Americans in millions of pounds t years since 1990.

Solution

Because both $B(t)$ and $C(t)$ represent amounts of meat consumed by Americans and both are measured in millions of pounds, we can substitute the expression that $B(t)$ equals for $C(t)$ in the second equation:

$$B(t) = 336.17t + 23844$$
$$C(t) = 854.125t + 16720.25$$
$$B(t) = C(t)$$
$$336.17t + 23844 = 854.125t + 16720.25 \qquad \text{Set the two expressions equal to}$$
$$\underline{-854.125t \qquad\qquad -854.125t} \qquad\qquad \text{each other.}$$
$$\qquad\qquad\qquad\qquad\qquad\qquad\qquad\qquad\qquad \text{Solve for } t.$$
$$-517.955t + 23844 = 16720.25$$
$$\underline{\qquad\qquad -23844 \quad -23844}$$
$$-517.955t = -7123.75$$

$$\frac{-517.955t}{-517.955} = \frac{-7123.75}{-517.955}$$

$$t \approx 13.7536$$

Now that we have the value of t, we need to find the amounts of beef and chicken consumed that year. We can do that by substituting this value for t in either equation.

$$B(13.7536) = 336.17(13.7536) + 23844$$
$$B(13.7536) \approx 28467.548$$

or

$$C(13.7536) = 854.125(13.7536) + 16720.25$$
$$C(13.7536) \approx 28467.544$$

Since both equations came to the same approximate answer, we have found the solution to this system. But because t represents a year, we need to round it to the nearest whole number—in this case, 14. If we substitute 14 into the variable t for these equations, we will find that they are not exactly equal:

First Equation

$$B(14) = 336.17(14) + 23844$$
$$B(14) = 28550.38$$

Second Equation

$$C(14) = 854.125(14) + 16720.25$$
$$C(14) = 28678$$

This gives us the following approximate answer. Americans ate approximately 28,600 million pounds of both beef and chicken in 2004.

EXAMPLE ② PRACTICE PROBLEM

The Pharmaceutical Management Agency of New Zealand, PHARMAC, manages the expenditures on various medical conditions and drugs throughout New Zealand. The amount spent on treating diabetes in New Zealand can be modeled by

$$T(t) = 1.1t + 6.84$$

where $T(t)$ represents the amount in millions of New Zealand dollars spent on treating diabetes t years since 1990. Also, the amount spent on diabetes research in New Zealand can be modeled by

$$R(t) = 1.3t + 0.52$$

where $R(t)$ represents the amount in millions of New Zealand dollars spent on diabetes research in New Zealand t years since 1990.

Source: PHARMAC.

Find the year in which the amount spent in New Zealand on treating diabetes will be the same as the amount spent in New Zealand on diabetes research.

EXAMPLE 3 SOLVING SYSTEMS USING THE SUBSTITUTION METHOD

Solve the following systems using the substitution method.

a.
$$p = 2m - 12$$
$$p - 3m = -17$$

b.
$$4c = 2d + 10$$
$$6c + 2d = -5$$

Solution

a. p is isolated in the first equation, so the expression that p is equal to can be substituted into p for the second equation:

$$p = 2m - 12$$
$$p - 3m = -17$$

Substitute and solve for m:

$$(2m - 12) - 3m = -17$$
$$-1m - 12 = -17$$
$$-1m = -5$$
$$m = 5$$

We know that $m = 5$, so we can substitute 5 for m in either of the equations to find p:

$$p = 2(5) - 12$$
$$p = -2$$

Therefore, we believe that the solution to this system is $(5, -2)$. We will check this solution in both equations to be sure that it is valid:

First Equation	Second Equation
$-2 = 2(5) - 12$	$-2 - 3(5) = -17$
$-2 = -2$	$-17 = -17$

b. Neither variable is already isolated, so we can solve the first equation for c, and then that expression can be substituted into c for the second equation:

$$4c = 2d + 10$$
$$\frac{4c}{4} = \frac{2d + 10}{4}$$
$$c = 0.5d + 2.5$$

Substitute this expression into c and solve for d:

$$6(0.5d + 2.5) + 2d = -5 \qquad \text{It is important to use parentheses around the}$$
$$3d + 15 + 2d = -5 \qquad \text{expression you are substituting into the variable.}$$
$$5d + 15 = -5$$
$$5d = -20$$
$$d = -4$$

We know that $d = -4$, so we can substitute -4 for d in either of the equations to find c:

$$4c = 2(-4) + 10$$
$$4c = -8 + 10$$
$$4c = 2$$
$$c = 0.5$$

Therefore the solution to this system is $c = 0.5$ and $d = -4$. You should check this solution in both equations to be sure that it is valid:

First Equation	Second Equation
$4(0.5) = 2(-4) + 10$	$6(0.5) + 2(-4) = -5$
$2 = 2$	$-5 = -5$

EXAMPLE ③ PRACTICE PROBLEM

Solve the following systems using the substitution method.

a.
$$y = 3x + 12$$
$$y = 20x - 158$$

b.
$$n = 4m + 3$$
$$n - 3m = 8.5$$

EXAMPLE ④ SOLVING APPLICATIONS USING SYSTEMS OF EQUATIONS

Clear Sign Designs, a local sign manufacturer, wants to give its sales force incentives to make bigger sales. To do this, the company executives are changing the salary structure. The old system was a base salary of $1000 per month and 8% commission on sales made. The new salary structure will consist of a base salary of $500 per month with a 10% commission on sales made.

a.   Find equations to represent the new and old salary structures.

b.   Find what sales amount will result in the same monthly salary for the sales force.

Solution

a.   Define the variables:

N = Salary in dollars for the new salary structure.

O = Salary in dollars for the old salary structure.

s = Sales made during the month in dollars.

Changing the commission percentages into decimals, we get the following two equations:

$$N(s) = 500 + 0.10s$$
$$O(s) = 1000 + 0.08s$$

b.   To find the sales amount that will result in the same salary, we need the new salary $N(s)$ to be equal to the old salary $O(s)$. So we can substitute the expression $500 + 0.10s$ into the second equation for $O(s)$. Solving this will give us the sales amount that will result in the same monthly salary. Using that sales amount, we can find the actual monthly salary earned.

$$N(s) = 500 + 0.10s$$
$$O(s) = 1000 + 0.08s$$

$$N(s) = O(s)$$

$500 + 0.10s = 1000 + 0.08s$ Set the two expressions equal to each other.

$\underline{-500 - 0.08s \quad -500 - 0.08s}$

$0.02s = 500$ Solve for s.

$$\frac{0.02s}{0.02} = \frac{500}{0.02}$$

$$s = 25000$$

so

$$N(25000) = 500 + 0.10(25000) \qquad \text{Find } N \text{ when } s = 25000$$
$$N(25000) = 3000$$

Check the solution in the table.

Therefore, if a salesperson has monthly sales of $25,000, the salary will be the same, $3000, on the new and old salary structures.

EXAMPLE 5 SOLVING INVESTMENT PROBLEMS USING THE SUBSTITUTION METHOD

Fay Clark is retired and has $500,000 to invest. She wants her investments to earn $24,000 per year in interest so she can live on the interest. She is considering depositing most of the funds in a very safe bank account that pays 3.5% simple interest per year and the rest in a more risky account that pays 10% simple interest per year.

a. Write a system of equations that will help you to find the amount in each investment.

b. How much should she invest in each account to earn the $24,000 she needs each year?

(*Note:* Some financial accounts, such as bank deposits, are federally insured, while others such as mutual funds and brokerage accounts are not federally insured and are therefore considered more risky.)

Solution

a. There are two main factors controlling this situation: the amount of money Fay can invest and the total amount of interest income she needs to earn each year. If we set up two equations using these facts, we can solve the system for the amounts in each account. We start by defining the variables:

A = The amount invested in dollars in the account paying 3.5% simple interest

B = The amount invested in dollars in the account paying 10% simple interest

With these definitions and the conditions we were given, we get the following two equations:

$$A + B = 500000 \qquad \text{The two accounts total \$500,0000.}$$
$$0.035A + 0.10B = 24000 \qquad \text{The total interest needs to be \$24,000.}$$

b. To find the amount that she should deposit in each account, we solve the system. To use the substitution method, we will isolate either variable. For this systems, we will isolate the variable A in the first equation and substitute the expression it equals into the second equation.

Say What?

Simple Interest:
Simple interest for one year is calculated as the amount invested multiplied by the interest rate (as a decimal).

If we have $5000 invested at 4.5% simple interest for one year, the interest will be calculated as

$$5000(0.045) = 225$$

The simple interest is $225.

Remember to change the interest rate into a decimal by multiplying by $\frac{1}{100}$ or 0.01.

$$A + B = 500000$$
$$0.035A + 0.10B = 24000$$

Subtract B from both sides to isolate A.

$$A = 500000 - B$$
$$0.035A + 0.10B = 24000$$

$$0.035(500000 - B) + 0.10B = 24000$$
$$17500 - 0.035B + 0.10B = 24000$$
$$17500 + 0.065B = 24000$$
$$\underline{-17500 \qquad\qquad -17500}$$

Substitute 500000 - B for A in the second equation.
Distribute the 0.035 through the parentheses.
Solve for B.

$$0.065B = 6500$$
$$\frac{0.065B}{0.065} = \frac{6500}{0.065}$$
$$B = 100000$$

Now that we found B, we can substitute \$100,000 into the B in the first equation and solve for A.

$$A + 100000 = 500000 \qquad \text{Solve for } A.$$
$$A = 400000$$

Check the solution in both equations:

First Equation	Second Equation
$400000 + 100000 = 500000$	$0.035(400000) + 0.10(100000) = 24000$
$500000 = 500000$	$24000 = 24000$

This solution tells us that Fay should deposit \$400,000 into the account paying 3.5% interest and \$100,000 into the more risky account paying 10% interest. This will give her a total of \$24,000 in interest each year.

EXAMPLE ⑤ PRACTICE PROBLEM

George is retiring and has \$350,000 to invest. George needs \$18,000 per year more than his Social Security to live on. George has two investments that he plans to use to earn the interest he needs to live on. One investment is conservative and pays 3.6% simple interest annually; the other is less conservative and pays 6% simple interest annually. Determine how much George should deposit into each account to make the interest he needs to live on each year. Note that George must invest the entire \$350,000.

DEPENDENT AND INCONSISTENT SYSTEMS

Recall from Section 2.1 that there are three types of solutions to linear systems. Although most systems will have a single solution, there are times when a system may be dependent or inconsistent and thus have an infinite number of solutions or no solutions, respectively. In solving a system of equations algebraically, finding out when you have a dependent or inconsistent system happens rather suddenly. Keeping an eye open for systems of parallel lines (inconsistent systems) or systems with multiples of the same equation (dependent systems) is one way to notice these situations. But these patterns are not always clear, since the equations can look very different yet be remarkably similar or the same.

While solving dependent or inconsistent systems, you will notice that all the variables will be eliminated, and you will be left with two numbers set equal to each other. If these remaining numbers are actually equal, then the equations must have been the same and you have a dependent system and an infinite number of solutions. If the remaining numbers are not equal, then the equations were not the same so the system is inconsistent and there is no solution.

Concept Connection

Consistent System: A system with one solution.

Inconsistent System: A system with no solutions. These systems have two parallel lines. The equations have the same slopes but different y-intercepts.

Dependent System: A system with infinitely many solutions. These systems are also considered consistent and occur when one equation is a multiple of the other equation. Once simplified, the two equations are the same.

E X A M P L E SOLVING SYSTEMS AND DETERMINING THE SYSTEM TYPE

Solve the following systems and label them as consistent, dependent, or inconsistent.

a. $r = 4.1t + 6.3$
$3r - 12.3t = 18.9$

b. $y = 4x + 9$
$y = 4x - 9$

c. $S = 1500 + 0.05c$
$S = 0.05c + 750$

Solution

a. This system is set up well for substitution, because one variable is already isolated:

$$r = 4.1t + 6.3$$
$$3r - 12.3t = 18.9$$

$$3(4.1t + 6.3) - 12.3t = 18.9 \quad \text{Substitute for } r.$$
$$12.3t + 18.9 - 12.3t = 18.9$$
$$18.9 = 18.9 \quad \text{These numbers are equal.}$$

Because both variables were eliminated and the remaining numbers are equal, this system must be dependent, so all inputs and outputs that solve one equation will solve the other. Thus, there are an infinite number of solutions and any solution to the equation $r = 4.1t + 6.3$ is a solution to the system.

This system is a set of two equations that are in fact the same linear function. If you put both of these equations into slope-intercept form, they are exactly the same.

First Equation	Second Equation
$r = 4.1t + 6.3$	$3r - 12.3t = 18.9$
	$3r = 12.3t + 18.9$
	$\dfrac{3r}{3} = \dfrac{12.3t + 18.9}{3}$
	$r = 4.1t + 6.3$

b. This system has two equations with the same slope but different vertical intercepts, so they are parallel lines. Therefore, this system is inconsistent and has no solution.

If you try the substitution method, both variables will be eliminated but the remaining numbers will not be equal:

$$4x + 9 = 4x - 9 \qquad \text{Substitute for } y.$$

$$\underline{-4x \qquad -4x}$$

$$9 \neq -9 \qquad \text{These numbers are not equal.}$$

c. This system of equations is set up for substitution, because the variable S is isolated in both equations:

$$S = 1500 + 0.05c$$

$$S = 0.05c + 750$$

_____

$$1500 + 0.05c = 0.05c + 750 \qquad \text{Substitute for } S.$$

$$\underline{-0.05c \qquad -0.05c}$$

$$1500 \neq 750 \qquad \text{These numbers are not equal.}$$

Because the variables were both eliminated but the remaining numbers are not equal, we have an inconsistent system with no solutions.

This system is a set of two parallel lines. If you put the first equation into slope-intercept form, you can see that both equations have the same slope.

EXAMPLE ⑥ PRACTICE PROBLEM

Solve the following systems and label them as consistent, dependent, or inconsistent.

a. $a = 1.25b + 4$

$8.75b - 7a = 14$

b. $x = -\dfrac{7}{6}y + \dfrac{2}{3}$

$-3x - 3.5y = -2$

2.2 Exercises

For Exercises 1 through 10, solve the systems using the substitution method. Label each system as consistent, inconsistent, or dependent.

1. $y = x + 7$
$\quad y = 2x + 3$

2. $y = 6x - 20$
$\quad y = 5x - 15$

3. $P = -5t + 20$
$\quad P = 2t - 8$

4. $B = 4c + 7$
$\quad B = -3c - 7$

5. $y = 2x - 5$
$\quad 3x + y = 10$

6. $W = 7p - 2$
$\quad -3p + W = -14$

7. $H = 3k + 6$
$\quad 2k + 4H = -4$

8. $C = 2r + 9$
$\quad 5r + 2C = -9$

9. $G = 4a - 5$
$\quad 5a - 2G = 1$

10. $y = x + 3$
$\quad 6x - 4y = 8$

11. Hope's Pottery from Exercise 13 in Section 2.1 has the revenue and cost functions

$$R(v) = 155v$$
$$C(v) = 5000 + 65v$$

where $R(v)$ represents the monthly revenue in dollars for selling v vases and $C(v)$ represents the monthly costs in dollars for producing and selling v vases. Use the substitution method to algebraically find the break-even point for Hope's Pottery.

12. Jim's Carburetors from Exercise 14 in Section 2.1 has revenue and cost functions:

$$R(c) = 65c$$
$$C(c) = 2300 + 17c$$

where $R(c)$ represents the monthly revenue in dollars for rebuilding c carburetors and $C(c)$ represents the monthly costs in dollars for rebuilding c carburetors. Use the substitution method to algebraically find the break-even point for Jim's Carburetors.

13. Optimum Traveling Detail comes to you and details your car at work or home. The owner has determined the following functions for the monthly revenue and cost:

$$R(a) = 175a$$
$$C(a) = 25a + 7800$$

where $R(a)$ represents the monthly revenue in dollars for detailing a automobiles and $C(a)$ represents the monthly costs in dollars for detailing a automobiles. Use the substitution method to algebraically find the break-even point for Optimum Traveling Detail.

14. A Table Affair sells European tablecloths at local farmers markets and can model their weekly revenue and cost using the following functions.

$$R(t) = 120t$$
$$C(t) = 45t + 800$$

where $R(t)$ represents the weekly revenue in dollars for selling t tablecloths and $C(t)$ represents the weekly costs in dollars for selling t tablecloths. Use the substitution method to algebraically find the break-even point for A Table Affair.

15. The number of associate's and master's degrees conferred at U.S. colleges can be modeled by the functions:

$$A(t) = 8.61t + 641.25$$
$$M(t) = 12.50t + 543.07$$

where $A(t)$ represents the number of associate's degrees in thousands conferred during the school year ending t years after 2000 and $M(t)$ represents the number of master's degrees in thousands conferred during the school year ending t years after 2000. Find what year the number of associate's degrees conferred will be same as the number of master's degrees.

Source: Models derived from data in the Statistical Abstract 2008.

16. The number of black and Hispanic students enrolled in U.S. schools can be modeled by the following functions:

$$B(t) = 107.7t + 10227.8$$
$$H(t) = 337.4t + 10424.2$$

where $B(t)$ represents the number of blacks in thousands enrolled in U.S. schools t years since 1980 and $H(t)$ represents the number of Hispanics in thousands enrolled in U.S. schools t years since 2000. Find what year the number of blacks equaled the number of Hispanics enrolled in U.S. schools.

Source: Models derived from data in the Statistical Abstract 2008.

17. The student-teacher ratio for public and private schools in the United States is given in the chart.

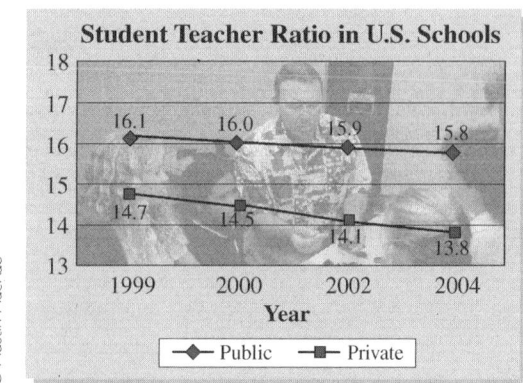

Source: *Statistical Abstract 2008.*

a. Find models for the student-teacher ratios at public and private schools in the United States

b. Determine when the student-teacher ratio will be the same in public and private schools.

18. The number of males and females who participated in high school athletic programs for various school years is given in the table.

School Year	Males	Females
03–04	4,039,253	2,865,299
04–05	4,110,319	2,908,390
05–06	4,206,549	2,953,355
06–07	4,321,103	3,021,807

Source: *www.infoplease.com.*

a. Find models for the data. (Hint: define the school year in years since school year 1999–2000; that is, 5 represents school year 2004 – 2005)

b. Determine the school year in which the same number of males and females will participate in high school athletic programs.

19. The percentages of Hispanics and African Americans 25 years old or older who have a college degree are given in the bar chart.

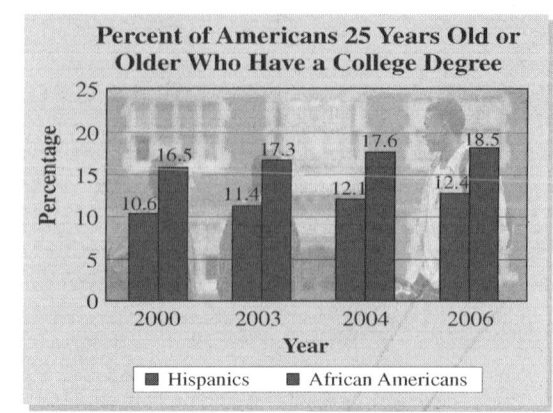

Source: *Statistical Abstract 2008.*

a. Find a model for the percentage of Hispanics who have a college degree.

b. How fast is this percentage growing each year?

c. Find a model for the percentage of African Americans who have a college degree.

d. How fast is this percentage growing each year?

e. Determine when the percentages of Hispanics and African Americans who have a college degree will be the same.

20. The number of American women of each age per thousand who are expected to die in the next 10 years from colon cancer or breast cancer is given in the table.

Age	Colon Cancer	Breast Cancer
60	4	7
65	6	9
70	8	10
75	11	11
80	14	12

Source: *Journal of the National Cancer Institute.*

a. Find models for these data.

b. Determine the age at which the risk of dying from colon cancer will be the same as the risk of dying from breast cancer.

21. Damian is investing $150,000 in two accounts to help support his daughter at college. Damian's daughter needs about $9600 each year to supplement her income from her part-time job. Damian decides to invest part of the money in an account paying 5% simple interest and the rest in another account paying 7.2% simple interest.

   a. Write a system of equations that will help you to find the amount that should be invested in each account.

   b. How much does Damian need to invest in each account to earn enough interest to help support his daughter?

22. Juan invests $225,000 in two accounts. Juan would like to earn $7900 a year to pay some expenses. Juan invests part of the money in an account paying 3% simple interest and the rest in an account paying 4% simple interest.

   a. Write a system of equations that will help you to find the amount that should be invested in each account.

   b. How much does Juan need to invest in each account to earn $9600 per year?

23. Henry was injured on the job and can no longer work. He received a settlement check from his company for $1.5 million that he will need to live on for the rest of his life. Henry is going to pay off his debts of $125,000 and invest the rest in two accounts. One account pays 5% simple interest, and the other pays 8% simple interest. Henry needs $87,500 a year in interest to continue to live at his current level.

   a. Write a system of equations that will help you to find the amount that should be invested in each account.

   b. How much should Henry invest in each account to earn the $87,500 he wants?

24. Mona settled a law suit for $2.5 million and wants to use the money to pay for a new home and invest the rest to earn enough interest to live on. The house Mona bought cost $450,000 and she wants $135,000 per year to live on. Mona invests the money in two accounts, one paying 6% simple interest and another paying 7.5% simple interest.

   a. Write a system of equations that will help you to find the amount that should be invested in each account.

   b. How much should Mona invest in each account to earn the $135,000 she wants?

25. Joan is retiring and has $175,000 to invest. Joan needs to earn $12,000 in interest each year to supplement her Social Security and pension income. Joan plans to invest part of her money in an account that pays 9% simple interest and the rest in a safer account that pays only 5% simple interest. Determine how much Joan should invest in each account to earn the money she needs.

26. Nikki has $600,000 to invest and decides to invest it in two accounts, one paying 5% simple interest and the other paying 6.5% simple interest. Determine how much Nikki should invest in each account if she needs $33,600 in interest per year.

27. Truong invests $40,000 in stocks and bonds and had a total return of $4180 in one year. If his stock investments returned 11% and his bond investments returned 9%, then how much did he invest in each?

28. Mike invests $70,000 in stocks and bonds and had a total return of $6305 in one year. If his stock investments returned 9.5% and his bond investments returned 8%, then how much did he invest in each?

29. When comparing two tour companies, you find that one charges a $500 base fee plus $25 per person who goes on the tour. The other tour company charges a $700 base fee plus $20 per person who goes on the tour. Find how many people would result in the same cost from either tour company.

30. A t-shirt printer charges a $150 set-up fee and $4.50 per shirt for a t-shirt with a company logo printed on it. Another company charges a $230 set-up fee and $4 per shirt for the same t-shirt and logo. Find what size order would result in the same cost from both companies.

For Exercises 31 through 52, solve the systems using the substitution method. Label each system as consistent, inconsistent, or dependent.

31. $P = \dfrac{2}{3}t + 5$

   $t + 3P = 21$

32. $H = \dfrac{3}{5}n + 8$

   $2n + 10H = 56$

33. $3x + 4y = 21$
$x = 6y - 4$

34. $5x - 8y = -48$
$x = 2y - 11$

35. $7d + 4r = 8.17$
$2r = 8d + 0.98$

36. $1.4h - 2.6t = -10.7$
$3h = 6.9t - 25.32$

37. $p = \frac{1}{4}r + \frac{7}{4}$
$p = 0.25r + 1.75$

38. $n = \frac{3}{8}m - \frac{7}{8}$
$n = 0.375m - 0.875$

39. $b = \frac{3}{10}a + \frac{4}{5}$
$b = 0.3a + 2.8$

40. $g = -\frac{3}{4}h - \frac{1}{2}$
$g = -0.75h - 2$

41. $y = 4(x - 7) + 2$
$y = -3(x + 2) + 5$

42. $f = 7(g + 2) - 8$
$f = 5(g - 6) + 13$

43. $w = 2(t + 5) - 12$
$w = 2(t - 3) + 4$

44. $d = -6(c - 4) - 10$
$d = -6(c + 2) + 26$

45. $6x - 15y = 24$
$y = \frac{2}{5}x - \frac{8}{5}$

46. $8w + 24y = 14$
$y = -\frac{1}{3}w + \frac{7}{12}$

47. $10x - 4y = 30$
$y = \frac{5}{2}x - 14$

48. $7x + 17.5y = 10.5$
$y = -0.4x + 1.8$

49. $g + 5h = 37$
$3g + 4h = 34$

50. $q - 2.5z = 23.25$
$6q + 5z = 49.5$

51. $\frac{2}{3}x + \frac{1}{2}y = -\frac{15}{2}$
$\frac{3}{4}x - y = \frac{5}{2}$

52. $\frac{3}{7}k - \frac{4}{5}m = \frac{9}{70}$
$k + \frac{2}{7}m = -\frac{71}{140}$

Solving Systems of Equations Using the Elimination Method

ELIMINATION METHOD

Although most of the models we find in this book will be in $y = mx + b$ form, we will sometimes be given a linear model or equation not in this form. There are also situations that may be best modeled by using another form. When this is the case, we can isolate a variable in one equation and then use the substitution method. A system that does not have a variable isolated in at least one of the equations, might be better suited for the elimination method.

In the elimination method, you will multiply one or both equations by constant(s) to make one of the variables add to zero (eliminate) when the two equations are added together. Recall that the multiplication property of equality states that we can multiply both sides of an equation by any non-zero constant. The addition property of equality states that we can add any expression to both sides of an equation. Your goal will be to get the coefficients of one variable to be the same but opposite signs in each equation. This will allow the variable to be eliminated when the two equations are added together.

ELIMINATION METHOD FOR SOLVING SYSTEMS OF EQUATIONS

- Best used when both equations are in general form, or no variable is already isolated or easily isolated.
- Multiply one or more equations by a number to make the coefficients of one variable opposite in signs but of the same value.
- Add the two equations to eliminate the variable, then solve.
- Remember to find the values for both variables.
- Check the solution in both equations.

EXAMPLE 1 SOLVE A SYSTEM OF EQUATIONS USING THE ELIMINATION METHOD

Solve the following system using the elimination method.

$$3x + 2y = 10$$
$$5x - 8y = 28$$

Solution

Because the signs of the coefficients of y are already opposites, we can eliminate the variable y by multiplying the first equation by 4:

$$4(3x + 2y) = 4(10) \qquad \longrightarrow \qquad 12x + 8y = 40$$
$$5x - 8y = 28 \qquad\qquad \longrightarrow \qquad 5x - 8y = 28$$

Now we can add the two equations together to eliminate y:

$$12x + 8y = 40$$
$$\underline{5x - 8y = 28}$$
$$17x = 68$$
$$\frac{17x}{17} = \frac{68}{17}$$
$$x = 4$$

Substitute 4 into x in either equation to find y.

$$3(4) + 2y = 10$$
$$12 + 2y = 10$$
$$2y = -2$$
$$y = -1$$

We found that the solution to this system is $(4, -1)$. We will check this solution in both equations to be sure that it is correct:

First Equation	Second Equation
$3(4) + 2(-1) = 10$	$5(4) - 8(-1) = 28$
$10 = 10$	$28 = 28$

In Section 2.2, we worked on applications that involved investments in two accounts with a goal of earning a certain amount of interest per year. Another set of applications that are set up using systems of equations involve mixtures. When mixing two chemicals or two other substances, we are trying to create a desired amount of the mixture that has a desired quality. For example, we may want 20ml of a 30% saline solution or 80 pounds of granola that cost $2 per pound. In these cases, we have an amount we want, which is similar to the total amount to be invested, and a desired percent for the mixture, which is similar to the interest needed each year.

EXAMPLE   2   SOLVING A MIXTURE APPLICATION USING THE ELIMINATION METHOD

Sally, a college chemistry student, needs 90 ml of a 50% saline solution to do her experiment for lab today. Sally can find only the two saline solutions shown below. How much of each of these solutions does Sally have to combine to get 90 ml of 50% saline solution for her lab experiment?

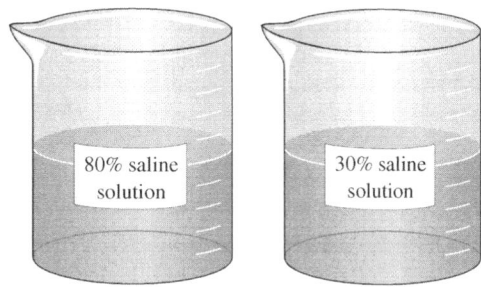

Solution

Sally has two constraints. She needs 90 ml of the solution, and she needs 50% of that to be saline. We define the following variables:

A = The amount of 30% saline solution used (in ml)

B = The amount of 80% saline solution used (in ml)

Using these variables and the two constraints given results in the following two equations:

$$A + B = 90 \qquad \text{The total mixture should be 90 ml.}$$
$$0.30A + 0.80B = 0.50(90) \quad \text{The saline should be 50\% of the total.}$$

Solving this system of equations will give us the amounts of each solution that Sally should use. First we will simplify the right side of the second equation.

$$A + B = 90$$
$$0.30A + 0.80B = 45$$

Now we will multiply the first equation by -0.80 to get the coefficients of B to be opposites.

$$-0.80(A + B) = -0.80(90) \quad\longrightarrow\quad -0.80A - 0.80B = -72$$
$$0.30A + 0.80B = 45 \quad\longrightarrow\quad 0.30A + 0.80B = 45$$

Add the two equations together to eliminate the B's.

$$-0.80A - 0.80B = -72$$
$$\underline{0.30A + 0.80B = 45}$$
$$-0.50A = -27 \qquad \text{Solve for } A.$$
$$\frac{-0.50A}{-0.50} = \frac{-27}{-0.50}$$
$$A = 54$$

so

$$54 + B = 90 \qquad \text{Solve for } B.$$
$$B = 36$$

Check the solution in both equations:

First Equation	Second Equation
$54 + 36 = 90$	$0.30(54) + 0.80(36) = 0.50(90)$
$90 = 90$	$45 = 45$

Sally should use 54 ml of the 30% saline solution and 36 ml of the 80% saline solution to get 90 ml of 50% saline solution.

EXAMPLE ② PRACTICE PROBLEM

Jim Johnson is a local veterinarian who has prescribed a diet of 24% protein for a client's Great Dane. Jim has the two types of dog food shown below, but neither is 24% protein.

The client wants 225 pounds of food to feed their Great Dane until the next vet appointment. How much of each type of food should Jim sell his client?

EXAMPLE ③ SOLVING SYSTEMS OF EQUATIONS USING THE ELIMINATION METHOD

Solve the following systems of equations using the elimination method.

a.
$$2x + 5y = 24$$
$$3x - 5y = 11$$

b.
$$3g + 5h = 27$$
$$6g + 7h = 39$$

Solution

a. The coefficients of y are already opposite numbers, so we can eliminate the y's by adding the two equations together and then we can solve for x:

$$2x + 5y = 24$$
$$\underline{3x - 5y = 11}$$
$$5x = 35$$
$$\frac{5x}{5} = \frac{35}{5}$$
$$x = 7$$

We need to solve for y:

$$2(7) + 5y = 24$$
$$14 + 5y = 24$$
$$5y = 10$$
$$y = 2$$

The solution to this system is (7, 2). You should check this solution in both equations to be sure that it is valid:

First Equation	Second Equation
$2(7) + 5(2) = 24$	$3(7) - 5(2) = 11$
$24 = 24$	$11 = 11$

b. Neither of the coefficients is already opposites, so we will eliminate the h's by multiplying the first equation by 7 and the second by -5 to make the coefficients 35 and -35 respectively.

$$7(3g + 5h) = 7(27) \quad \longrightarrow \quad 21g + 35h = 189$$
$$-5(6g + 7h) = -5(39) \quad \longrightarrow \quad -30g - 35h = -195$$

Add the two equations together to eliminate the h terms.

$$21g + 35h = 189$$
$$\underline{-30g - 35h = -195}$$
$$-9g = -6 \qquad \text{Solve for } g.$$
$$\frac{-9g}{-9} = \frac{-6}{-9}$$
$$g = \frac{2}{3}$$

We can use this value of g to find h.

$$3\left(\frac{2}{3}\right) + 5h = 27$$
$$2 + 5h = 27$$
$$5h = 25$$
$$h = 5$$

The solution to this system is $g = \frac{2}{3}$ and $h = 5$, $\left(\frac{2}{3}, 5\right)$. You should check this solution in both equations to be sure that it is valid:

First Equation	Second Equation
$3\left(\frac{2}{3}\right) + 5(5) = 27$	$6\left(\frac{2}{3}\right) + 7(5) = 39$
$27 = 27$	$39 = 39$

EXAMPLE ③ PRACTICE PROBLEM

Solve the following systems of equations using the elimination method.

a. $\quad x + 8y = -35$
$\quad -3x + 5y = -11$

b. $\quad 24d + 36g = 180$
$\quad 14d + 20g = 96$

EXAMPLE ④ SOLVING SYSTEMS OF EQUATIONS USING THE ELIMINATION METHOD

Solve the following systems of equations using the elimination method.

a. $\quad 5p + 2w = 4$

b. $\quad -3x + 2y = 7$

b. $\quad -4p - w = -3$
$\quad 12x - 8y = -28$

Solution

a. This system is set up well for elimination, because neither equation has a variable isolated:

$$5p + 2w = 4 \qquad\longrightarrow\qquad 5p + 2w = 4$$
$$2(-4p - w) = 2(-3) \qquad\longrightarrow\qquad -8p - 2w = -6$$

Now add the two equations together to eliminate the w terms.

$$
\begin{aligned}
5p + 2w &= 4 \\
-8p - 2w &= -6 \\
\hline
-3p &= -2 \\
\frac{-3p}{-3} &= \frac{-2}{-3} \\
p &= \frac{2}{3}
\end{aligned}
$$

We can use this value of p to find w, or we can redo the elimination process and eliminate the p's this time so we can solve for w.

$$4(5p + 2w) = 4(4) \qquad\longrightarrow\qquad 20p + 8w = 16$$
$$5(-4p - w) = 5(-3) \qquad\longrightarrow\qquad -20p - 5w = -15$$

Now add the two equations together to eliminate the p terms.

$$
\begin{aligned}
20p + 8w &= 16 \\
-20p - 5w &= -15 \\
\hline
3w &= 1 \\
w &= \frac{1}{3}
\end{aligned}
$$

This system has one solution at the point $p = \frac{2}{3}$ and $w = \frac{1}{3}$, $\left(\frac{2}{3}, \frac{1}{3}\right)$, so it is a consistent system. We can check this solution in both equations.

First Equation	Second Equation
$5\left(\frac{2}{3}\right) + 2\left(\frac{1}{3}\right) = 4$	$-4\left(\frac{2}{3}\right) - \frac{1}{3} = -3$
$\frac{10}{3} + \frac{2}{3} = 4$	$-\frac{8}{3} - \frac{1}{3} = -3$
$\frac{12}{3} = 4$	$-\frac{9}{3} = -3$
$4 = 4$	$-3 = -3$

b. This system is set up well for elimination, because no variable is already isolated and the variables are already lined up to be eliminated:

$$4(-3x + 2y) = 4(7) \qquad\longrightarrow\qquad -12x + 8y = 28$$
$$12x - 8y = -28 \qquad\longrightarrow\qquad 12x - 8y = -28$$

Now add the two equations together.

$$-12x + 8y = 28$$
$$\underline{12x - 8y = -28}$$
$$0 = 0$$

All the variable terms eliminated and these numbers are equal.

Because the variables were both eliminated and the remaining numbers are equal, this system is dependent. The two equations are multiples of each other. If you divide the second equation by -4, it will reduce to be the same as the first equation. Thus, there are an infinite number of solutions that solve the equations $-3x + 2y = 7$ and $12x - 8y = -28$ simultaneously.

EXAMPLE 4 PRACTICE PROBLEM

Solve the following systems of equations using the elimination method.

a.
$$15x + 14y = 16$$
$$-3x + 21y = 7$$

b.
$$\frac{5}{2}x - \frac{3}{2}y = 6$$
$$10x - 6y = 8$$

2.3 Exercises

For Exercises 1 through 10 solve the systems using the elimination method. Label each system as consistent, inconsistent, or dependent.

1. $x + 4y = 11$
 $5x - 4y = 7$

2. $6x + 2y = 6$
 $-6x - 5y = 21$

3. $2g - 3h = 9$
 $-2g - 4h = 26$

4. $7f + g = 20$
 $3f - g = 20$

5. $8x - 2y = 36$
 $3x + 4y = 23$

6. $p + 4W = -82$
 $-3p + W = -14$

7. $7k - 4H = 16$
 $2k + 8H = 32$

8. $2G - 3F = -20$
 $5G + 9F = -17$

9. $3t + 8k = 20$
 $5t + 6k = 26$

10. $4x - 12y = -8$
 $10x - 8y = 46$

11. A chemistry student needs to make 20 ml of 15% HCl solution but has only the solutions shown below to work with.

5% HCl solution   50% HCl solution

How much of each solution should this student use to get the 20 ml of 15% HCl solution?

12. Fred needs 45 ml of 12% sucrose solution to do his science experiment. He needs to mix a 5% sucrose solution with some 30% sucrose solution to get the 12% solution he wants. How much of each should Fred use?

13. Kristy is doing her chemistry lab, but the student in front of her used the last of the 5% NaCl solution she needed. Kristy will have to make more 5% solu-tion with the 2% and 10% NaCl solutions left in the lab. If Kristy needs 25 ml for her experiment, how much of each solution should she use?

14. Camren needs 30 ml of 10% HCl solution but only has 5% HCl and 30% HCl solutions. How much of each solution should Camren use to make the 30 ml of 10% HCL solution?

15. The local fair offers two options for admission and ride tickets. For option 1, you pay $22 for admission and 50 cents per ride ticket. For option 2, you pay $15 for admission and 75 cents per ride ticket.

a. Use a system of equations to find the number of rides that will result in the same cost for both options.

b. If you want to ride a large number of rides which option should you take?

16. Two car rental agencies advertised a weekend rate in the newspaper. U-rent advertised a weekend rate of $29 plus 15 cents per mile driven. Car Galaxy advertised a weekend rate of $40 plus 10 cents per mile driven.

a. Use a system of equations to find the number of miles that will result in the same charges from both companies.

b. If you plan to drive about 200 miles, which company should you rent from?

17. While working on a home project, a man purchased two orders of wood. The first order was for 150 planks and 8 4X4's for a total cost of $222.92. The second order was for 45 planks and 2 4X4's for a total cost of $87.53. Use a system of equations to find the cost for a plank and a 4X4.

18. A golf resort offers two getaway options. The Mini Escape includes 2 nights' stay and 2 rounds of golf, and the Grand Escape includes 3 nights' stay and 4 rounds of golf. The Mini Escape costs $398 and the Grand Escape costs $647. Use a system of equations to find the cost for one nights stay and the cost for one round of golf for these options.

19. In one month, Ana earned $1420 for 165 hours of work. Ana gets paid $8 per hour for her regular time and $12 per hour for any overtime hours. Find the number of regular hours and overtime hours Ana worked that month.

20. In a month, Jerrell earned $3430, for 230 hours worked. Jerrell earns $14 per hour for regular hours and $21 per hour for overtime. Find the number of regular hours and overtime hours Jerrell worked that month.

For Exercises 21 through 52, solve each system by elimination or substitution. Label each system as consistent, inconsistent, or dependent.

21. $w + 8s = -42$
    $-9w - s = -48$

22. $-3x + 7y = -1$
    $3x + 9y = -2$

23. $3.7x + 3.5y = 5.3$
    $y = 3.57x + 4.21$

24. $4.1w + 3.7t = 5.1$
    $t = 4.43w + 4.63$

25. $\frac{5}{3}d - \frac{3}{5}g = 52$
    $\frac{3}{5}d + \frac{5}{3}g = -66$

26. $\frac{2}{5}P + \frac{5}{2}T = 26$
    $\frac{5}{2}P - \frac{2}{5}T = 40.7$

27. $-2x + 11y = 2$
    $11(x - y) = -2$

28. $-8x + 7y = -3$
    $7(x - y) = 3$

29. $-3.5x + y = -16.95$
    $y = -2.4x + 4.88$

30. $4.5c + 2.1b = 4.7$
    $b = 3.51c + 3.43$

31. $W = \frac{3}{7}d + 6$
    $W = \frac{2}{7}d + 8.8$

32. $M = \frac{5}{9}r + 8$
    $M = \frac{7}{9}r - 2.5$

33. $T = 0.5g + 8.5$
    $g - 2T = -17$

34. $B = \frac{2}{8}k - 4$
    $k - 4B = 16$

35. $-9c - 7d = 8$
    $c - 7d = 8$

36. $-4r + t = -1$
    $-3r + 8t = 5$

37. $3x + 4y = -15$
    $y = x - 9$

38. $3s - 2t = -2$
    $t = -s - 9$

39. $2.5x + y = 4$
    $x + 0.4y = -5$

40. $7.5p + k = 3$
    $15p + 2k = 7$

41. $W = -3.2c + 4.1$
    $W = 2.4c - 8.8$

42. $H = 2.6t + 4.8$
    $H = -4.6t + 3.5$

43. $p = \frac{7}{4}t + 9$
    $4p - 7t = 36$

44. $y = \frac{7}{2}x + 2$
    $2x - 7y = 8$

45. $\frac{4}{3}x + y = 13$
    $x - \frac{7}{6}y = 4$

46. $\frac{7}{4}x + y = -1$
    $x - \frac{5}{3}y = 3$

47. $-8.4x - 2.8y = -58.8$
    $3x + y = 21$

48. $2.3p + 3.5t = 7.8$
    $11.27p + 17.15t = 38.22$

49. $3c - b = -17$
    $b = 3c + 18$

50. $8x - y = 20$
    $y = 8x + 4$

51. $-2.4x + y = 2.5$
    $4.1x + y = -2$

52. $7r - 2c = -8$
    $-2r - 6c = 7$

Solving Linear Inequalities

In life, we often want to know when one quantity is less than or greater than another quantity. We may want to know how many questions we can miss and still earn the grade we want on a test. For example, if we get more than 20 questions of 25 on a test, we will pass. We would like to know how many things we can buy and stay within a budget. For example, if we spend less than $300 on groceries each month, we will stay below our budget. The desire to compare quantities leads us to mathematical inequalities. We will work with the four basic inequalities, less than, greater than, less than or equal to, and greater than or equal to.

INTRODUCTION TO INEQUALITIES

When working with inequalities in algebra, you should be cautious about which operations you perform. In most cases, using the operations to solve an inequality is exactly the same as using them to solve an equation except that the answer will be expressed differently and will have a different interpretation.

CONCEPT INVESTIGATION I

WHAT CHANGES AN INEQUALITY?

For each inequality, perform the given operations and determine whether the inequality remains true.

a. $20 > 15$

i.	Add 12 to both sides.	$20 + 12 > 15 + 12$	$32 > 27$	True
ii.	Subtract 4 from both sides.	$20 - 4 > 15 - 4$	$16 > 11$	True
iii.	Multiply both sides by 3.	$3(20) > 3(15)$	$60 > 45$	True
iv.	Divide both sides by 5.	$\dfrac{20}{5} > \dfrac{15}{5}$	$4 > 3$	?

b. $45 \geq 39$

i.   Add -7 to both sides.

ii.   Subtract -9 from both sides.

iii.   Multiply both sides by -2.

iv.   Divide both sides by -3.

c. $36 > -12$

i.   Add 10 to both sides.

ii.   Subtract 15 from both sides.

iii.   Multiply both sides by 9.

iv.   Divide both sides by 4.

d. $-15 \leq 7$

i.   Add -5 to both sides.

Say What?

Inequality Symbols
* Less than: $<$

    $5 < 20$
* Greater than: $>$

    $15 > 8$
* Less than or equal to: $\leq$

    $11 \leq 73$
* Greater than or equal to: $\geq$

    $45 \geq 3$

    ii. Subtract -6 from both sides.

    iii. Multiply both sides by -6.5.

    iv. Divide both sides by -4.

1. Which operations cause the inequality to become false?

2. Write in your own words what operations to be cautious with in working with inequalities.

3. Check your statement with other students in your class or your instructor.

SOLVING INEQUALITIES

From the concept investigation above, we can see that when working with inequalities we need to watch for operations that will reverse the inequality relationship. When solving inequalities, we will reverse the inequality symbol whenever we multiply or divide both sides of the inequality by a negative number.

In general it is best to keep the variable on the left side of an inequality symbol so that the interpretation of the solution will be easier. To check the solution to an inequality, we must check that the number we find will make both sides equal to one another and then also check that the inequality is facing the correct direction. We can check the direction of the inequality symbol by picking a value in the solution set and plugging it into the original inequality to check if it satisfies the inequality.

EXAMPLE   **1** SOLVING INEQUALITIES ALGEBRAICALLY

Solve the following inequalities.

a. $-5x + 7 \geq 22$

b. $12w + 15 > 3w - 6$

c. $\dfrac{b}{-5} \leq -20$

Solution

a.   We will solve the inequality the same way we would solve an equation. Note, though, that we have to reverse the inequality symbol any time that we multiply or divide both sides by a negative number.

$$-5x + 7 \geq 22$$
$$\underline{\quad -7 \quad\; -7 \quad}$$
$$-5x \geq 15$$
$$\frac{-5x}{-5} \leq \frac{15}{-5} \qquad \text{Divide both sides by -5 and reverse the inequality symbol.}$$
$$x \leq -3$$

Therefore, any number less than or equal to -3 will solve this inequality. To check this solution, we need to check the -3 to be sure it makes the sides equal to one another and then we need to pick a value less than -3. We will test -5:

Check the number -3 for equality.	Check -5 to confirm the direction of the inequality.
$-5(-3) + 7 = 22$	$-5(-5) + 7 \geq 22$
$22 = 22$	$32 \geq 22$

The equality and resulting inequality are true, so the solution is correct.

b.

$$12w + 15 > 3w - 6$$

$$\underline{-3w \qquad\quad -3w}$$

$$9w + 15 > -6$$

$$\underline{\quad -15 \ -15}$$

$$9w > -21$$

$$\frac{9w}{9} > \frac{-21}{9}$$

$$w > \frac{-21}{9}$$

Divide both sides by 9. Because we are dividing by a positive number, we will not reverse the inequality symbol.

$\frac{-21}{9}$ is approximately -2.33 so to check this solution we will test -2.33 for equality and a number greater than -2.33 to test the direction of the inequality. We will test $w = 0$:

Check the number -2.33 for equality.	Check 0 to confirm the direction of the inequality.
$12(-2.33) + 15 = 3(-2.33) - 6$	$12(0) + 15 > 3(0) - 6$
$-12.96 \approx -12.99$	$15 > -6$

The equality is approximate because we rounded our solution. The equality and resulting inequality are true, so our solution is correct.

c.

$$\frac{b}{-5} \leq -20$$

$$-5\left(\frac{b}{-5}\right) \geq -5(-20)$$

Multiply both sides by -5 and reverse the inequality symbol.

$$b \geq 100$$

Note that we had to reverse the inequality symbol because we multiplied both sides by a negative number. To test this solution, we will check $b = 100$ and pick a number greater than 100, we will use $b = 110$:

Check the number 100 for equality.	Check 110 to confirm the direction of the inequality.
$\frac{100}{-5} = -20$	$\frac{110}{-5} \leq -20$
$-20 = -20$	$-22 \leq -20$

The equality and the resulting inequality are true, so the solution is correct.

EXAMPLE ⓘ PRACTICE PROBLEM

Solve the following inequalities.

a. $4x + 12 > 10$

b. $8h + 9 \leq 10h + 1$

c. $\frac{m}{-3} - 8 < 10$

EXAMPLE **2** FINDING ERRORS IN SOLUTIONS OF INEQUALITIES

George and Martha both found the same solution to the inequality

$$7v + 10 < 5(2v - 7).$$

Compare their solutions and determine whether they made any mistakes. If they have, explain what they did wrong.

George's Solution	Step #	Martha's Solution	Step #
$7v + 10 < 5(2v - 7)$	1	$7v + 10 < 5(2v - 7)$	1
$7v + 10 < 10v - 35$	2	$7v + 10 < (10v - 35)$	2
$\underline{-10 \qquad\quad -10}$		$\underline{-10 \qquad\quad -10}$	
$7v > 10v - 45$	3	$7v < 10v - 45$	3
$\underline{-10v \quad -10v}$		$\underline{-10v \quad -10v}$	
$-3v < -45$	4	$-3v < -45$	4
$\dfrac{-3v}{-3} > \dfrac{-45}{-3}$	5	$\dfrac{-3v}{-3} > \dfrac{-45}{-3}$	5
$v > 15$	6	$v > 15$	6

Solution

George made mistakes on both lines 3 and 4. He should not have reversed the inequality symbol when subtracting from both sides. By making both of these errors, he actually ended with the correct solution but not the correct method. Martha did the problem correctly.

EXAMPLE **2** PRACTICE PROBLEM

Tom and Shannon found different answers to the inequality .

$$2.5n + 6.5 > -3.2n - 4.3$$

Can you find which one made a mistake and explain what he or she did wrong?

Tom's Solution	Step #	Shannon's Solution	Step #
$2.5n + 6.5 > -3.2n - 4.3$	1	$2.5n + 6.5 > -3.2n - 4.3$	1
$2.5n + 6.5 > -3.2n - 4.3$	2	$2.5n + 6.5 > -3.2n - 4.3$	2
$\underline{3.2\,n \qquad\qquad 3.2\,n}$		$\underline{3.2\,n \qquad\qquad 3.2\,n}$	
$5.7n + 6.5 > -4.3$	3	$5.7n + 6.5 > -4.3$	3
$\underline{-6.5 \quad -6.5}$		$\underline{-6.5 \quad -6.5}$	
$5.7n > -10.8$	4	$5.7n > -10.8$	4
$\dfrac{5.7n}{5.7} > \dfrac{-10.8}{5.7}$	5	$\dfrac{5.7n}{5.7} < \dfrac{-10.8}{5.7}$	5
$n > -1.895$	6	$n < -1.895$	6

SYSTEMS AS INEQUALITIES

Many times, when we work with systems of equations, we are not interested in finding when the two equations are equal. Instead, we are interested in finding when one equation is less than or greater than the other. Going back to the example of U-Haul and Bob's Truck Rental from Section 2.1, we are not concerned with when charges from these companies are equal but when one company's plan is less expensive than the other.

EXAMPLE   3   SOLVING SYSTEMS AS INEQUALITIES

If we reconsider the U-Haul and Bob's Truck Rental example from Section 2.1, we get the following:

$U(m)$ = The total cost in dollars to rent a 10-foot truck from U-Haul

$B(m)$ = The total cost in dollars to rent a 10-foot truck from Bob's Truck Rental

m = The number of miles traveled in rented truck

$$U(m) = 19.95 + 0.79m$$
$$B(m) = 24.95 + 0.75m$$

Find the distances that you can drive that will result in Bob's being less expensive than U-Haul.

Solution

Because we want Bob's to be less expensive than U-Haul, we can set the expression for the cost at Bob's less than the expression for the cost at U-Haul:

$$B(m) < U(m)$$
$$24.95 + 0.75m < 19.95 + 0.79m$$
$$\underline{-0.79m \qquad\quad -0.79m}$$
$$24.95 - 0.04m < 19.95$$
$$\underline{-24.95 \qquad\quad -24.95}$$
$$-0.04m < -5$$
$$\frac{-0.04m}{-0.04} > \frac{-5}{-0.04}$$
$$m > 125$$

Reverse the inequality symbol when dividing by a negative number.

Thus, Bob's will be less expensive than U-Haul for any distances more than 125 miles. We can easily check this solution and the direction of the symbol using the following table.

X	Y1	Y2
110	107.45	106.85
120	114.95	114.75
125	118.7	118.7
130	122.45	122.65
140	129.95	130.55
150	137.45	138.45

X=110

We can see that for distances greater than 125, Bob's (Y1) costs less than U-Haul (Y2). Therefore, our solution is correct.

You should notice in this example that the solution to the system is not one set mileage but a range of miles that makes Bob's less expensive than U-Haul. In solving a system as an inequality, the answer will be an interval of values that make the situation true.

EXAMPLE **4** SOLVING SYSTEMS AS INEQUALITIES

In Exercise 18 of Section 2.2, we looked at the number of males and females who participated in high school athletic programs for various school years. The data are given again in the table.

School Year	Males	Females
03–04	4,039,253	2,865,299
04–05	4,110,319	2,908,390
05–06	4,206,549	2,953,355
06–07	4,321,103	3,021,807

Source: www.infoplease.com.

a. Find models for these sets of data.

b. Use the models to approximate when there will be more females than males participating in high school athletic programs.

Solution

a. Define variables:

S = School year in years since school year 1999–2000; that is, 5 represents school year ending in 2005 (which would mean 2004 – 2005).

M = The number of males in millions participating in high school athletic programs.

F = The number of females in millions participating in high school athletic programs.

For the model of the number of males, we will use the first and last points:

$$(4, 4.04) \quad \text{and} \quad (7, 4.32)$$

$$m = \frac{4.32 - 4.04}{7 - 4} = 0.093$$

$$M = 0.093S + b$$
$$4.04 = 0.093(4) + b$$
$$4.04 = 0.372 + b$$
$$3.668 = b$$
$$M = 0.093S + 3.668$$

This model should be adjusted down a little to make it fit the data better.

$$M(S) = 0.093S + 3.66$$

The adjusted model.

This first model has both points it missed below the graph.

$M = 0.093S + 3.668$

By adjusting the value of b, we can shift the graph down a little to make it fit the data better.

$M = 0.093S + 3.66$

This first model has both points it missed below the graph.

$$F = 0.05S + 2.67$$

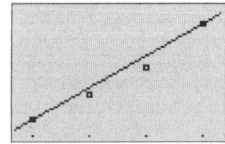

By adjusting the value of b, we can shift the graph down a little to make it fit the data better.

$$F = 0.05S + 2.66$$

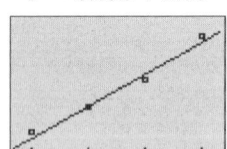

Now for the model of the female data, we will use the first and last points:

$$(4, 2.87) \quad \text{and} \quad (7, 3.02)$$

$$m = \frac{3.02 - 2.87}{7 - 4} = 0.05$$

$$F = 0.05S + b$$

$$2.87 = 0.05(4) + b$$

$$2.87 = 0.2 + b$$

$$2.67 = b$$

$$F = 0.05S + 2.67$$

This model should be adjusted down a little to make it fit the data better.

$$F(S) = 0.05S + 2.66$$

The adjusted model.

b. Because we want to know when the number of females will be larger than the number of males, we can set up the following inequality:

$$\text{Females} > \text{Males}$$

$$0.05S\ 1 + 2.66 > 0.093S + 3.66$$

$$\underline{-0.093S \qquad\qquad -0.093S}$$

$$-0.043S + 2.66 > 3.66$$

$$\underline{-2.66 \quad -2.66}$$

$$-0.043S > 1$$

$$\frac{-0.043S}{-0.043} < \frac{1}{-0.043}$$

$$S < -23.26$$

This solution can be stated different ways.

The number of females participating in high school athletic programs was greater than the number of males before the school year 1976–1977.

or

The number of females participating in high school athletic programs was greater than the number of males during the school year 1975–1976 and prior.

Either statement interprets the inequality correctly, because the first one does not include the year 1976–1977 and the second ends at the previous year, 1975–1976. Although the solution says that more females participated in high school athletic programs in the past, this is model breakdown since in the 1970's not nearly as many girls participated in high school athletics.

When solving an inequality, you might find it easiest to keep the variable on the left side of the inequality symbol. Doing so will make the solution easier to interpret. If you do keep the variable on the left side, you may run into places where you must multiply or divide by a negative number, requiring you to reverse the direction of the inequality symbol. Follow the simple rule of reversing the inequality symbol when multiplying or dividing by a negative number, and you will solve inequalities using the same skills you use to solve equations.

EXAMPLE ④ PRACTICE PROBLEM

The percentage of college freshmen who were male and the percentage of college freshmen who were female for various years are given in the chart.

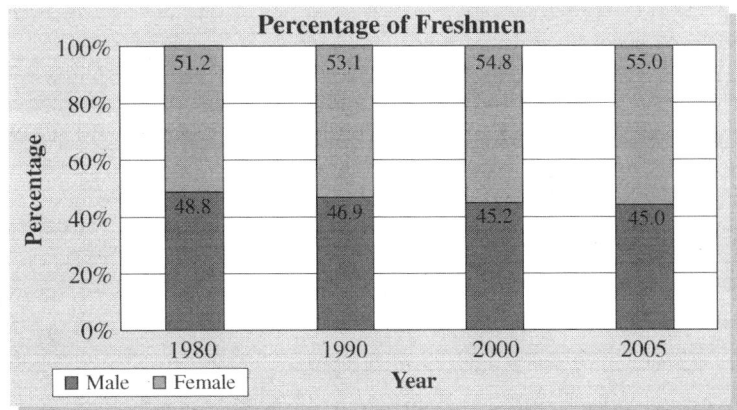

Source: Statistical Abstract 2008.

a.   Find models for the data.

b.   Determine the years for which the number of male freshmen was greater than or equal to the number of female freshman.

SOLVING INEQUALITIES NUMERICALLY AND GRAPHICALLY

We can use a table to estimate the solutions to inequalities by looking for the input value(s) that make the left side of the inequality equal to the right side of the inequality. Once we have the input value(s) that make the sides equal, we will compare the output value(s) on either side of those inputs to see when one is less than or greater than the other.

Start by putting the left side into Y1 and the right side into Y2 of the Y = screen of your calculator. When looking for the input value(s) that make the two sides equal, notice when one side changes from being smaller than the other side to being larger. The value where these two sides are equal must be between these input values. Once you have found the input value that makes the two sides of the inequality equal, look for the inequality relationship you want to be true. This inequality relationship will be when the left side is less than or greater than the right side.

EXAMPLE ⑤ SOLVING INEQUALITIES NUMERICALLY

Solve the following inequalities numerically:

a. $5x + 7 > 2x + 31$

b. $4x + 9 \le 6x - 1$

Solution

a. We will first put the two sides of the inequality into the calculator and then use the table to estimate a solution.

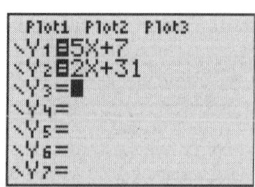 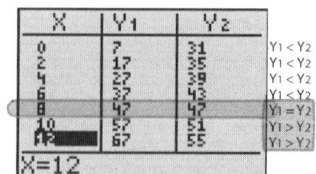

From the table, we can see that the two sides of the inequality are equal at $x = 8$ and that the left side (Y1) is greater than the right side (Y2) when $x > 8$. The solution to this inequality is $x > 8$.

b. We start by putting the two sides of the inequality into the calculator and use the table.

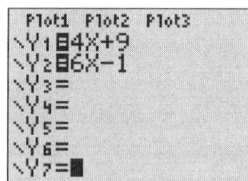 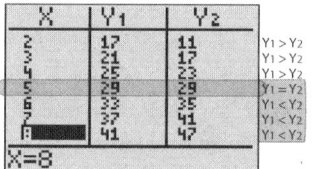

We can see on the table that the two sides are equal when $x = 5$ and the left side (Y1) is less than the right side (Y2) when $x > 5$. The solution to this inequality is $x \geq 5$.

EXAMPLE ⑤ PRACTICE PROBLEM

Solve the following inequality numerically.

$$-3x + 8 > x - 4$$

We can also use a graph of the two sides of the inequality to estimate a solution. On a graph, the curve that is vertically below the other curve is "less than" the other for all x (input) values that result in the curve staying below the other. If the two curves intersect, the x-value of the intersection will be the endpoint of the solution interval.

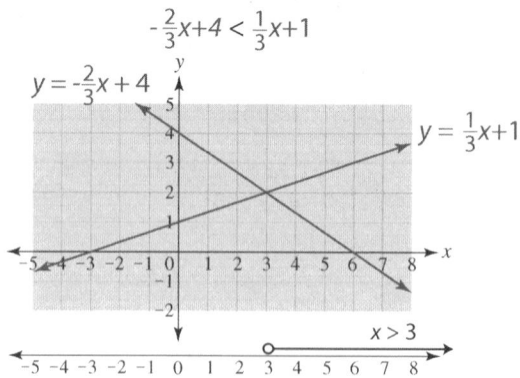

In this graph we can see that the blue line is below the red line when x is greater than 3. The solution to the inequality is $x > 3$.

We can use the calculator to solve graphically by putting one side of the inequality into Y1 and the other side into Y2 and then graphing. Once you find the intersection, you will need to determine the values of x that make the graph you want less than or greater than the other. The end of the solution interval will always be determined by the places where the graphs intersect. The typical solutions are either less than the x-value or greater than the x-value of the intersection.

EXAMPLE **6** SOLVING INEQUALITIES GRAPHICALLY

Solve the following inequalities graphically.

a. $-2x + 9 \leq 4x - 33$

b. $2.5x + 4 < 3.5x + 6$

Solution

a. We will put the two sides of the inequality into the calculator and graph them both on the same window. Watch the calculator graph the lines, so you know which line is the left side of the inequality and which is the right side of the inequality.

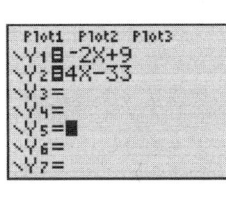 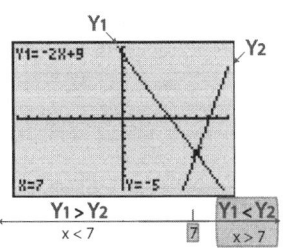

From the graph, we can see that the two sides of the inequality are equal at $x = 7$. The graph of the $-2x + 9$ (Y1) is lower (less than) than the graph of the $4x - 33$ for values of x greater than 7. Therefore the solution to this inequality is $x \geq 7$. The "greater than or equal to" symbol is used because the original inequality had the "equal to" part included in the inequality symbol.

b. Put each side of the inequality into the graphing calculator and graph them both on the same window.

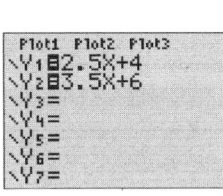 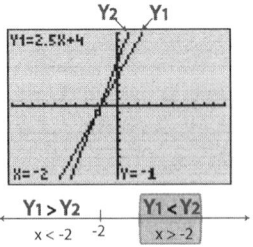

From the graph, we can see that the intersection is at $(-2, -1)$, so the two sides are equal at $x = -2$. The left side (Y1) is less than the right side (Y2) when $x > -2$. The solution to this inequality is $x > -2$.

EXAMPLE **6** PRACTICE PROBLEM

Solve the following inequality graphically.

$$1.25x - 2 \geq 2x - 5$$

Exercises

For Exercises 1 through 10 solve the given inequalities.

1. $5x + 7 > 37$

2. $4x - 10 < 2$

3. $\dfrac{P}{4} \geq 2$

4. $\dfrac{M}{6} < 4.5$

5. $8v + 4 < 5v - 20$

6. $3x + 7 \leq x - 5$

7. $2t + 12 < 5t + 39$

8. $4p - 8 > 15p - 41$

9. $\dfrac{K}{-4} + 7 \leq 21$

10. $\dfrac{V}{-5} - 4 \geq 3$

11. The numbers of full-time and part-time faculty in U.S. higher education institutions are given in the graph.

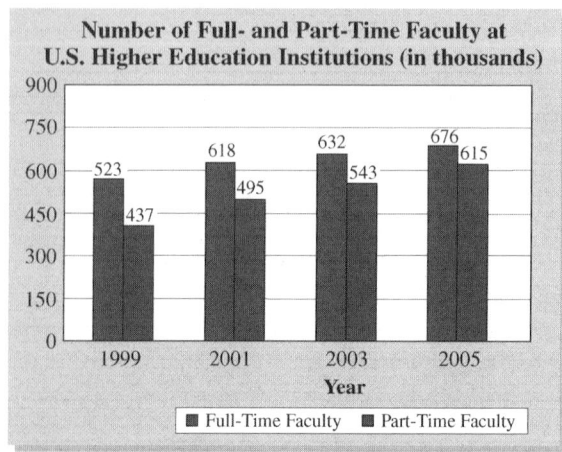

Source: Statistical Abstract 2008.

a. Find models for the data.

b. Find when the number of part-time faculty will be greater than the number of full-time faculty.

12. The populations of Afghanistan and Algeria are given below.

Year	Afghanistan Population (in thousands)	Algeria Population (in thousands)
1990	14669	25093
1995	21489	28364
2000	23898	30409
2006	31057	32930

Source: Statistical Abstract 2008.

a. Find models for the populations of Afghanistan and Algeria.

b. Determine the years for which Afghanistan will have more people than Algeria.

c. Is your prediction in part b reasonable?

13. The population on the continent of Africa has been growing rapidly over the past several decades. The populations of Africa and Europe are given in the charts.

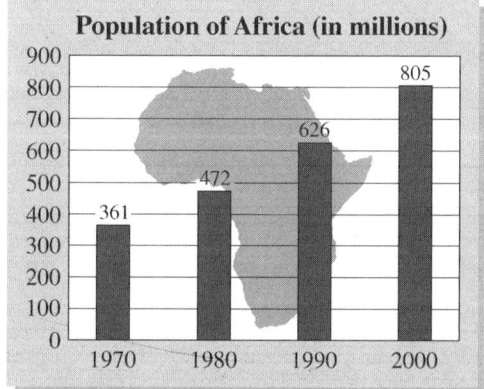

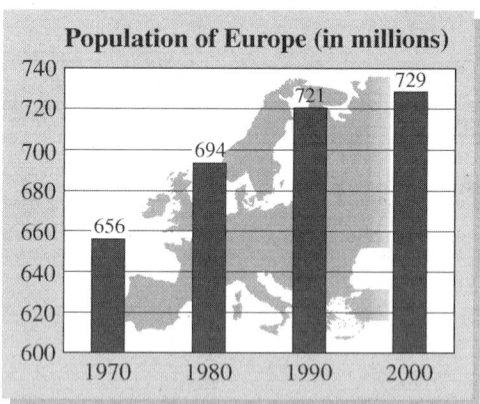

Source: Statistical Abstract 2001.

a. Find models for the populations of Africa and Europe.

b. Estimate when Africa will have a greater population than Europe.

14. The percentages of births to teenage mothers in the United States for various states are given below.

Year	Idaho	Utah
2002	10.0	7.3
2003	9.6	6.7
2004	9.2	6.4
2005	8.8	6.0

Source: Statistical Abstract 2008.

  a. Find a model for the percentage of births to teenage mothers in Idaho.

  b. What does the slope for the model in part a mean in this context?

  c. Find a model for the percentage of births to teenage mothers in Utah.

  d. Estimate when the percentage of births to teenage mothers in Utah will be greater than the percentage in Idaho.

15. The revenue and cost functions for a local cabinet manufacturer are

$$R(c) = 450c$$
$$C(c) = 280c + 20000$$

where $R(c)$ represents the monthly revenue in dollars from selling c cabinets and $C(c)$ represents the monthly cost in dollars to manufacture and sell c cabinets. Find the number of cabinets this company must sell each month to break even or make a profit.

16. Dan's Mobile Window Tinting tints car windows. Dan has determined his company's weekly revenue and cost functions as follows:

$$R(w) = 35w + 10$$
$$C(w) = 30w + 350$$

where $R(w)$ represents the weekly revenue in dollars from tinting w windows and $C(w)$ represents the weekly cost in dollars to tint w windows. Find the number of windows Dan must tint each week to break even or make a profit.

17. Optimum Traveling Detail from Exercise 13 in Section 2.2 has the revenue and cost functions:

$$R(a) = 175a$$
$$C(a) = 25a + 7800$$

where $R(a)$ represents the monthly revenue in dollars for detailing a automobiles and $C(a)$ represents

the monthly costs in dollars for detailing a automobiles. Find the number of automobiles Optimum Traveling Detail must detail to break-even or make a profit.

18. A Table Affair from Exercise 14 in Section 2.2 has the following weekly revenue and cost functions.

$$R(t) = 120t$$
$$C(t) = 45t + 800$$

where $R(t)$ represents the weekly revenue in dollars for selling t tablecloths and $C(t)$ represents the weekly costs in dollars for selling t tablecloths. Find the number of tablecloths A Table Affair must sell to break-even or make a profit.

19. Several prepaid cell phone services are advertised in the flyer below.

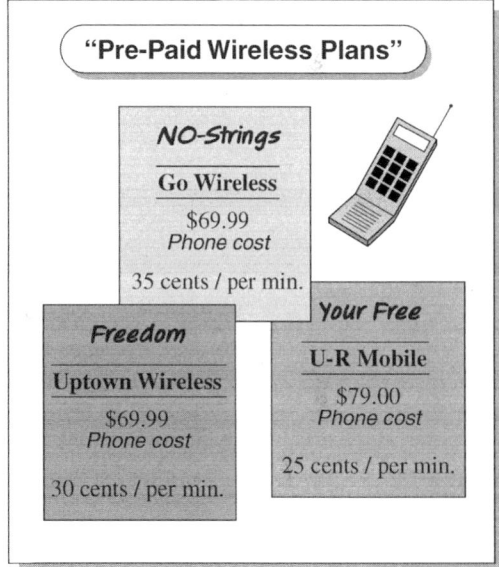

  a. Find models for the cost of these three plans if a person uses m minutes.

  b. For what number of minutes will U-R Mobile have the cheapest plan?

20. Several equipment rental services have the following charges for a dump truck rental.

Company	Base Charges ($)	Hourly Rate ($)
Pauley's Equipment	100	65
You Haul It	80	70
Big Red's Equipment	80	75

a. Find models for the cost of these three companies rentals if a person rents the dump truck for h hours.

b. For what number of hours will You Haul it be the cheapest?

21. The percentage of births to unmarried women in the United States can be modeled by

$$U(t) = 0.71t + 37.1$$

where $U(t)$ represents the percentage of births to unmarried women in the United States t years since 2005. The percentage of births to unmarried women in the United Kingdom t years since 2005 can be modeled by

$$K(t) = 1.3t + 44.7$$

Source: Models derived from data in the Statistical Abstract 2008.

Determine when the percentage of births to unmarried women in the United States was greater than the percentage in the United Kingdom.

22. The marriage rate per 1,000 population aged 15-64 years in the United States can be modeled by

$$U(t) = -0.19t + 11.48$$

where $U(t)$ represents the marriage rate in the United States t years since 2005. The marriage rate in the United Kingdom t years since 2005 can be modeled by

$$K(t) = -0.15t + 7.66$$

Source: Models derived from data in the Statistical Abstract 2008.

Determine when the marriage rate in the United States will be less than the marriage rate in the United Kingdom.

23. A salesperson earns a base salary of $600 per month plus 21% commission on all sales. Find the minimum amount a salesperson would need to sell to earn at least $1225 for one month.

24. A salesperson earns a base salary of $725 per month plus 17% commission on all sales. Find the minimum amount a salesperson would need to sell to earn at least $1225 for one month.

25. Use the table to find when Y1 < Y2.

X	Y₁	Y₂
0	5	15
1	8	13
2	11	11
4	17	7
5	20	5
6	23	3

X=

26. Use the table to find when Y1 > Y2.

X	Y₁	Y₂
-6	-29	-8
-5	-24	-10
-4	-19	-12
-3	-14	-14
-2	-9	-16
-1	-4	-18

X=

27. Use the table to find when Y1 > Y2.

X	Y₁	Y₂
-10	11	-7
-9	9	-3
-8	7	1
-7	5	5
-6	3	9
-5	1	13

X=

28. Use the table to find when Y1 > Y2.

X	Y₁	Y₂
5	-10	50
10	0	30
15	10	10
20	20	-10
25	30	-30
30	40	-50

X=

For Exercises 29 through 36, solve each inequality numerically.

29. $8x + 1 > -2x + 51$

30. $4x - 8 < -x + 22$

31. $4.5x - 8 \le 7.5x - 20$

32. $0.25x + 12 \ge 0.75x + 2$

33. $-1.5x + 18 \ge 2.5x + 4$

34. $-4.25x + 5.125 \le 3.6x + 45.25$

35. $\frac{1}{4}x - 2 < -\frac{1}{2}x - 4.5$

36. $\frac{2}{3}x + 4 > \frac{1}{3}x + 6$

For Exercises 37 through 40, use the graph to solve the given inequality.

37. Use the graph to find when $f(x) < g(x)$.

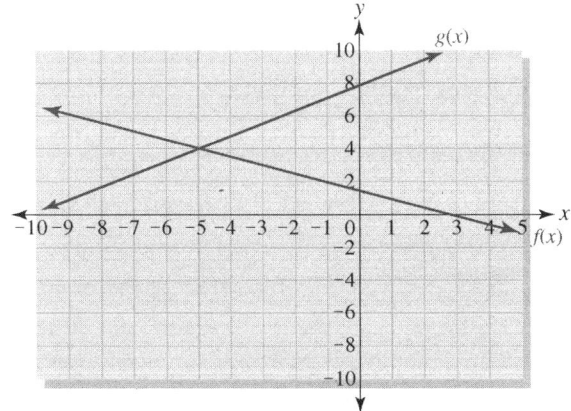

38. Use the graph to find when $f(x) < g(x)$.

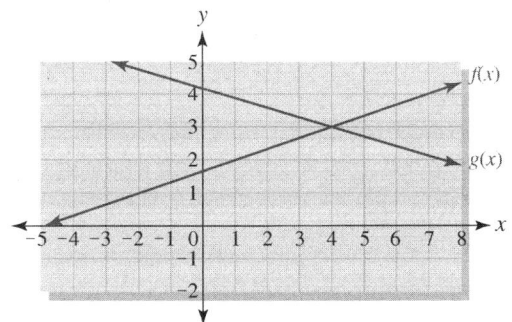

39. Use the graph to find when $f(x) \geq g(x)$.

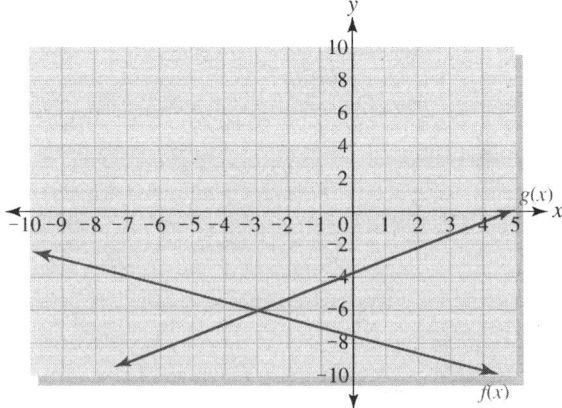

40. Use the graph to find when $f(x) \geq g(x)$.

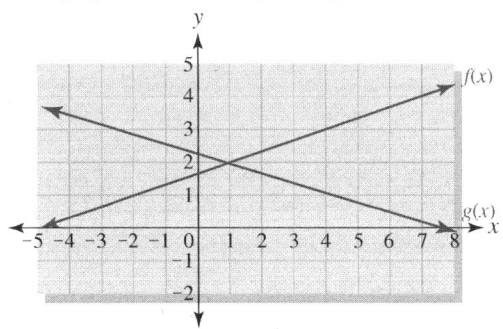

For Exercises 41 through 44, solve the inequality graphically.

41. $5x + 3 > -2x + 17$

42. $4.5x - 8 < 7.5x - 20$

43. $\frac{1}{2}x - 6 < -\frac{1}{3}x + 4$

44. $\frac{2}{5}x + 11 > \frac{4}{5}x + 3$

For Exercises 45 through 58, solve the inequality algebraically.

45. $5 + \frac{3x}{2} \leq -3$

46. $7 + \frac{4x}{3} \geq -15$

47. $5x + 3 \geq 3(x - 2)$

48. $4t + 3 > 2(t + 1)$

49. $\frac{-7d}{3} + 5 < 3$

50. $\frac{-2g}{9} + 12 > 4$

51. $2.7v + 3.69 > 1.5v - 6.5$

52. $3.4b + 2.45 < 0.3b - 8.5$

53. $3.2 + 2.7(1.5k - 3.1) \geq 9.43k - 17.5$

54. $8.7 - 1.4(8.2m + 6) \geq -2.3(7.1m - 4.3)$

55. $\frac{2}{5}(w - 20) \leq -\frac{3}{7}(4w - 9)$

56. $\frac{2}{3}(P + 4) < -\frac{5}{7}(2P - 12)$

57. $2.35x + 7.42 < 1.3x - 4.75$

58. $3.74x - 5.87 > 7.28x + 3.25$

Absolute Value Equations and Inequalities

ABSOLUTE VALUE EQUATIONS

The absolute value can be defined in several ways. One way is to use the concept of distance on a number line. The absolute value of a number is defined as the distance from 0 to the number on a real number line. Using distance in the definition may help you to see why $|6| = 6$ and $|-6| = 6$. Both 6 and -6 are 6 units from 0 on the real number line.

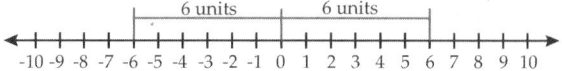

Since distance is always positive, the absolute value will always have a positive result.

DEFINITION

Absolute Value: The absolute value of a real number n, $|n|$, is the distance from zero to n on a real number line.

Using the distance definition of absolute value guides us when solving equations that contain absolute values. The equation $|x| = 8$ is asking what number(s), x, are 8 units from zero on the number line. If we look at the number line, we can see that both 8 and -8 are 8 units from zero.

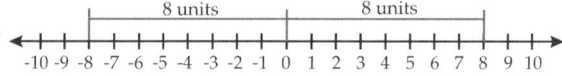

The absolute value equation becomes two equations, one for the positive side of zero and the other for the negative side of zero.

$$|x| = 8$$
$$x = 8 \quad \text{or} \quad x = -8$$

DEFINITION

Absolute Value Equation: If n is a positive real number and u is any algebraic expression ,

$$|u| = n$$
$$u = n \quad \text{or} \quad u = -n$$

If n is negative then the equation $|u| = n$ has no real solution.

When solving an equation that contains an absolute value, you will isolate the expression containing the absolute value and then write two equations, one with a positive value and one with a negative value.

SOLVING ABSOLUTE VALUE EQUATIONS

- Isolate the expression containing the absolute value.
- Rewrite the equation as two equations .

$$|u| = n$$

$$u = n \quad \text{or} \quad u = -n$$

- Solve each equation.
- Check the solutions in the original absolute value equation.

EXAMPLE ☐ SOLVING ABSOLUTE VALUE EQUATIONS

Solve the following equations.

a. $|x + 5| = 12$ **b.** $2|x - 4| + 3 = 9$ **c.** $|h + 3| + 10 = 4$

Solution

a. The absolute value is already isolated, so we rewrite it as two equations and solve.

$$|x + 5| = 12 \qquad \text{Rewrite into two equations.}$$

$$x + 5 = 12 \quad \text{or} \quad x + 5 = -12$$

$$x = 7 \quad \text{or} \qquad x = -17 \qquad \begin{array}{l}\text{Solve by subtracting 5}\\\text{from both sides.}\end{array}$$

We have two solutions for this equation. We check them both in the original absolute value equation.

$x = 7$	$x = -17$
$\|7 + 5\| = 12$	$\|-17 + 5\| = 12$
$\|12\| = 12$	$\|-12\| = 12$
$12 = 12$	$12 = 12$

Both solutions work. We can see the two solutions by looking at the graph of both sides of the equation and looking for where they interesect.

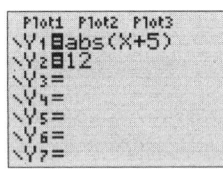 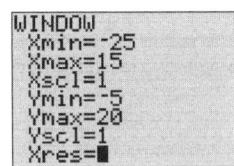 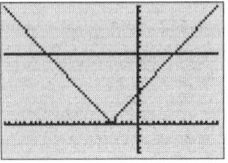

b. First, we need to isolate the absolute value.

$$2|x - 4| + 3 = 9 \qquad \begin{array}{l}\text{Subtract 3 from both sides of the}\\\text{equation.}\end{array}$$

$$\underline{ -3 \quad -3}$$

$$2|x - 4| = 6$$

$$\frac{2|x - 4|}{2} = \frac{6}{2} \qquad \begin{array}{l}\text{Divide both sides of the equation}\\\text{by 2.}\end{array}$$

$$|x - 4| = 3$$

Using Your TI Graphing Calculator

To take the absolute value of a number or expression on your graphing calculator, you need to use the abs function found in the math menu. To get to the abs function, press [MATH] and [▶] to get to the NUM submenu.

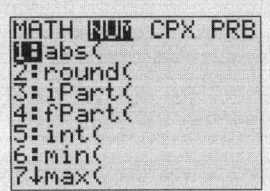

The first option on this sub-menu is abs(. Pressing 1 or [ENTER] will bring abs(into your equation or calculation. Remember to close the parenthesis at the end of the expression you are taking the absolute value of.

Now, we rewrite into two equations and solve.

$$|x - 4| = 3$$

$x - 4 = 3$ or $x - 4 = -3$ **Solve by adding 4 to both**
$x = 7$ or $x = 1$ **sides of the equation.**

We check both equations in the original absolute value equation.

$x = 7$	$x = 1$
$2|7 - 4| + 3 = 9$	$2|1 - 4| + 3 = 9$
$2|3| + 3 = 9$	$2|-3| + 3 = 9$
$9 = 9$	$9 = 9$

Both solutions work.

c. First, we will isolate the absolute value.

$$\begin{aligned} |h + 3| + 10 &= 4 \\ \underline{-10 \quad -10} & \\ |h + 3| &= -6 \end{aligned}$$

Subtract 10 from both sides of the equation.

The absolute value equals a negative number, which is not possible. There is no real solution to this equation. A good way to check this result is to graph both sides of the equation and see if they intersect.

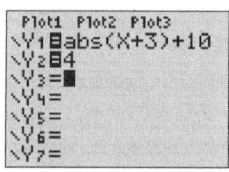 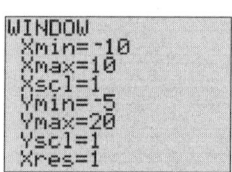 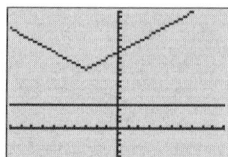

From the graph, we can see that the absolute value curve does not intersect the line. This confirms that our result of no real solutions is correct.

EXAMPLE ① PRACTICE PROBLEM

Solve the following equations.

a. $|x - 2| = 7$ **b.** $|d - 12| - 9 = -15$ **c.** $-3|w + 1| + 12 = -3$

EXAMPLE 2 AN APPLICATION OF ABSOLUTE VALUE EQUATIONS

Rebecca is taking a trip across the country and is concerned about gasoline costs. Her most efficient mileage expressed as miles per gallon for Rebecca's car can be modeled by the function

$$m(s) = -\frac{1}{3}|s - 60| + 35$$

where m is the gas mileage for Rebecca's car when she is driving at an average speed of s miles per hour, for speeds between 40 and 80 miles per hour. At what speed should Rebecca drive to get gas mileage of 30 miles per gallon during her trip?

Solution

Rebecca wants to get 30 miles per gallon so $m = 30$. We will substitute 30 into m and solve.

$$m(s) = -\frac{1}{3}|s - 60| + 35$$

$$30 = -\frac{1}{3}|s - 60| + 35 \qquad \text{Isolate the absolute value by subtracting 35 from both sides.}$$

$$\underline{-35 \qquad\qquad\qquad -35}$$

$$-5 = -\frac{1}{3}|s - 60|$$

$$-3(-5) = -3\left(-\frac{1}{3}|s - 60|\right) \qquad \text{Multiply both sides by -3 to eliminate the fraction.}$$

$$15 = |s - 60|$$

$$s - 60 = 15 \quad \text{or} \quad s - 60 = -15 \qquad \text{Rewrite into two equations and}$$
$$s = 75 \quad \text{or} \qquad\quad s = 45 \qquad \text{solve.}$$

We check these solutions in the original equation using the table on the calculator.

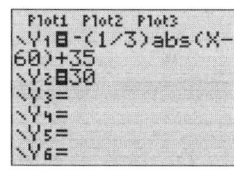 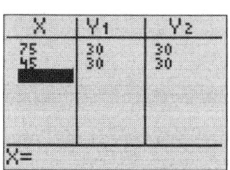

Rebecca can drive either 75 miles per hour or 45 miles per hour on her trip to get gas mileage of 30 miles per gallon.

EXAMPLE ② PRACTICE PROBLEM

Svetlana is driving from San Antonio to Houston, Texas, on Interstate 10 and will pass through Columbus, Texas, on the way. Svetlana's distance from Columbus can be modeled by the function

$$D(t) = |125 - 65t|$$

where D is Svetlana's distance in miles from Columbus, Texas, after driving for t hours. Find the time when Svetlana will be 25 miles away from Columbus.

ABSOLUTE VALUE INEQUALITIES INVOLVING LESS THAN OR LESS THAN OR EQUAL TO

Inequalities that contain absolute values can also be interpreted using the distance definition of absolute value. The inequality $|x| < 3$ is asking for all values of x that are less than 3 units from zero on the number line.

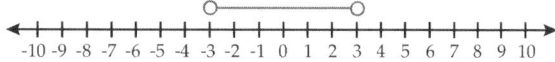

If we look at the graph of each side of the inequality, we can see the solution set.

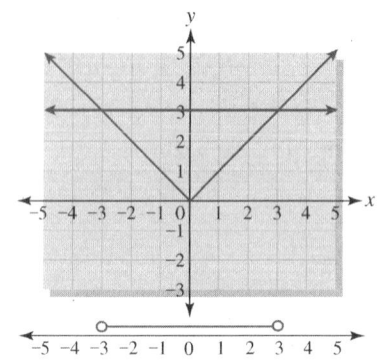

We can see that the red graph of the absolute value is less than 3 when it is below the blue graph of $y = 3$ when x is between -3 and 3.

When solving inequalities with absolute values, it is important to keep in mind what the inequality is asking. The absolute value inequality $|x| < 3$ is asking for x values that are closer than 3 units away from zero. From the above graph, we can see that this interpretation implies that x must be between -3 and 3. Writing this interval with inequalities is similar to the way we rewrote absolute value equations into two equations but the inequality symbols must be in the correct direction and order.

$$|x| < 3$$
$$-3 < x < 3$$

The last inequality says that x must be between -3 and 3, as seen as the interval on the above graph. When solving absolute value inequalities with a less-than symbol, you write a compound inequality to indicate that the expression must be between the positive number and the opposite of the number. Remember that an absolute value cannot be less than a negative number in the real number system.

SOLVING ABSOLUTE VALUE INEQUALITIES INVOLVING THE SYMBOLS LESS THAN OR LESS THAN OR EQUAL TO

* Isolate the expression containing the absolute value to the left side of the inequality.
* Rewrite the absolute value inequality as a compound inequality .

$$|u| < n \qquad\qquad |u| \leq n$$
$$-n < u < n \qquad\qquad -n \leq u \leq n$$

* Solve the compound inequality. (Isolate the variable in the middle.)
* Check the solution set in the original absolute value inequality.

EXAMPLE **3** SOLVING ABSOLUTE VALUE INEQUALITIES WITH LESS THAN

Solve the following inequalities. Give the solution as an inequality and graph the solution set on a number line.

a. $|x| < 4$ **b.** $|x - 2| \leq 5$ **c.** $|b + 5| + 15 < 20$

Solution

a.   The absolute value is isolated on the left side and it is a less than symbol so we will rewrite the absolute value inequality as a compound inequality.

$$|x| < 4$$

$$-4 < x < 4$$

Therefore, x must be between -4 and 4.

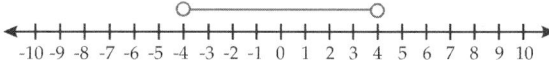

The solution set is the interval from -4 to 4. To confirm this solution set, we check the endpoints of the interval and pick a number inside the interval to check the direction of the inequalities. Check each of these values in the original absolute value inequality. For the point inside the interval, we can pick any value we want. We will pick $x = 3$.

Left endpoint.	Right endpoint.	Point inside the interval.
$|-4| = 4$	$|4| = 4$	$|3| < 4$
$4 = 4$	$4 = 4$	$3 < 4$

The equations and the inequality are true, so the solution set is correct.

b.   The absolute value is isolated on the left side. It uses a less than or equal to symbol, so we rewrite the absolute value inequality as a compound inequality and solve.

$$|x - 2| \le 5$$ Rewrite as a compound inequality.

$$-5 \le x - 2 \le 5$$

$$\underline{+2 \qquad +2 +2}$$ Add 2 to all sides of the inequality.

$$-3 \le x \le 7$$

Therefore, x must be between or equal to -3 and 7.

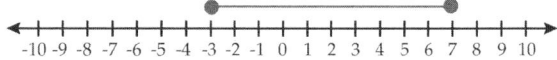

To confirm the solution set, we will check the endpoints of the interval and pick a number inside the interval to check the direction of the inequality symbols. If we pick $x = 4$ and check all of these values in the original absolute value inequality, we get:

Left endpoint.	Right endpoint.	Point inside the interval.
$|-3 - 2| = 5$	$|7 - 2| = 5$	$|4 - 2| \le 5$
$|-5| = 5$	$|5| = 5$	$|2| \le 5$
$5 = 5$	$5 = 5$	$2 \le 5$

The equations for the endpoints and the inequality are true, so the solution set is correct.

c.   First, we isolate the absolute value and then rewrite it as a compound inequality.

$$|b + 5| + 15 < 20$$

Isolate the absolute value.

$$\underline{\quad -15 \quad -15\quad}$$

$$|b + 5| < 5$$

Rewrite into a compound inequality and solve.

$$-5 < b + 5 < 5$$

$$\underline{-5 \quad\quad -5 \;\; -5\quad}$$

Subtract 5 from all sides.

$$-10 < b < 0$$

Therefore, b must be between or equal to -10 and 0.

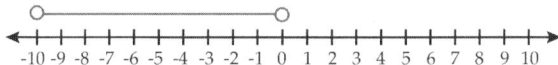

We will confirm the solution set by graphing the left and right side of the inequality on the calculator and comparing the graphs.

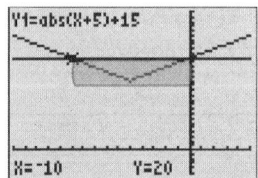

We can see that the absolute value graph is less than 20 when x is between -10 and 0.

EXAMPLE ③ PRACTICE PROBLEM

Solve the following inequalities. Give the solution as an inequality, and graph the solution set on a number line.

a. $|x - 8| < 2$ **b.** $|d + 1| - 3 \leq 5$

EXAMPLE ④ AN APPLICATION OF ABSOLUTE VALUE INEQUALITIES WITH THE SYMBOL FOR LESS THAN

When a part is made for a machine or other precise application, the accuracy of the size of each part is very important for that part to function properly. The amount the size of a particular part can be larger or smaller than the ideal size is called the tolerance. When a 20 millimeter bolt is made for an engine, the actual size must be within 0.02 mm of 20mm. If L is the actual length of the bolt in millimeters, the acceptable lengths of the bolt can be modeled by.

$$|L - 20| \leq 0.02$$

Find the acceptable lengths of these bolts.

Solution

We will solve the absolute value inequality.

$$|L - 20| \leq 0.02$$

$$-0.02 \leq L - 20 \leq 0.02$$

$$19.98 \leq L \leq 20.02$$

We will confirm this solution set by checking the endpoints of the interval in the original absolute value inequality. We know that the ideal length of 20mm is inside the interval, so the inequality symbols are ok.

Left endpoint.	Right endpoint.
$\lvert 19.98 - 20\rvert = 0.02$	$\lvert 20.02 - 20\rvert = 0.02$
$\lvert -0.02\rvert = 0.02$	$\lvert 0.02\rvert = 0.02$
$0.02 = 0.02$	$0.02 = 0.02$

The equations are true, so the bolts must be between 19.98mm and 20.02mm long.

EXAMPLE ④ PRACTICE PROBLEM

A rod in an engine is supposed to be 4.125 inches long. The tolerance for the rod is 0.0005 inches. The acceptable lengths of the rod can be modeled by

$$\lvert L - 4.125\rvert \le 0.0005$$

where L is the actual length of the rod in inches. Find the acceptable lengths for this rod.

ABSOLUTE VALUE INEQUALITIES INVOLVING GREATER THAN OR GREATER THAN OR EQUAL TO

Where the less than symbol asks for numbers closer to zero, the greater than symbol asks for numbers farther away from zero. The inequality $\lvert x\rvert > 5$ is asking for the values of x that are more than 5 units away from zero on the number line.

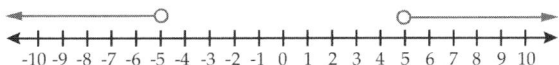

If we look at the graph of each side of the inequality, we can see the solution set.

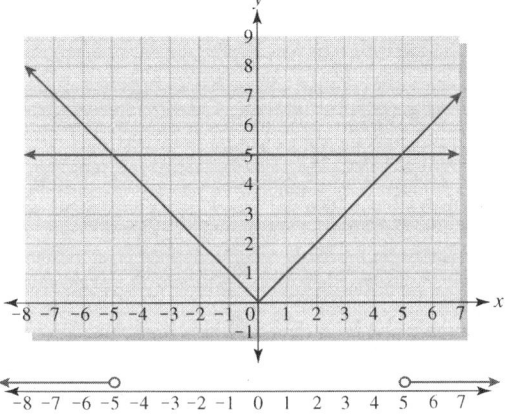

When x is less than -5 and when x is greater than 5, we see that the red graph of the absolute value is greater than, above, the blue graph of the line $y = 5$.

To represent values that are farther away from zero requires an interval on the right and another interval on the left to account for distances in both directions. Therefore, we will write two inequalities to accommodate the absolute value and will have one inequality going left (<) and another going right (>).

$$\lvert x\rvert > 5$$
$$x < -5 \quad \text{or} \quad x > 5$$

SOLVING ABSOLUTE VALUE INEQUALITIES INVOLVING GREATER THAN OR GREATER THAN OR EQUAL TO

* Isolate the expression containing the absolute value to the left side of the inequality.

* Rewrite the absolute value inequality into two inequalities, one less than and one greater than. .

$$|u| > n \qquad\qquad\qquad |u| \geq n$$

$$u < -n \quad \text{or} \quad u > n \qquad\qquad u \leq -n \text{ or } \quad u \geq n$$

* Solve the two inequalities.

* Check the solution set in the original absolute value inequality.

E X A M P L E **5** SOLVING ABSOLUTE VALUE INEQUALITIES WITH GREATER THAN

Solve the following inequalities. Give the solution as an inequality and graph the solution set on a number line.

a. $|x| > 2$ **b.** $|p + 4| > 3$ **c.** $|x - 2| + 7 \geq 10$

Solution

a. The absolute value is already isolated, so we will rewrite the absolute value inequality as two inequalities.

$$|x| > 2$$

$$x < -2 \quad \text{or} \quad x > 2$$

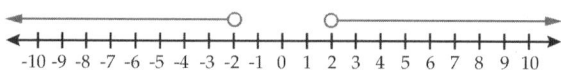

To check this solution, we will check the endpoints and a point from each of the intervals. We will check -6 for the left interval and 6 for the right interval.

Left interval endpoint.	Right interval endpoint.	Left interval test point.	Right interval test point.
$\|-2\| = 2$	$\|2\| = 2$	$\|-6\| > 2$	$\|6\| > 2$
$2 = 2$	$2 = 2$	$6 > 2$	$6 > 2$

Each of these points satisfies the original absolute value inequality, so the solution set is correct.

b. The absolute value is already isolated, so we will rewrite the absolute value inequality as two inequalities and solve.

$$|p + 4| > 3$$

$$p + 4 < -3 \quad \text{or} \quad p + 4 > 3$$

$$p < -7 \quad \text{or} \qquad p > -1$$

Rewrite as two inequalities and solve.
Subtract 4 from both sides.

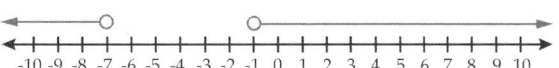

We will check this solution set using the graph of each side on the calculator.

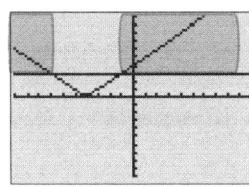

We can see that the absolute value graph is greater than 3 when x is less than -7 and greater than -1.

c. First, we will isolate the absolute value on the left side of the inequality and then rewrite into two inequalities and solve.

$$|x - 2| + 7 \geq 10$$

Isolate the absolute value.

$$\underline{\quad -7 \quad -7 \quad}$$

$$|x - 2| \geq 3$$

Rewrite into two inequalities and solve.

$$x - 2 \leq -3 \quad \text{or} \quad x - 2 \geq 3$$

Add 2 to both sides.

$$x \leq -1 \quad \text{or} \quad x \geq 5$$

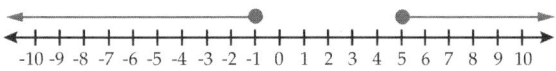

We will check this solution set using the graph of each side of the inequality on the calculator.

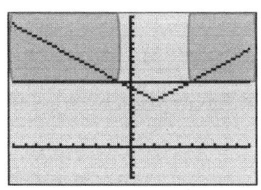

We can see that the absolute value graph is greater than 10 when x is less than -1 and greater than 5.

EXAMPLE ⑤ PRACTICE PROBLEM

Solve the following inequalities. Give the solution as an inequality and graph the solution set on a number line.

a. $|x + 2| > 7$

b. $|g - 3| - 4 \geq 1$

EXAMPLE ⑥ AN APPLICATION OF ABSOLUTE VALUE INEQUALITIES WITH GREATER THAN

A baby's birth weight is considered unusually high or low if it satisfies the inequality

$$|b - 118| > 30$$

where b is a baby's birth weight in ounces. Find the birth weights that would be considered unusually low or high.

Solution

We will solve the inequality by rewriting it into two inequalities.

$$|b - 118| > 30$$

$$b - 118 < -30 \quad \text{or} \quad b - 118 > 30$$

$$b < 88 \quad \text{or} \quad b > 148$$

We will check this solution set using the endpoints of the intervals and points within each interval.

Left interval endpoint.	Right interval endpoint.	Left interval test point.	Right interval test point.
$|88 - 118| = 30$	$|148 - 118| = 30$	$|70 - 118| > 30$	$|160 - 118| > 30$
$30 = 30$	$30 = 30$	$48 > 30$	$42 > 30$

These checks are true, so the solution set is correct. A baby is considered to have an unusually low birth weight when it weighs less than 88 ounces, and unusually high birth weight when it weighs more than 148 ounces.

EXAMPLE ⑥ PRACTICE PROBLEM

A recent poll asked people how much TV they watched in an average week. If the number of hours h a person watches TV satisfies the inequality

$$\left|\frac{h - 32}{4.5}\right| > 1.96$$

it would be considered unusual. Find the number of hours that would be considered unusual.

2.5 Exercises

For Exercises 1 through 10, write the solution set using inequality notation for the given number line.

1.

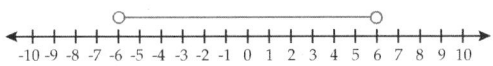

2.

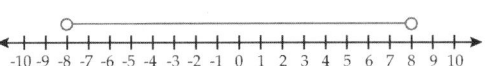

3.

4.

5.

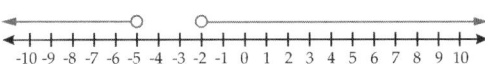

6.

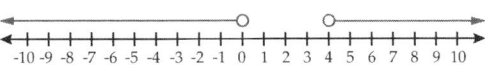

7.

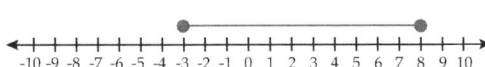

8.

9.

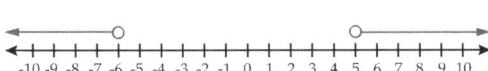

10.

For Exercises 11 through 16, write the solution set using inequality notation for the given graph and inequality.

11.

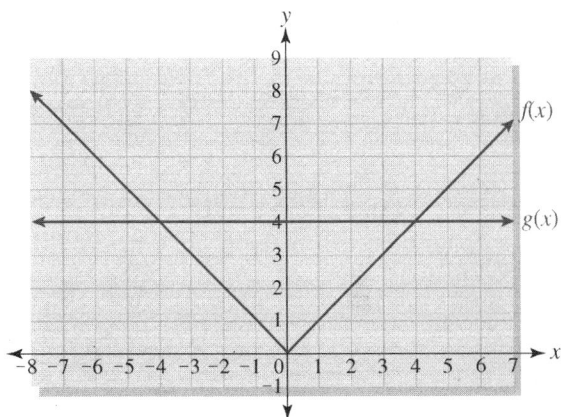

a. $f(x) < g(x)$ **b.** $f(x) > g(x)$

12.

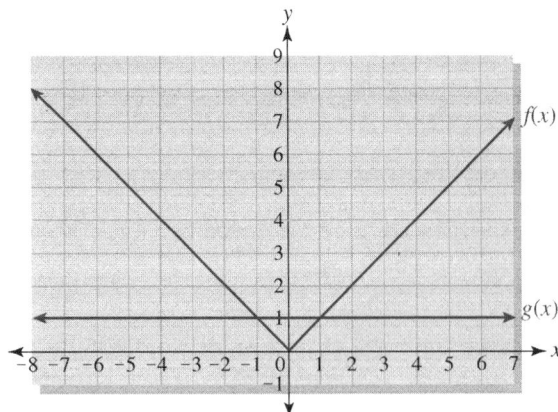

a. $f(x) < g(x)$ **b.** $f(x) > g(x)$

13.

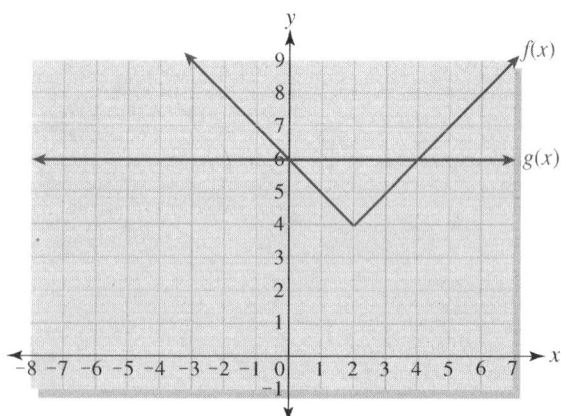

a. $f(x) \le g(x)$ **b.** $f(x) \ge g(x)$

14.

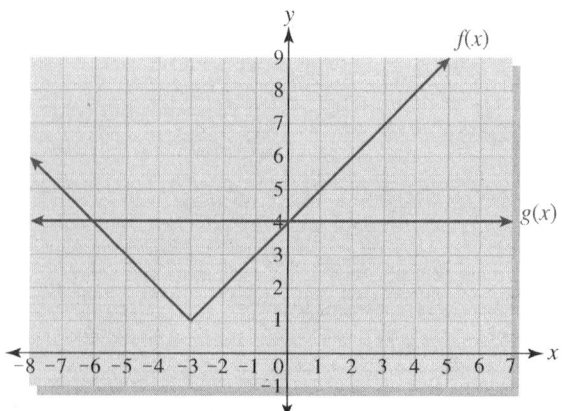

a. $f(x) \le g(x)$ **b.** $f(x) \ge g(x)$

15.

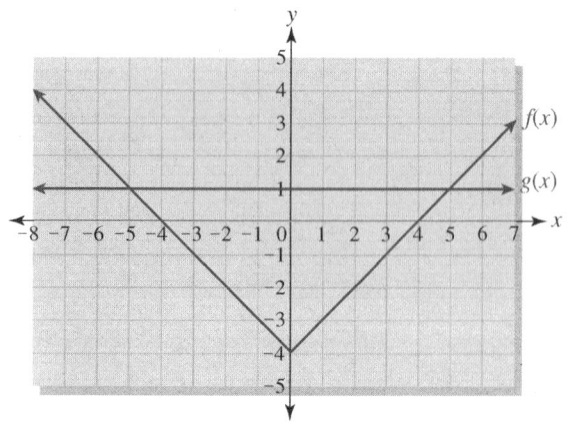

a. $f(x) < g(x)$ **b.** $f(x) > g(x)$

16.

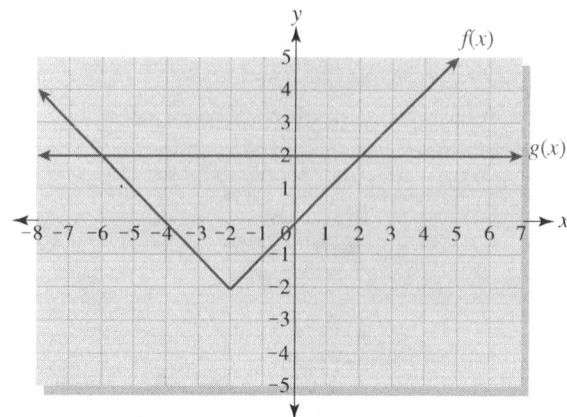

a. $f(x) \le g(x)$ **b.** $f(x) \ge g(x)$

For Exercises 17 through 32, solve the following equations.

17. $|x| = 12$

18. $|x| = 20$

19. $|h + 7| = 15$

20. $|m + 9| = 11$

21. $|b - 12| = 8$

22. $|w - 8| = 30$

23. $|k + 10| = -3$

24. $|s - 4| = -15$

25. $2|r + 11| = 36$

26. $8|h - 3| = 4$

27. $|g - 9| + 12 = 8$

28. $|x + 4| + 6 = 1$

29. $-2|d - 8| + 1 = 11$

30. $-4|m + 3| + 7 = 3$

31. $-3|x - 12| = -5$

32. $-5|y + 14| = -36$

33. Ricardo is taking a trip from Arizona to Yosemite National Park in California. The gas mileage in miles per gallon for Ricardo's truck can be modeled by the function

$$m(s) = -\frac{1}{3}|s - 60| + 25$$

where m is the gas mileage for Ricardo's truck when he is driving at an average speed of s miles per hour, for speeds between 40 and 80 miles per hour. At what speed should Ricardo drive to get gas mileage of 20 miles per gallon during his trip?

34. William is driving from Yakima, Washington, to Salt Lake City, Utah. The gas mileage in miles per gallon for William's car can be modeled by the function

$$m(s) = -\frac{1}{2}|s - 55| + 24$$

where m is the gas mileage for William's car when he is driving at an average speed of s miles per hour, for speeds between 40 and 80 miles per hour. At what speed should William drive to get gas mileage of 22 miles per gallon during his trip?

35. On Ricardo's trip from Arizona to Yosemite National Park, he will pass through Fresno, CA. Ricardo's distance from Fresno can be modeled by the function

$$D(t) = |700 - 60t|$$

where D is Ricardo's distance in miles from Fresno after driving for t hours. Find the time when Ricardo will be 50 miles away from Fresno.

36. On William's trip from Yakima, Washington, to Salt Lake City, Utah, he will pass through Boise, Idaho. William's distance from Boise can be modeled by the function

$$D(t) = |365 - 55t|$$

where D is William's distance in miles from Boise after driving for t hours. Find the time when William will be 30 miles away from Boise.

For Exercises 37 through 42, solve the following inequalities. Give the solution as an inequality, and graph the solution set on a number line.

37. $|x| < 5$

38. $|x| < 7$

39. $|h - 3| \le 4$

40. $|k + 5| \le 2$

41. $|b - 1| < 6$

42. $|p + 3| < 3$

43. A coupling for a satellite fuel line is designed to be 3.00 centimeters long. The tolerance for this part is 0.001 centimeters. The acceptable lengths of the coupler can be modeled by

$$|L - 3.00| \le 0.001$$

where L is the actual length of the coupler in centimeters. Find the acceptable lengths for this part.

44. A cylinder for a small engine is designed to be 3.625 inches in diameter. The tolerance for this part is 0.0005 inches. The acceptable diameters of the cylinder can be modeled by

$$|D - 3.625| \le 0.001$$

where D is the actual diameter of the cylinder in inches. Find the acceptable diameters for this part.

45. A particular pressure sensor measures the pressure in a gas chamber. The accuracy of this sensor is 0.5 pounds per square inch, psi. For an experiment, the technician requires the gas chamber to be at a pressure of 20psi. The acceptable pressures for this experiment will satisfy

$$|P - 20| \le 0.5$$

where P is the measured pressure in psi. Find the acceptable pressures for this experiment.

46. A particular thermocouple measures the temperature during a chemical reaction. The accuracy of this thermocouple, is 0.05°C. For an experiment the technician wants to keep the temperature at 22°C. The acceptable temperature readings for this reaction will satisfy

$$|T - 22| \le 0.05$$

where T is the temperature reading in degrees Celcius. Find the acceptable temperature readings for this chemical reaction.

For Exercises 47 through 52, solve the following inequalities. Give the solution as an inequality and graph the solution set on a number line.

47. $|x| > 3$

48. $|x| > 7$

49. $|p - 4| \ge 2$

50. $|m - 3| \ge 1$

51. $|y + 6| > 3$

52. $|w + 2| > 5$

53. Men's heights have a mean of 69 in. A height h that satisfies

$$\left|\frac{h - 63.6}{2.5}\right| > 2$$

is considered unusual. Find the heights that are considered unusual. (The word mean is a statistics word for the average value.)

54. Women's heights have a mean of 63.6 in. A height h that satisfies

$$\left|\frac{h - 63.6}{2.5}\right| > 2$$

is considered unusual. Find the heights that are considered unusual.

55. The normal mean score of the Stanford Binet IQ test is 100. An IQ score S that satisfies

$$\left|\frac{S - 100}{16}\right| > 2$$

is considered unusual.

a. Find the IQ scores that are considered unusual.

b. Albert Einstein was said to have an IQ of 160. Was Einstein's IQ unusual?

56. Human body temperatures have a mean of $98.20°$F. A body temperature that satisfies

$$\left|\frac{T - 98.20}{0.62}\right| > 2$$

is considered unusual.

a. Find the body temperatures that are considered unusual.

b. If a person has a body temperature of $99°$F, would it be considered unusual?

For Exercises 57 through 70 solve the following inequalities. Give the solution as an inequality and graph the solution set on a number line.

57. $2|r + 6| < 8$

58. $5|t + 3| < 20$

59. $|d - 5| + 10 \leq 7$

60. $|x + 3| + 15 < 4$

61. $|m + 1| - 2 > 6$

62. $-3|s - 2| + 2 > -4$

63. $-6|p - 4| + 3 \leq 16$

64. $-4|x + 3| + 9 < 20$

65. $-|x - 4| \geq -2$

66. $-|k + 11| \geq -36$

67. $5 + 2|2x - 4| > 11$

68. $4 - 5|4x + 8| > -6$

69. $12 + 8|3d - 8| \leq 10$

70. $20 + 3|5x - 2| \leq 11$

Solving Systems of Linear Inequalities

GRAPHING LINEAR INEQUALITIES WITH TWO VARIABLES

Many things in life are limited by our circumstances. When we go shopping, we are limited by the amount of money we have as well as by how much of something we need or want. Businesses are often limited by the production in a plant or by the costs associated with production. Dieticians must keep their clients within a certain range of nutrients and, therefore, must balance the types of foods and supplements that they prescribe for someone.

All of these situations can best be looked at as inequalities because a company or person might not need to use the maximum amount of something but may be able to use less. The use of inequalities allows for a set of solutions that will work within the given limitations or constraints. When two or more constraints are considered, a system of inequalities emerges.

Before we consider a system of inequalities, we will investigate an inequality that has two variables and look at its set of solutions. A graph is used to visualize the set of solutions to an inequality with two variables.

GRAPHING A LINEAR INEQUALITY WITH TWO VARIABLES

i. Graph the line as if it were an equation. Use a dashed line if the inequality does not include an "equal to" symbol.

ii. Pick a point on one side of the line and test the inequality.
(If not on the line. the point (0,0) is a very easy point to use as a test point.)

iii. If the point satisfies the inequality, shade that side of the line; if it does not satisfy the inequality, shade the other side of the line.

EXAMPLE   **1** AN APPLICATION OF LINEAR INEQUALITIES WITH TWO VARIABLES

A bicycle plant can produce both beach cruisers and mountain bikes. The plant needs 2 hours to build one beach cruiser and 3.5 hours to produce a mountain bike. If the plant operates up to 60 hours per week, what combinations of bikes can they produce in a week?

© Schwinn Bicycles   Photo provided by Trek Bicycle Corporation. All rights reserved.

Solution

First, we will define some variables:

C = number of beach cruisers built in a week

M = number of mountain bikes built in a week

Because each cruiser takes 2 hours to build, $2C$ will represent the number of hours it takes the plant to build cruisers that week. It takes 3.5 hours to build a mountain bike, so $3.5M$ represents the number of hours that plant spends building mountain bikes that week. The maximum number of hours the plant operates in a week is 60, so we have the following inequality:

$$2C + 3.5M \leq 60$$

This inequality can be graphed. You can solve for C to get

$$2C + 3.5M \leq 60$$
$$2C \leq -3.5M + 60$$
$$C \leq -1.75M + 30$$

To graph this inequality, we first graph it as if it were an equation, giving us the line below:

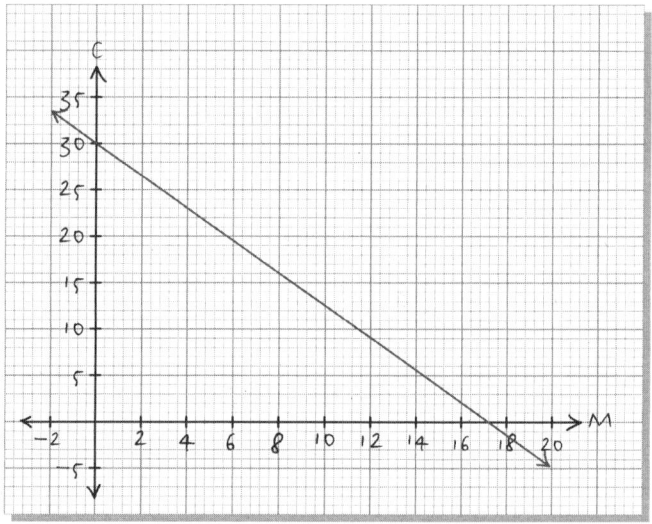

Because this situation requires that the factory work 60 hours or less a week, there will be many possible solutions. To represent all possible solutions, we will shade the side of the line that gives us 60 hours or less each week. To determine which side of the line to shade, test a point on either side of the line to see whether it is a solution to the inequality. If it is a solution, you shade the side of the line that the test point is on. If we test the point $(0, 0)$,

$$2(0) + 3.5(0) \leq 60$$
$$0 \leq 60$$

it satisfies the inequality indicating that the lower portion of the graph should be shaded.

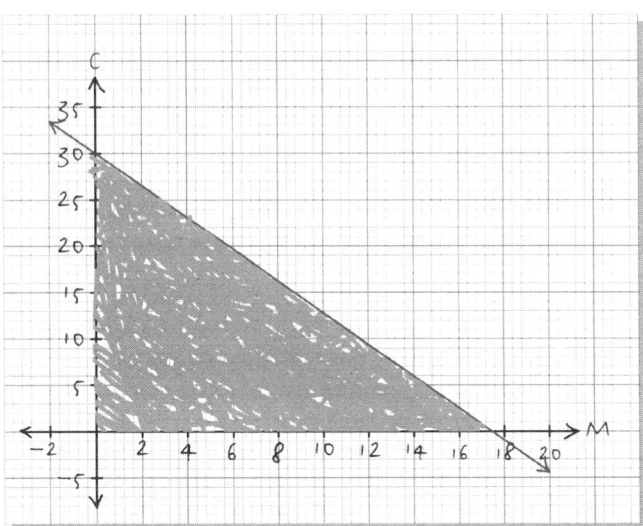

Any point within the shaded region is a solution to this inequality, meaning that the plant can build anywhere from no bikes at all to a combination of bikes ranging from no cruisers and 17 mountain bikes to no mountain bikes but 30 cruisers. Any point on the line itself would mean that exactly 60 hours are being used a week. Any point in the shaded region means that less than 60 hours are being used, and any point outside the shaded region implies that more than 60 would be needed to do the work.

As you can see from the example, there is a large set of solutions for this situation. Working with inequalities most often means finding not a single solution but a set of possible solutions. When graphing an inequality that has two variables, you can graph the inequality as if it were an equation and then decide which side of the line to shade. Deciding which side of a line to shade can be done by using any point not on the line as a test point. If that point is a solution to the inequality, then shade the side that the point is on. If the point is not a solution to the inequality, then test a point on the other side of the line.

In this example, the inequality included things equal to and less than, so the line was drawn as a solid line. If the inequality you are graphing does not have an "equal to" symbol, then the line is drawn as a dashed line, indicating that it is not a part of the solution set.

E X A M P L E   **2**   GRAPHING LINEAR INEQUALITIES WITH TWO VARI-ABLES

Graph the following inequalities by hand:

a. $y < 2x - 7$ **b.** $y \le -\dfrac{3}{5}x + 6$ **c.** $-3x - y < -7$

Solution

a. First we will graph the line as if it is an equation. Because the inequality is a "less than" and does not have an "equal to" part, we will use a dashed line.

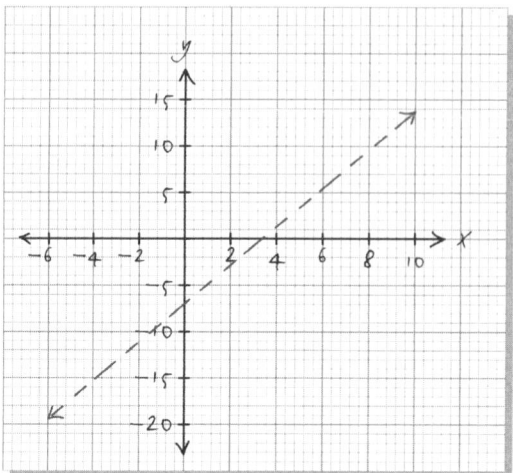

Now that we have drawn the line, we want to do the shading to show the solution set. We can choose the point $(0, 0)$ as our test point.

$$y < 2x - 7$$

Plug in the test point $(0, 0)$.

$$0 < 2(0) - 7$$

$$0 \not< -7$$

This inequality is not true so the test point is not a solution.

The test point $(0, 0)$ does not satisfy the inequality because we get $0 < -7$, which is not true. So we should shade the opposite side of the graph, the lower portion of the graph. When y is isolated on the left side of the inequality, the "less than" symbol implies that you will shade the lower portion of the graph. Shading gives us the following graph:

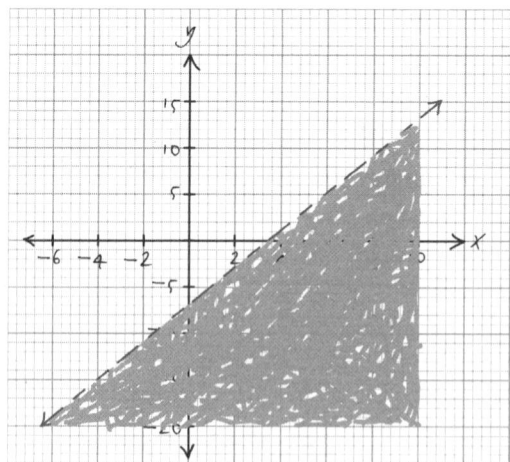

b. Again, we will graph the inequality as if it were an equation. This time, the line will be solid, since the symbol does include an "equal to."

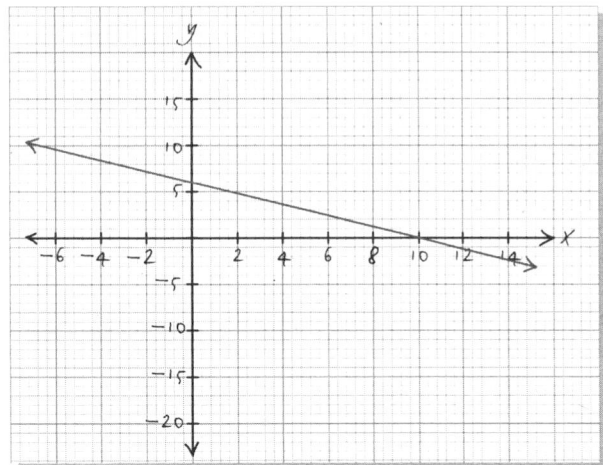

Now if we test the point (0, 0), we find that it satisfies the inequality, so we shade the lower portion of the graph.

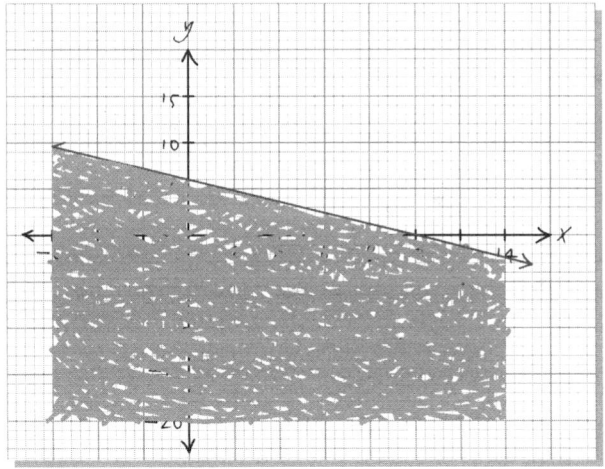

c. We will start by putting the equation into slope-intercept form.

$$-3x - y < -7$$

$$-y < 3x - 7$$

$$\frac{-y}{-1} > \frac{3x - 7}{-1}$$

Remember to reverse the inequality symbol when dividing by a negative number.

$$y > -3x + 7$$

This inequality does not have an "equal to" part, so we will graph the line using dashes to indicate that it is not part of the solution set but only a boundary for it.

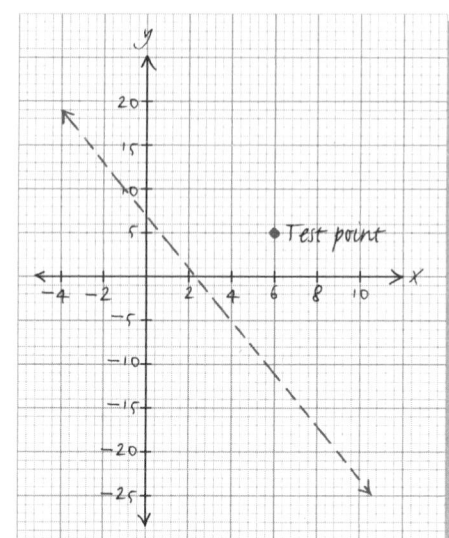

$$y > -3x + 7$$

Now we will test the point (6, 5), which is clearly above the line.

$$-3x - y < -7$$
$$-3(6) - (5) < -7$$
$$-23 < -7$$

Plug the test point (6, 5) into the original inequality.

This inequality is true so the test point is a solution.

We see that the test point (6, 5) does satisfy the inequality, so we will shade on that side of the line.

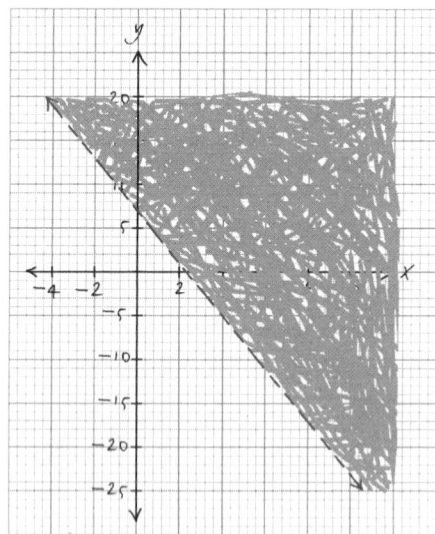

Note that if we pick a point not in the solutions set, it will not satisfy the inequality.

If we test the point (0, 0) that is clearly below the line we get:

$$-3x - y < -7$$
$$-3(0) - (0) < -7$$
$$0 < -7$$

This inequality is not true, so the point (0, 0) is not a solution and thus not in the shaded region.

EXAMPLE ② PRACTICE PROBLEM

Graph the following inequalities by hand.

a. $y > 2x - 8$

b. $y \le -\frac{3}{4}x + 5$

SOLVING SYSTEMS OF LINEAR INEQUALITIES

Now that we can graph an inequality with two variables, we can consider a system of inequalities. Most situations have more than one constraint. The bicycle plant from Example 1 might also have a constraint that they must keep the plant in production at least 40 hours per week. With more than one constraint, the situation calls for a system of inequalities, not just one inequality. When solving a system of inequalities, we will graph each inequality, shading according to the inequality symbol of each line. The area where all the shaded regions overlap will be the solution set to the system of equations. If the shaded regions do not overlap, there is no solution to the system.

GRAPHING A SYSTEM OF LINEAR INEQUALITIES

1. Graph both inequalities on the same set of axes, shading according to the inequality symbol of each line.

2. The solution set is the intersection of the shaded regions of all the inequalities in the system.

3. If the shaded regions do not intersect, there is no solution.

E X A M P L E   **3** | AN APPLICATION OF SYSTEMS OF LINEAR INEQUALITIES

Example 1 discussed a bicycle plant that had the time constraint that they could not exceed 60 hours of production per week. The same plant must stay in production at least 40 hours per week.

a. Create a system of inequalities to model this situation.

b. Graph the solution set for this system.

c. Can the plant produce 10 mountain bikes and 5 cruisers in a week?

d. Can the plant produce 5 mountain bikes and 8 cruisers a week?

Solution

a. First, we will define the variables as we did in Example 1:

C = number of cruisers built in a week

M = number of mountain bikes built in a week

The constraint of not exceeding 60 hours per week results in the inequality from Example 1:

$$2C + 3.5M \leq 60$$

The new constraint of not working less than 40 hours a week results in the inequality

$$2C + 3.5M \geq 40$$

This gives us the following system of inequalities:

$$2C + 3.5M \leq 60$$
$$2C + 3.5M \geq 40$$

b. We need to graph both of these inequalities at the same time, being careful to shade the correct sides of each line.

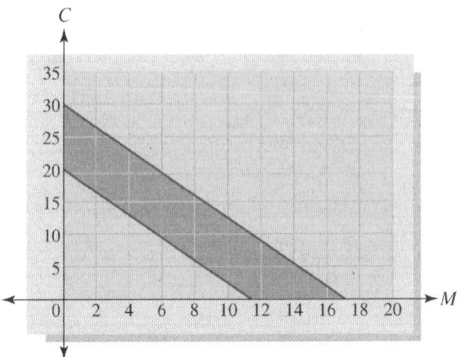

The *pink* area shows the solution set for the first equation.

The *blue* area shows the solution set for the second equation.

The *purple* area (overlap) is the solution set for the system as a whole.

c. 10 mountain bikes and 5 cruisers would be the point (10, 5) on the graph. This point is inside the area of the graph where the solution sets of the two inequalities overlap. Therefore this combination would be a reasonable amount of each type of bike to produce a week.

d. 5 mountain bikes and 8 cruisers would be the point (5, 8) and would lie outside the overlapping section of this system. Therefore producing this combination of bikes per week would not meet the constraints.

In Example 3, we needed to graph only the area covered by positive numbers of bikes being produced. Using only positive values is typical of many real-life situations and actually adds two additional inequalities to the system: $M \geq 0$ and $C \geq 0$. We did not write these inequalities this time, but we could have. Common sense told us that a negative number of bikes could not be produced.

EXAMPLE ③ PRACTICE PROBLEM

A small appliance manufacturer has a plant that produces toaster ovens and electric griddles. The plant must keep its weekly costs below $40,000 and cannot be in operation more than 80 hours a week. It takes 1.25 hours and costs the plant $850 to produce one hundred toaster ovens. It takes 2 hours and costs the plant $920 to produce a hundred electric griddles.

a. Create a system of inequalities to model this situation.

b. Graph the solution set for this system.

c. Can the plant produce 1,600 toaster ovens and 2,000 electric griddles a week?

d. Can the plant produce 2,000 toaster ovens and 3,000 electric griddles a week?

Another type of application that we can consider as a system of inequalities is investment questions like those we did in Section 2.2.

EXAMPLE 4 AN INVESTMENT PROBLEM AS A SYSTEM OF INEQUALITIES

Mary is retiring and has $500,000 to invest in two accounts. One account pays 5% annual interest, and the other pays 3.5% annual interest. How much should she invest in each account to earn at least $21,000 in interest each year?

Solution

We define variables as follows:

A = the amount in the account paying 5% annual interest

B = the amount in the account paying 3.5% annual interest

Because Mary can invest up to $500,000 and must earn at least $21,000 per year in interest, we can write the following system of inequalities:

$$A + B \leq 500000$$

$$0.05A + 0.035B \geq 21000$$

If we graph this system, we can find the possible solutions to this situation.

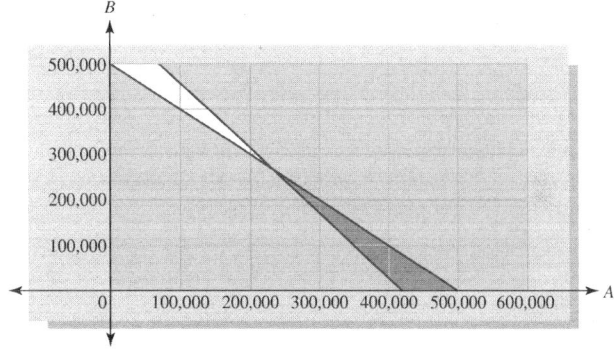

From this graph, we can see that Mary can have many different investment amounts that would earn her the minimum amount of interest she needs and still not use all of her $500,000. Any combination of investments from the overlapping section of the graph can be used to meet Mary's investment goal. An investment of $450,000 in the account paying 5% and $50,000 in the account paying 3.5% will satisfy the needs Mary has. An investment of $100,000 in the account paying 5% and $400,000 in the account paying 3.5% will not earn enough interest to meet Mary's needs.

In many systems of equations or inequalities the points of intersection for the lines of the system are critical values of interest. In the case of Mary's investments, that point of intersection is the investment that meets her needs and invests the total $500,000. This is also the investment with the least amount invested in the more risky account paying 5% and the most money she can invest in the safer account paying 3.5% and still meet her minimum interest needs.

EXAMPLE ④ PRACTICE PROBLEM

Don is retiring and has $800,000 to invest in two accounts. One account pays 6.5% annual interest, and the other pays 4.5% annual interest. How much should he invest in each account to earn at least $40,500 in interest each year?

EXAMPLE ⑤ GRAPHING SYSTEMS OF EQUATIONS

Graph each system of inequalities by hand.

a. $2x + 5y < 15$
　　$3x - 2y > 6$

b. $y \geq -1.5x + 5$
　　$y \leq -1.5x + 2$

Solution

a. We first need to solve these inequalities for y, and then we can graph them.

$2x + 5y < 15$

$\qquad 5y < -2x + 15$ Solve the first inequality for y.

$\qquad y < -\dfrac{2}{5}x + 3$

$3x - 2y > 6$ Solve the second inequality for y.

$\qquad -2y > -3x + 6$

$\qquad y < \dfrac{3}{2}x - 3$ Remember to reverse the inequality symbol when dividing by a negative number.

$\qquad y < -\dfrac{2}{5}x + 3$ Now graph the system of inequalities.

$\qquad y < \dfrac{3}{2}x - 3$

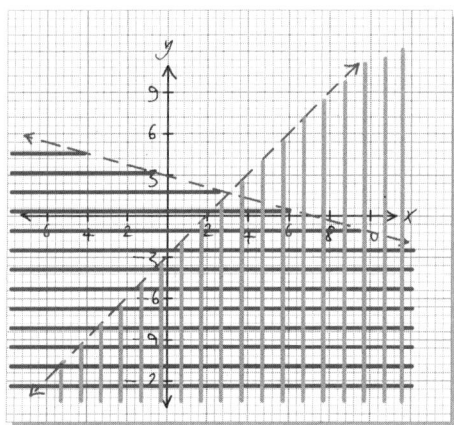

The section of the graph where the two shadings overlap is the solution set for this system of inequalities.

b. We start by graphing the two inequalities.

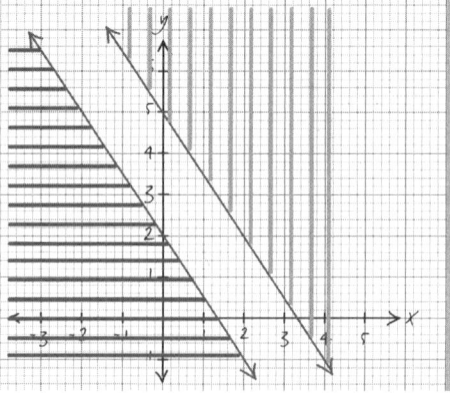

We can see from the graph that the solution sets for these two inequalities give us no intersection and therefore there is no solution to the system.

EXAMPLE PRACTICE PROBLEM

Graph the following system of inequalities by hand.

a. $y \geq x + 2$

$y \leq 2x - 3$

Graphing these systems can also be done using the graphing calculator. See the instructions in the Using Your Graphing Calculator box to the left or in Appendix C.

EXAMPLE **6** GRAPHING SYSTEMS OF INEQUALITIES ON A GRAPHING CALCULATOR

Graph the following system of inequalities using a graphing calculator.

$y < -2x + 5$

$y > 3x - 5$

Solution

a. Both inequalities have y isolated on the left side. The first inequality has a "less than" symbol, so we shade below the line. For the second inequality, we shade above the line because it has a "greater than" symbol. Both lines should be dashed because these inequalities do not include any "equal to" parts. The calculator does not show dashed lines when shading, something we have to remember when considering possible solutions.

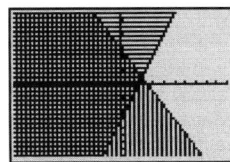 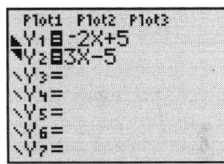

The two shaded regions intersect on the left side, giving us our solution set.

EXAMPLE PRACTICE PROBLEM

Graph the following system of inequalities using a graphing calculator.

a. $2x - 5y > 12$

$4y > 7x - 5$

Using Your TI Graphing Calculator

To graph inequalities on your calculator, you can have the calculator shade above or below the line. Use the cursor buttons to go to the left of the Y1 and press Enter until you get the shade above or below symbol you are looking for.

Note that the calculator cannot use shading with dashed lines, so any inequalities without "equal to" in the inequality symbol will graph with a solid line but should be interpreted as a dashed line. Any points on that line should not be considered part of the solution set.

2.6 Exercises

According to the National Institutes of Health, adults are considered underweight or overweight on the basis of a combination of their height and weight. The graph below shows this relationship as a system of inequalities. Use the graph to answer Exercises 1 through 8.

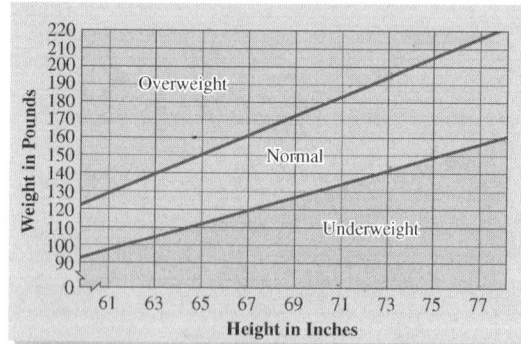

1. What category does a 5′1″ person weighing 105 pounds fall into?

2. What category does a 6′3″ person weighing 210 pounds fall into?

3. Above what weight is a 5′9″ person considered overweight?

4. Under what weight is a 6′ person considered underweight?

5. If a 150-pound person is to be considered in the normal weight range, how tall must the person be?

6. If a 130-pound person is to be considered in the normal weight range, how tall must the person be?

7. If a person is 67″ tall, what is the person's normal weight range?

8. If a person is 73″ tall, what is the person's normal weight range?

Bicycles Galore can manufacture up to 500 bikes a month in its Pittsburgh plant. Cruisers cost $75 to manufacture, and mountain bikes cost $135 to manufacture. Bicycles Galore must keep its total monthly costs below $50,000. Use the graph of this situation to answer Exercises 9 through 12.

9. What is the greatest number of mountain bikes the company can build per month?

10. What is the maximum number of cruisers the company can build per month?

11. Can Bicycles Galore build 400 cruisers and 100 mountain bikes per month? Explain.

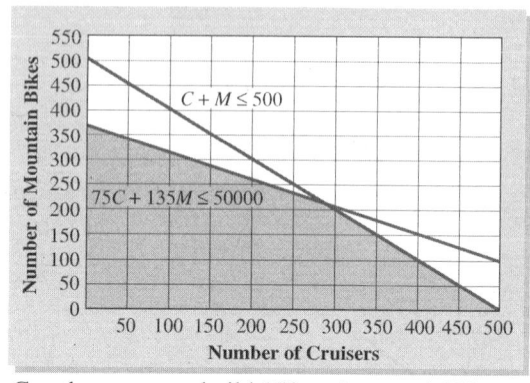

12. Can the company build 150 cruisers and 350 mountain bikes per month? Explain.

13. Bicycles Galore can produce up to 500 bikes per month in their Pittsburgh plant. They can make a profit of $65 per cruiser they build and $120 per mountain bike they build. The company's board of directors wants to make a profit of at least $40,000 per month.

    a. Write a system of inequalities to model this situation and graph it by hand.

    b. Can Bicycles Galore meet a demand by the board of directors for $40,000 in profit per month?

    c. If the company wants to maximize production at 500 bikes, how many of each bike should the plant produce to make the $40,000 profit?

14. Theresa has started a small business creating jewelry to sell at boutiques. She can make a necklace in about 45 minutes and a bracelet in about 30 minutes. Theresa can work only up to 25 hours a month making the jewelry. She can make $20 profit from a necklace and $18 profit from bracelets. Theresa wants to earn at least $800 profit.

    a. Write a system of inequalities to model this situation and graph it by hand.

    b. Should Theresa make more necklaces or bracelets?

    c. If Theresa knows that most of her customers want a bracelet and necklace combination, how many of each should she make to meet her profit goal?

15. Juanita is retiring and has $750,000 to invest in two accounts. One account pays 4.5% simple interest, and the other pays 3.75% simple interest. Juanita wants to know how much she should invest in each account to earn at least $30,000 in interest each year. Write a system of inequalities and graph it by hand to show the possibilities.

16. Casey is retiring and has $2.5 million to invest in stocks and bonds. Casey estimates that the stocks will have an average return of 9.5% and the bonds will pay 7% simple interest. Casey wants $210,000 per year to cover his living and travel expenses each year. Write a system of inequalities and graph it by hand to show the possible amounts he could invest in each account.

17. In preparing meals to eat during an all-day bike race, the team dietician determines that each athlete must consume at least 2000 calories and at least 350 grams of carbohydrates to sustain the energy needed to compete at peak performance. The dietician is providing each athlete with a combination of power bars and sports drinks for the race. Each power bar has 240 calories and 30 grams of carbs. The sports drinks contain 300 calories and 70 grams of carbs.

    a. Write a system of inequalities for these dietary needs and graph it by hand.

    b. What is the minimum number of power bars and sports drinks that an athlete should consume during the race?

    c. If racers can carry only four drinks, what is the minimum number of power bars they should eat?

18. The same dietician from Exercise 17 had found that he should increase the calories and carbohydrates for the team to improve their performance. The dietician decides that each athlete must consume at least 2500 calories and at least 420 grams of carbohydrates to sustain the energy needed to compete at peak performance. The dietician is providing each athlete with a combination of power bars and sports drinks for the race. Each power bar has 240 calories and 30 grams of carbs. The sports drinks contain 300 calories and 70 grams of carbs.

    a. Write a system of inequalities for these dietary needs and graph it by hand.

    b. What is the minimum number of power bars and sports drinks that an athlete should consume during the race?

    c. If racers can carry only four drinks, what is the minimum number of power bars they should eat?

For Exercises 19 through 30, graph the inequalities by hand.

19. $y > 3x + 2$ **20.** $y < -4x + 7$

21. $y > \dfrac{5}{4}x - 2$ **22.** $y < \dfrac{3}{7}x + 1$

23. $y \geq \dfrac{2}{3}x + 6$ **24.** $y \geq \dfrac{4}{5}x - 5$

25. $y \leq -2x + 6$ **26.** $y \leq x + 5$

27. $2x + 5y > 10$ **28.** $3x + 4y < 8$

29. $12x - 4y > 8$ **30.** $10x - 5y < 20$

For Exercises 31 through 34, find the inequality for the given graph.

31.

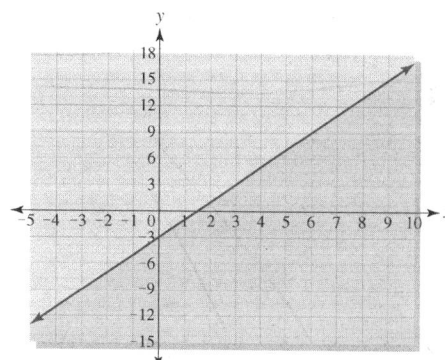

32.

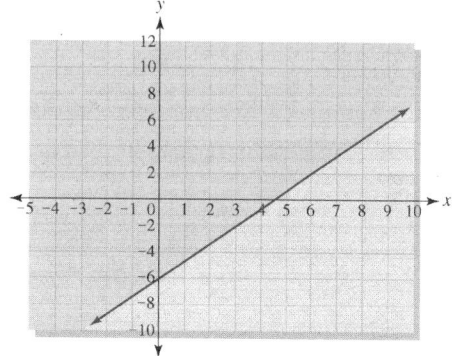

33.

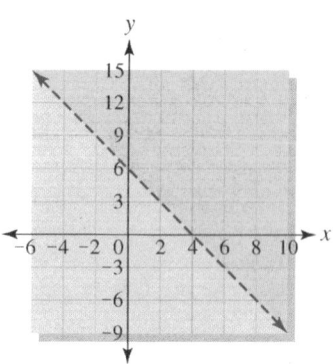

34.

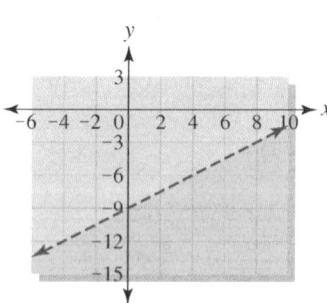

35.

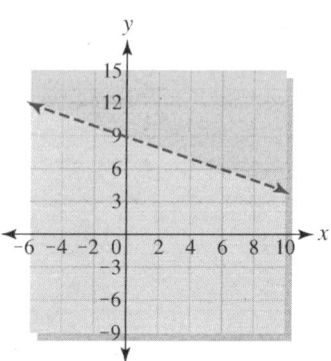

36.

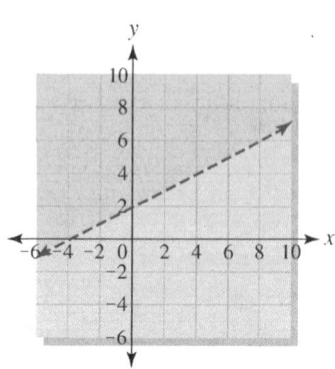

37.

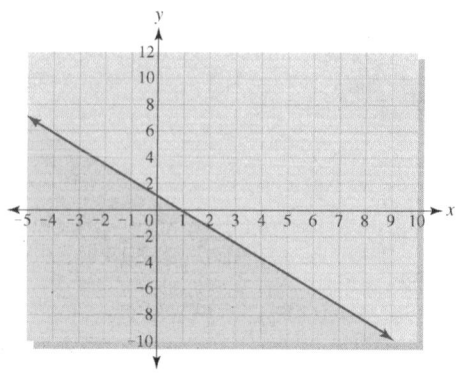

38.

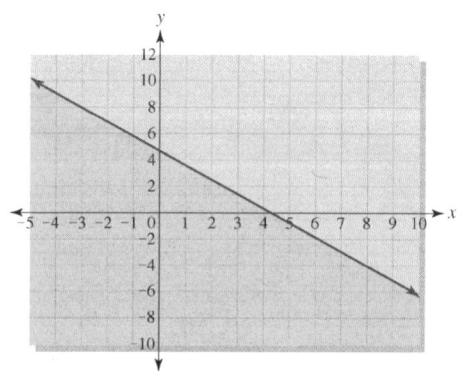

For Exercises 39 through 50, graph the systems of inequalities by hand or on the calculator.

39. $y > 2x + 4$
$y < -3x + 7$

40. $y < 4x + 5$
$y > -x + 1$

41. $y < \dfrac{2}{5}x - 3$

$y < -\dfrac{3}{4}x + 6$

42. $y \geq \dfrac{1}{3}x + 2$

$y \leq -\dfrac{2}{3}x + 7$

43. $y \leq 2x + 5$
$y \geq 2x + 8$

44. $y < -4x + 10$
$y > -4x + 2$

45. $2x + 4y \leq 5$
$2x - 4y \leq 5$

46. $-4x - 3y \geq 11$
$4x + 3y \geq 5$

47. $5x + 3y > 7$
$4x + 2y < 10$

48. $4x + 5y < 10$
$8x + 10y > 10$

49. $y > \dfrac{2}{3}x - 12$

$6x - 4y > 12$

50. $1.5x + 4.6y \leq 1.8$

$y \geq \dfrac{2}{17}x + 1.2$

Chapter 2 Summary

Section 2.1 Introduction to Systems

- A **system of equations** is a set of two or more equations for which you are seeking a common solution.
- Two ways in which systems can be solved are **graphically** and **numerically.**
- The solution to a system of equations can be seen on a graph where all the graphed lines **intersect** each other.
- Numerically, a system can be solved by looking for input values that give you the same output value for every equation in the system.
- There are three possible types of systems of equations:

   1. **Consistent systems** have a finite number of solutions.
   2. **Dependent systems** have infinitely many solutions and occur when the equations are the same graph.
   3. **Inconsistent systems** have no solution and occur when the graphs do not intersect, such as with parallel lines.

EXAMPLE

Solve the system by graphing:

$$y = 2x - 5$$
$$y = -\frac{1}{2}x + \frac{5}{2}$$

Solution

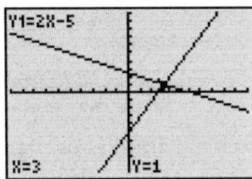

Therefore this system has the solution (3, 1).

EXAMPLE

Solve the system numerically:

$$y = -3x + 8$$
$$y = 4.5x - 10.75$$

Solution

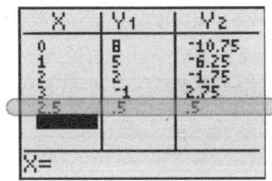

Therefore, this system has the solution (2.5, 0.5).

EXAMPLE

Determine whether the following system is consistent, dependent, or inconsistent:

$$y = 2.75x - 3.5$$
$$11x - 4y = 20$$

Solution

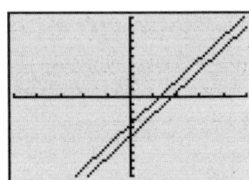

This system is inconsistent and has no solutions because the lines are parallel.

Section 2.2 Solving Systems of Equations Using the Substitution Method

- **Substitution Method**
  - **i.** Best used when a variable is already isolated in at least one equation.
  - **ii.** Substitute the expression representing the isolated variable from one equation in place of that variable in the other equation.
  - **iii.** Remember to find the values for both variables.
  - **iv.** Check the solution in both equations.

- In solving an **inconsistent system** algebraically, all the variables will be eliminated, and the remaining sides of the equation will not be equal. This means that there will be no solutions to the system.

- In solving a **dependent system** algebraically, all the variables will be eliminated, but the remaining sides of the equation will be equal. This means that any solution to one equation will be a solution to the other, giving infinitely many solutions.

EXAMPLE 4

Solve the following system by substitution:

$$y = 7x - 15$$
$$3x + 2y = 45$$

Solution

$$y = 7x - 15$$
$$3x + 2y = 45$$

$$3x + 2(7x - 15) = 45$$
$$3x + 14x - 30 = 45$$

$$17x - 30 = 45$$
$$17x = 75$$
$$x = 4.412$$

$$y = 7(4.412) - 15$$ Remember to find the value of the other variable.
$$y = 15.88$$

This system has a solution of (4.412, 15.88). We check this solution in both equations.

First Equation.

$$15.88 = 7(4.412) - 15$$
$$15.88 \approx 15.884$$

Second Equation.

$$3(4.412) + 2(15.88) = 45$$
$$44.996 \approx 45$$

EXAMPLE 5

Solve the following system:

$$y = 3x + 8$$
$$-6x + 2y = 20$$

Solution

$$y = 3x + 8$$
$$-6x + 2y = 20$$

$$-6x + 2(3x + 8) = 20$$
$$-6x + 6x + 16 = 20$$
$$16 \neq 20$$

All the variables were eliminated and the remaining numbers are not equal, so this is an inconsistent system and has no solutions.

Section 2.3 Solving Systems of Equations Using the Elimination Method

- **Elimination Method**
  - **i.** Best used when neither variable is already isolated.
  - **ii.** Multiply one or more equations by a number to make the coefficients of one variable opposite in signs but of the same size.
  - **iii.** Add the two equations to eliminate the variable, then solve.

iv. Remember to find the values for both variables.

v. Check the solution in both equations.

EXAMPLE 6

Solve the following system by elimination:

$$3x + 5y = 41$$
$$10x - 4y = -8$$

Solution

$$4(3x + 5y) = 4(41) \longrightarrow 12x + 20y = 164$$
$$5(10x - 4y) = 5(-8) \longrightarrow 50x - 20y = -40$$

$$12x + 20y = 164$$
$$\underline{50x - 20y = -40}$$
$$62x = 124$$
$$x = 2$$

$$3(2) + 5y = 41$$ **Remember to find the value**
$$y = 7$$ **of the other variable.**

Therefore, this system has a solution of $(2, 7)$. We check this equation using the graph.

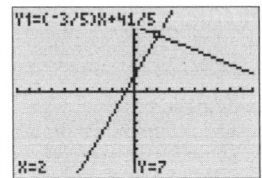

Section 2.4 Solving Linear Inequalities

- When solving **inequalities,** use the same methods as for solving equations, but remember to reverse the inequality symbol whenever you multiply or divide both sides of the inequality by a negative number.

- Many systems of equations can be solved as inequalities when the context of the problem implies a "less than or greater than" relationship.

- Use caution when **interpreting the results** from an inequality problem. The solution will include more than one input, so your wording will need to indicate that.

- When solving an inequality **numerically,** look for the input value that makes both sides equal and then whether a number greater than or less than that value will make the inequality relationship you want true.

- When solving an inequality **graphically,** look for the intersection of the graphs and determine on which side of the intersection the inequality relationship you want is true.

EXAMPLE 7

The percentage of people in the United States who believe that the amount of violence on TV does damage to children can be modeled by $V(t) = -2.4t + 73$, where t is time in years since 2000. The percentage of people in the United States who believe that the amount of sex on TV does damage to children can be modeled by $S(t) = 1.7t + 19.8$, where t is time in years since 2000. Find when the percentage of people who believe that violence is damaging will be less than those who believe that the amount of sex on TV is damaging.

Solution

$$-2.4t + 73 < 1.7t + 19.8$$
$$-4.1t < -53.2$$
$$t > 12.98$$

In about 2013 and beyond, more people will believe that the amount of violence on TV does damage to children than will believe that the amount of sex on TV does damage to children.

EXAMPLE

Use the table to find when Y1 < Y2.

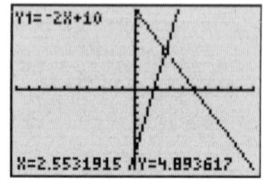

Solution

Y1 = Y2 at x = −1.5 and is less than Y2 when x < −1.5, so the solution is x < −1.5.

EXAMPLE 9

Solve the inequality $-2x + 10 < 5x - 8$ graphically.

Solution

These lines intersect when x = 2.55 and $-2x + 10 < 5x - 8$ to the right of the intersection, so the solution is x > 2.55.

Section 2.5 Absolute Value Equations and Inequalities

- **Absolute Value:** The absolute value of a real number n, $|n|$, is the distance from zero to n on a real number line.

- **Absolute Value Equation:** If n is a positive real number and u is any algebraic expression ,
$$|u| = n$$
$$u = n \quad \text{or} \quad u = -n$$

- If n is negative, then the equation $|u| = n$ has no real solution.

- **Solving absolute value equations:**

    1. Isolate the expression containing the absolute value.

    2. Rewrite the equation into two equations .
    $$|u| = n$$
    $$u = n \quad \text{or} \quad u = -n$$

    3. Solve each equation.

    4. Check the solutions in the original absolute value equation.

- **Solving absolute value inequalities involving less than or less than or equal to:**

    1. Isolate the expression containing the absolute value to the left side of the inequality.

    2. Rewrite the absolute value inequality into a compound inequality .

$|u| < n$	$|u| \leq n$
$-n < u < n$	$-n \leq u \leq n$

3. Solve the compound inequality.

4. Check the solution set in the original absolute value inequality.

- **Solving absolute value inequalities involving greater than or greater than or equal to:**

    1. Isolate the expression containing the absolute value to the left side of the inequality.

    2. Rewrite the absolute value inequality into two inequalities, one less than and one greater than.

$|u| > n$	$|u| \geq n$
$u < -n \quad \text{or} \quad u > n$	$u \leq -n \text{ or } u \geq n$

3. Solve the two inequalities.

4. Check the solution set in the original absolute value inequality.

EXAMPLE 10

Solve the equation $|x + 5| = 11$.

Solution

$$|x + 5| = 11$$
$$x + 5 = -11 \quad \text{or} \quad x + 5 = 11$$
$$x = -16 \quad \text{or} \quad x = 6$$

We check these solutions in the original equation.

x = -16	x = 6
$|-16 + 5| = 11$	$|6 + 5| = 11$
$11 = 11$	$11 = 11$

EXAMPLE

Solve the given inequality. Give the solution as an inequality and graph the solution set on a number line.
$$|x + 3| \leq 4$$

Solution

$$|x + 3| \leq 4$$
$$-4 \leq x + 3 \leq 4$$
$$-7 \leq x \leq 1$$

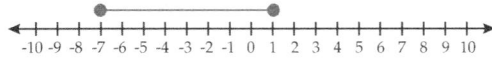

We check this solution using a graph.

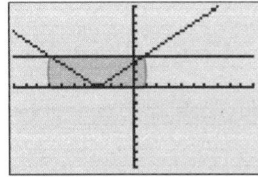

We can see that the absolute value graph is less than 4 when x is between -7 and 1.

EXAMPLE

Solve the given inequality. Give the solution as an inequality and graph the solution set on a number line.
$$|x - 2| - 1 > 4$$

Solution

$$|x - 2| - 1 > 4$$
$$|x - 2| > 5$$
$$x - 2 < -5 \quad \text{or} \quad x - 2 > 5$$
$$x < -3 \quad \text{or} \quad x > 7$$

We check this solution using a graph.

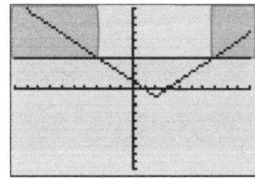

We can see that the absolute value graph is greater than 4 when x is less than -3 or greater than 7.

Section 2.6 Solving Systems of Inequalities

- To **graph a linear inequality with two variables,** graph the line as you would if it were an equation and shade the side of the line that makes the inequality true. Use a dashed line if the inequality does not include the symbol for equal to.

- To **solve a system of inequalities,** graph all the inequalities on the same set of axes. The region where all the shaded areas overlap is the solution set of the system.

EXAMPLE 13

Graph the inequality $y < 3x - 12$.

Solution

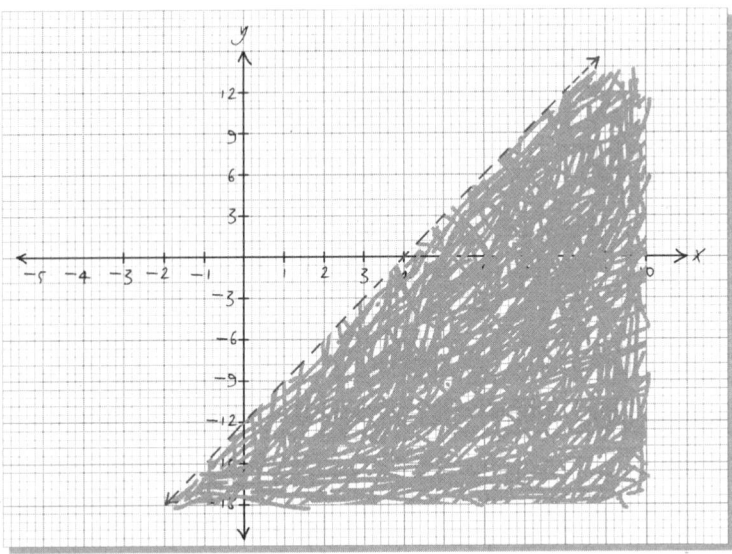

EXAMPLE 14

Graph the following system of inequalities by hand:

$$y \geq 2x - 6$$
$$y \leq -3x + 7$$

Solution

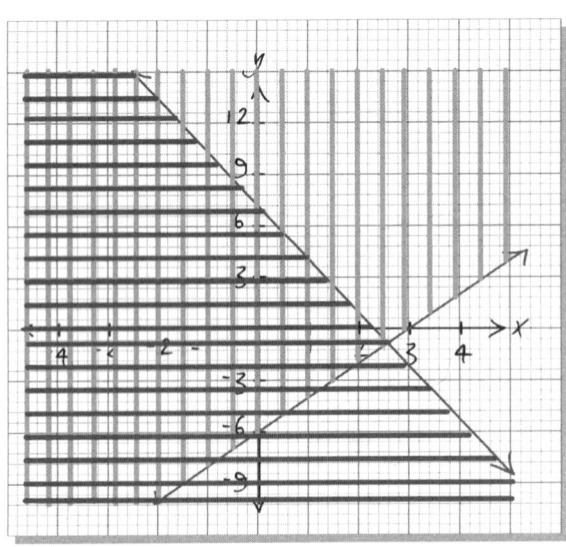

Chapter 2 Review Exercises

At the end of each exercise, you'll find the section number [in brackets] where the material is covered if you need help solving the problem.

1. Frank's Shoe Repair works out of a kiosk in the local mall and has determined the following models for the monthly revenue and costs associated with repairing shoes.

$$R(s) = 7.5s$$
$$C(s) = 235 + 2.25s$$

where R(s) represents the monthly revenue in dollars for repairing s shoes and C(s) is the monthly costs in dollars for repairing s shoes.

   a. Graph both functions on the same calculator screen.

   b. Determine the break-even point for Frank's Shoe Repair.   [2.1]

2. Joanne's Cosmetics Outlet has two salary options for their salespeople. Each salesperson can choose which option to base his or her salary on:

$$Q_1(s) = 350 + 0.07s$$
$$Q_2(s) = 0.065s + 450$$

where $Q_1(s)$ represents monthly salary option 1 in dollars when s dollars in sales are made per month and $Q_2(s)$ represents monthly salary option 2 in dollars when s dollars in sales are made per month.

   a. Graph both functions on the same calculator screen.

   b. Find what sales level would result in both salary options having the same monthly salary.   [2.1]

3. You decide to market your own custom computer software. You must invest $3232 on computer hardware and spend $3.95 to buy and package each disk. If each program sells for $9.00, how many copies must you sell to break even?   [2.2]

4. John has cashews that sell for $4.50 a pound and peanuts that sell for $2.00 a pound. How much of each must he mix to get 100 pounds of a mixture that sells for $3.00 per pound?   [2.3]

5. The residual values of cars make up a major part of the long-term value of a car. When considering two sports cars, Stephen found the following data in the 2004 *Kelley Blue Book.*

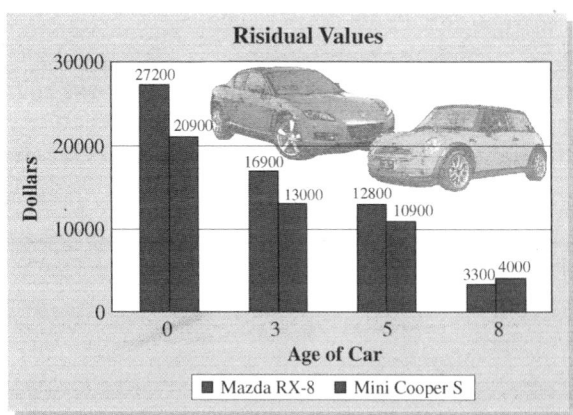

Source: November/December 2004 Kelley Blue Book Residual Values Guide.

   a. Find models for the residual values for each car given.

   b. Determine when these two cars will have the same value.   [2.2]

6. The percentage of births to unmarried women in Germany and the United Kingdom is given below.

Year	Germany	United Kingdom
2001	25.0	40.1
2002	26.0	40.6
2003	27.0	41.5
2004	27.9	42.3
2005	29.2	42.9

Source: Statistical Abstract 2008.

   a. Find a model for the percent of births to unmarried women in Germany.

b. What does the slope for the model in part a mean in this context?

c. Find a model for the percent of births to unmarried women in the United Kingdom.

d. Estimate when the percent of births to unmarried women in Germany will be greater than the percent in the United Kingdom. [2.4]

For Exercises 7 through 11, solve the systems by graphing. Label each system as consistent, inconsistent, or dependent.

7.

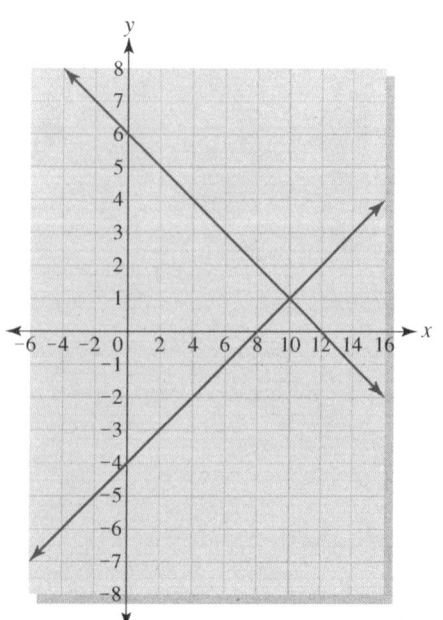

[2.1]

8. $x + 3y = -12$
$-5x + y = 15$ [2.1]

9. $w = 2d + 7$
$w = \frac{3}{7}d + \frac{1}{2}$ [2.1]

10. $p = 1.625t + 9$
$p = \frac{13}{8}t - 5$ [2.1]

11. $5x - 4y = 28$
$1.25x - y = -7$ [2.1]

12. The Palomar College Foundation has been given an endowment of $3 million to fund scholarships for outstanding math and science students. The foundation wants the endowment to earn enough interest each year to fund 130 $2000 scholarships. They plan to invest part of the money in a safe investment earning 7% simple interest and the rest in another account earning 11% simple interest. How much should they invest in each account to have enough interest to fund the 130 scholarships? [2.2]

13. Brian is an agriculture scientist who is testing different applications of insecticides on tomato plants. For part of the trial he is running now, he needs 150 gallons of a 12% solution of test chemical AX-14. Brian received only containers with 5% solution and 15% concentrations of test chemical AX-14. How much of each chemical should Brian mix to run his trial? [2.2]

For Exercises 14 through 24, solve each system by elimination or substitution. Label any systems that are dependent or inconsistent.

14. $3x + 4y = -26$
$y = x - 3$ [2.2]

15. $2w - 5t = -1$
$3w - 4t = 2$ [2.3]

16. $2.35d + 4.7c = 4.7$
$c = -7.05d - 21.15$ [2.2]

17. $\frac{5}{6}m + n = 25$
$m - \frac{4}{5}n = 5$ [2.3]

18. $y = 4.1x - 2.2$
$y = -2.9x - 7.1$ [2.2]

19. $-8x + 7y = -3$
$7(x - y) = 3$ [2.3]

20. $4.1w + 3.7t = 5.1$
$t = 4.43w + 4.63$ [2.2]

21. $3f + 2g = -22$
$g = -f - 9$ [2.2]

22. $-6x + 15y = 5$
$15(x - y) = -5$ [2.3]

23. $\frac{w}{2} + \frac{2z}{3} = 32$

$\frac{w}{4} - \frac{5z}{9} = 40$ [2.3]

24. $-2.7x + y = -13.61$
$y = -4.28x + 5.69$ [2.2]

25. The percentage of total births that are to teenage mothers in Delaware can be modeled by
$$D(t) = -0.46t + 12.35$$

where $D(t)$ represents the percent of total births that are to teenage mothers in Delaware t years since 2000. The percentage of total births that are to teenage mothers in Louisiana t years since 2000 can be modeled by
$$L(t) = -0.67t + 16.94$$

Source: Models derived from data in the Statistical Abstract 2008.

Determine when the percentage of total births that are to teenage mothers in Louisiana will be less than the percentage in Delaware. [2.4]

For Exercises 26 through 33, solve the inequality.

26. Use the graph to determine when $f(x) < g(x)$. [2.4]

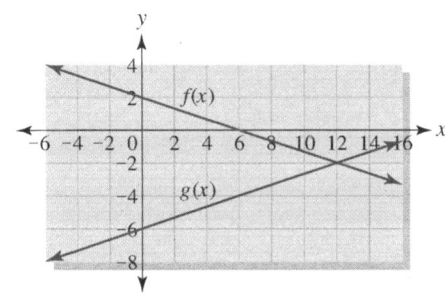

27. Use the table to find when Y1 > Y2. [2.4]

28. $7 - \dfrac{9x}{11} \le -8$ [2.4]

29. $-5x + 4 \ge 7(x - 1)$ [2.4]

30. $3t + 4 > -6(4t + 2)$ [2.4]

31. $\dfrac{3d}{5} + 7 < 4$ [2.4]

32. $-1.5v + 2.84 > -3.2v - 1.48$ [2.4]

33. $1.85 + 1.34(2.4k - 5.7) \ge 3.25k - 14.62$ [2.4]

For Exercises 34 through 37, solve the following equations.

34. $|x - 7| = 20$ [2.5]

35. $|2x + 3| - 10 = 40$ [2.5]

36. $|x + 19| = 30$ [2.5]

37. $5|x - 7| + 10 = 95$ [2.5]

For Exercises 38 through 43, solve the given inequality. Give the solution as an inequality and graph the solution set on a number line. [2.5]

38. $|x| < 8.5$ [2.5]

39. $|x| > 6.5$ [2.5]

40. $|x + 3| \ge 12$ [2.5]

41. $|x - 7| \le 3$ [2.5]

42. $2|x + 5| - 4 > 16$ [2.5]

43. $-3|x + 5| + 7 \le 4$ [2.5]

For Exercises 44 through 47, graph the inequalities by hand.

44. $y \le 4x - 10$ [2.6]

45. $y \ge 1.5x - 5$ [2.6]

46. $2x + 5y > 15$ [2.6]

47. $-3x - 4y > 12$ [2.6]

For Exercises 48 through 51, graph the system of inequalities.

48. $y > 2x + 5$
$y < -x + 7$ [2.6]

49. $y > 1.5x + 2$
$4.5x - 3y > 6$ [2.6]

50. $4x + 5y \le 12$
$2x + 8y \ge 5$ [2.6]

51. $y > \dfrac{2}{3}x - 7$
$y < \dfrac{2}{3}x + 5$ [2.6]

52. A college is selling tickets to a championship basketball game and has up to 12,000 tickets to sell. They can sell both regular admission tickets and student discount tickets. Student tickets cost $9, and regular admission tickets cost $15. The school wants to make at least $140,000 from selling tickets. Write a system of inequalities to determine the combinations of tickets they can sell to make the money they want.   [2.6]

53. Find the inequality for the graph.   [2.6]

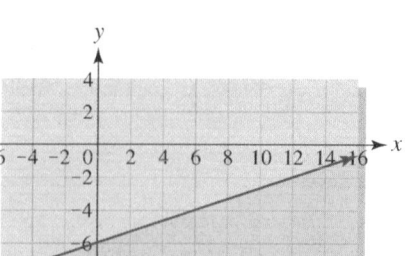

54. Find the inequality for the graph.   [2.6]

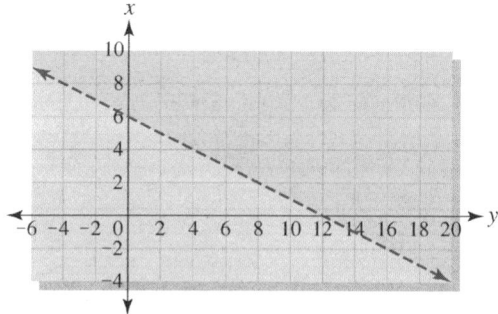

Chapter 2 Test

1. The average hourly earnings of production workers in manufacturing industries for California and Massachusetts are given in the table.

Year	California (in dollars per hour)	Massachusetts (in dollars per hour)
2003	15.04	16.53
2004	15.36	16.89
2005	15.70	17.66
2006	15.95	18.26

Source: Statistical Abstract 2008.

    a. Find models for the data.

    b. Find when the average hourly earnings were the same in both states.

2. Christine is ready to retire and has $500,000 to invest. Christine needs to earn $44,500 in interest each year to continue her current lifestyle. She plans to invest the money in two accounts: one paying 12% and a safer account paying 7%.

    a. Write a system of equations that will help you to find the amount Christine needs to invest in each account.

    b. How much should Christine invest in each account to earn the $44,500 she wants?

3. Georgia has up to 20 hours a week to study for her math and history classes. She needs to study at least 12 hours a week to pass her math class. Write a system of inequalities to describe the possible amounts of time she can study for each subject. Graph the system by hand.

4. Wendy is working in a forensics lab and needs 2 liters of an 8% HCl solution to test some evidence. Wendy has only a 5% HCl solution and a 20% HCl solution to work with. Because she knows that the solution must be 8% for the test to be valid, how much of the 5% solution and the 20% solution should she use?

For Exercises 5 through 7, solve the inequality.

5. $5x + 7 < 12x - 8$

6. $3.2m + 4.5 \geq 5.7(2m + 3.4)$

7. $-4.7 + 6.5(a + 2.5) \leq 2.43.1a - 5$

For Exercises 8 through 11, solve the given system.

8. $x + 7y = -2$
   $3x + y = 34$

9. $0.4375w + 4t = 22$
   $-2.4t = 0.2625w - 13.2$

10. $5c + 3d = -15$
    $d = -\dfrac{5}{3}c - 12$

11. $2.68g - 3.45f = 23.87$
    $4.75g + 6.9f = -12.47$

12. Use the graph to find when $f(x) \geq g(x)$.

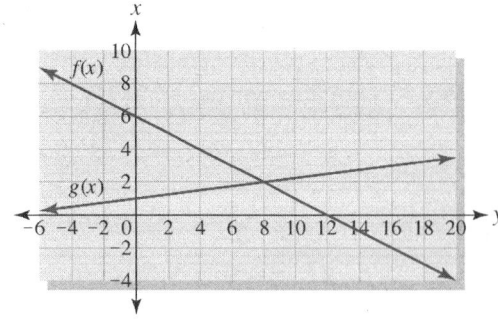

13. Scott has decided to market his custom-built hammock stands. He needs to invest $7500 on tools and spend $395 to buy and build each stand. If each hammock stand sells for $550, how many must Scott sell to break even?

14. Graph by hand: $y > -\dfrac{4}{5}x + 6$

15. The revenue for local telephone service providers can be modeled by

$$L(t) = 3804.32t + 84252.07$$

where $L(t)$ represents the revenue in millions of dollars for local telephone service providers t years since 1990. The revenue in millions of dollars for cellular telephone service providers t years since 1990 can be modeled by

$$C(t) = 6428.29t - 11553$$

Determine when the revenue for cellular telephone providers will be greater than that for local telephone providers.

16. Use the table to find when Y1 < Y2.

X	Y₁	Y₂
-6	-4	-2.5
-5	-3.5	-2.75
-4	-3	-3
-3	-2.5	-3.25
-2	-2	-3.5
-1	-1.5	-3.75
0	-1	-4

X=0

17. Solve $|x + 5| + 3 = 17$

18. Solve $|x - 4| - 8 < 5$. Give the solution as an inequality and graph the solution set on a number line.

For Exercises 19 and 20, graph the system of inequalities.

19. $y \geq 2x - 5$

$y \leq \frac{1}{3}x + 4$

20. $4x - 3y < 6$

$3x + y > 12$

Chapter 2 Projects

Research project

One or more people

What you will need:

- Prices for basic cell phone plans from two or more cell phone service providers

WHICH CELL PHONE PLAN IS CHEAPER?

It will be your job to compare two or more cell phone plans from various cell phone service providers. You will need to visit, phone, or investigate on the Internet two or more cell phone service providers and find their prices for the most basic service plan they offer. Then you will explore which of these companies is cheaper for a variety of users.

Write up

a. Describe how you collected the information. What companies are you comparing? Give a basic description of the plans you are considering.

b. Create a table of data for each plan that includes several different amounts of minutes used for a month and the cost for those minutes.

c. Draw a graph with all of the plans you are comparing that depicts the monthly cost for different amounts of minutes used. (This might not be a straight line.)

d. Determine when each plan will be the cheapest.

e. Is there a number of minutes when two or more plans cost the same amount? If so, find that amount and describe its significance.

f. Describe what else besides cost you might consider in choosing a cell phone service provider.

g. Which of the cell phone service providers would you choose and why?

Research project

One or more people

What you will need:

- Find data for two things to compare and analyze using a system of equations.
- You might want to use the Internet or library. Statistical abstracts and some journals and scientific articles are good resources for data.

FIND YOUR OWN COMPARISON

You are given the task of finding two real-world situations to compare and analyze using a system of equations. You will need to find two related sets of data that can be modeled by using linear functions. Using these two functions, you will need to solve the system of equations and explain the results. You may use the problems in this chapter to get ideas of things to investigate, but your data should not be discussed in this textbook.

Write up

a. Describe the data you found and where you found it. Cite any sources you used.

b. Create a scatter plot of your data.

c. Find models for your sets of data and write a system of equations. Graph your functions on the same graph. Be sure to show the intersection of the two functions.

d. Solve the system of equations.

e. Explain the meaning of the solution to the system of equations you found in part d.

f. Explain any restrictions you would place on this system of equations to avoid model breakdown.

Exponents, Polynomials and Functions

- Use the rules for exponents to simplify expressions.
- Recognize the relationship between rational exponents and radical expressions.
- Classify polynomials by name and degree.
- Combine polynomials using basic arithmetic operations.
- Combine functions using basic arithmetic operations.
- Combine functions using composition.
- Combine functions in a context to create a desired new function.
- Factor polynomials using the AC method.
- Factor polynomials using trial and error.
- Factor perfect square trinomials.
- Factor the difference of two squares.
- Factor the sum and difference of two cubes.

 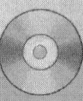 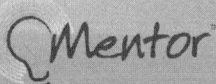 

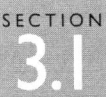

SECTION
3.1

Rules for Exponents

RULES FOR EXPONENTS

In order to work with functions other than linear functions, we need to have a good understanding of the basic rules for exponents and how to use them to simplify problems containing exponents. Many of these rules you already know and have used in previous classes and sections of this textbook. This section is a recap to prepare us for more applications in the rest of the textbook. In this section, assume that all variables are not equal to zero. This allows us to ignore the possibility of division by zero, which is not defined.

Recall that the basic concept of an exponent is repeated multiplication.

$$2 \cdot 2 \cdot 2 \cdot 2 \cdot 2 \cdot 2 = 2^6 = 64$$

$$xxxxx = x^5$$

$$xxxyy = x^3 y^2$$

$$3 \cdot 3 \cdot 3 \cdot 3 \cdot 7 \cdot 7 \cdot 7 \cdot 7 \cdot 7 \cdot 7 = 3^4 \cdot 7^6 = 81 \cdot 117649 = 9529569$$

As you can see, exponents allow us to write a long expression in a very compact way. When we work with exponents, there are two parts to an exponential expression: the **base** and the **exponent**. The base is the number or variable being raised to a power, and the exponent is the power to which the base is being raised.

$$5^{\overset{\text{exponent}}{3}}_{\text{base}}$$

One of the most common operations we do with exponential expressions is to multiply them together. When we multiply any exponential expressions with the same base the expressions can be combined into one exponential.

$$x^7 x^2 = xxxxxxx \cdot xx = x^9$$

In this example, we see that we had seven x's multiplied by two more x's, which gives us a total of nine x's multiplied together. Therefore, we can write a final simpler expression, x^9. Combining expressions with the same base leads us to the **product rule for exponents.**

THE PRODUCT RULE FOR EXPONENTS

$$x^m x^n = x^{m+n}$$

When multiplying exponential expressions that have the same base, you add exponents.

$$x^5 x^3 = x^8$$

When more than one base is included in an expression or multiplication problem, the associative and commutative properties can be used along with the product rule for exponents to simplify the expression or multiplication.

Using Your TI Graphing Calculator

To enter exponents on the graphing calculator, you can use either of two methods.

If you are squaring a term, you can use the

button on the left-hand side of the calculator. Enter the number or variable that you wish to square and press the button.

If you want to raise a variable or number to any other power, you should use the caret button

in the far right column of the calculator. In this case, you will need to enter the number or variable you want to raise to a power, then press the button,

followed by the power you wish to use.

For example, x^5 would be entered as

EXAMPLE ▮ USING THE PRODUCT RULE FOR EXPONENTS

Simplify the following expressions.

a. $x^5 x^2 x^3$ **b.** $(3f^4 g^5)(7f^2 g^7)$ **c.** $(a^2 b^5 c)(a^3 b^4 c^3)$

Concept Connection

Associative Property:
When multiplying more than two variables or constants together we can group them in any way we wish.

$$abc = (ab)c = a(bc)$$

Commutative Property:
When multiplying variables or constants together they can be multiplied in any order we wish.

$$ab = ba$$
$$abc = bac = cab = cba$$

Solution

a.

$$x^5 x^2 x^3 = x^{10}$$ Add the exponents.

b.

$$(3f^4 g^5)(7f^2 g^7)$$
$$3 \cdot 7 f^4 f^2 g^5 g^7$$ Use the commutative property to re-arrange the coefficients and bases.
$$21 f^6 g^{12}$$ Add the exponents of the like bases and multiply the coefficients.

c.

$$(a^2 b^5 c)(a^3 b^4 c^3)$$ Use the commutative property to re-arrange the bases.
$$a^2 a^3 b^5 b^4 c c^3$$
$$a^5 b^9 c^4$$ Add the exponents of the like bases.

EXAMPLE ① PRACTICE PROBLEM

Simplify the following expressions.

a. $m^3 m^4 m^2$ **b.** $(4w^6 x^2)(8wx^9)$

It is very important with the exponent rules that you notice what operation you are working with in the expression and what operation this results in for the exponents. Since the coefficients are not exponents, the exponent rules will not apply. Many students will forget that they are working with exponent rules and want to change multiplication of coefficients into addition.

$$2x^5 \cdot 3x^4$$
$$6x^5 x^4$$ Multiply the coefficients.
$$6x^9$$ Add the exponents.

$$2x^5 \cdot 3x^4 \neq 5x^9$$ Do not add the coefficients. They are not exponents.

The next rule is closely related to the product rule for exponents. What should we do with exponents when we are dividing two exponential expressions that have the same base?

$$\frac{x^5}{x^3} = \frac{xxxxx}{xxx} = \frac{xx}{1} = x^2$$

Because multiplication and division are inverse operations, we can eliminate any variables that are in both the numerator and denominator of this fraction. That means that we can cancel three x's from the top and bottom, leaving us with only two x's remaining on the top. When we multiplied exponential expressions with the same base, we added exponents; when we divide exponential expressions with the same base, we will subtract exponents. Working with exponents this way leads us to the **quotient rule for exponents.**

THE QUOTIENT RULE FOR EXPONENTS

$$\frac{x^m}{x^n} = x^{m-n}$$

When dividing exponential expressions that have the same base, you subtract exponents.

$$\frac{x^7}{x^3} = x^4$$

Remember that coefficients are not exponents so they will simply be reduced when part of a division problem. Do not divide the coefficients into a decimal form. Leaving the coefficients as a reduced fraction is the standard form these fractions should be left in.

EXAMPLE 2 USING THE QUOTIENT RULE FOR EXPONENTS

Simplify the following expressions.

a. $\dfrac{x^{12}}{x^5}$ b. $\dfrac{a^5 b^3 c^4}{a^3 b^2 c}$ c. $\dfrac{35 m^2 n^4}{7 mn}$

d. $\dfrac{10 k^7 p^3}{8 k^4 p}$

Solution

a.

$$\frac{x^{12}}{x^5} = x^7 \qquad \text{Subtract the exponents.}$$

b.

$$\frac{a^5 b^3 c^4}{a^3 b^2 c} = a^2 bc^3 \qquad \text{Subtract the exponents of like bases.}$$

c.

$$\frac{35 m^2 n^4}{7 mn} = 5 mn^3 \qquad \text{Subtract the exponents of like bases and reduce the coefficients.}$$

d.

$$\frac{10 k^7 p^3}{8 k^4 p} = \frac{5 k^3 p^2}{4} = \frac{5}{4} k^3 p^2 \qquad \text{Subtract the exponents of like bases and reduce the coefficients.}$$

EXAMPLE   ② PRACTICE PROBLEM

Simplify the following expressions.

a. $\dfrac{a^3 b^8 c}{a^2 b^2}$ b. $\dfrac{40 t^{11} w^{14}}{5 t^3 w^9}$ c. $\dfrac{24 b^{18} c^4}{14 b^{10} c^3}$

The next rule for exponents deals with what to do when an exponential expression is raised to another exponent. Changing the exponent to repeated multiplication gives us

$$(x^3)^2 = (x^3)(x^3) = x^6$$
$$(x^2)^5 = x^2 x^2 x^2 x^2 x^2 = x^{10}$$

Because raising to a power is simply repeated multiplication, the **power rule for exponents** follows from the product rule for exponents.

THE POWER RULE FOR EXPONENTS

$$(x^m)^n = x^{mn}$$

When raising an exponential expression to another power, you multiply the exponents.

$$(x^4)^7 = x^{28}$$

When more than one variable or constant is being raised to a power, we can use the rules for **powers of products and quotients**. When an exponential expression contains more than one variable or a numeric constant that is multiplied or divided, the outside power must be distributed to each constant or variable by using the power rule for exponents. It is very important to remember that exponents can only be distributed over multiplication and division. When a sum or difference is raised to a power, simply distributing the exponent will not work.

POWERS OF PRODUCTS AND QUOTIENTS

$$(xy)^m = x^m y^m \qquad \left(\frac{x}{y}\right)^m = \frac{x^m}{y^m}$$

In raising an expression to a power, that power can distribute over multiplication or division.

$$(xy)^5 = x^5 y^5 \qquad \left(\frac{x}{y}\right)^4 = \frac{x^4}{y^4}$$

Distribution of exponents does **NOT** work over addition or subtraction.

Incorrect: $(x+y)^2 \neq x^2 + y^2$

Correct: $(x+y)^2 = (x+y)(x+y) = x^2 + 2xy + y^2$

EXAMPLE **3** USING THE POWER RULE FOR EXPO-
NENTS AND DISTRIBUTING EXPONENTS

Simplify the following expressions.

a. $(w^3)^5$ **b.** $(a^3b^2c)^3$ **c.** $(5m^5n^3)^2$

d. $\left(\dfrac{m^2}{n^4}\right)^3$ **e.** $\left(\dfrac{2xy^2}{3z^5}\right)^4$

Solution

a.

$$(w^3)^5 = w^{15}$$ Using the power rule, multiply the exponents.

b.

$$(a^3b^2c)^3$$ Distribute the exponent through to each base.

$$(a^3)^3(b^2)^3(c)^3$$

$$a^9b^6c^3$$ Using the power rule, multiply the exponents.

c.

$$(5m^5n^3)^2$$ Distribute the exponent through to each base and the coefficient.

$$(5)^2(m^5)^2(n^3)^2$$

$$25m^{10}n^6$$ Using the power rule, multiply the exponents and raise the coefficient 5 to the second power.

d.

$$\left(\frac{m^2}{n^4}\right)^3$$ Distribute the exponent through to each base.

$$\frac{(m^2)^3}{(n^4)^3}$$

$$\frac{m^6}{n^{12}}$$ Using the power rule, multiply the exponents.

e.

$$\left(\frac{2xy^2}{3z^5}\right)^4$$ Distribute the exponent through to each base and the coefficients.

$$\frac{2^4x^4(y^2)^4}{3^4(z^5)^4}$$

$$\frac{16x^4y^8}{81z^{20}}$$ Using the power rule, multiply the exponents and raise the coefficients to the fourth power.

EXAMPLE PRACTICE PROBLEM

Simplify the following expressions.

a. $(3x^5y^2z)^3$

b. $\left(\dfrac{4m^4p^8}{5m^3p^5}\right)^2$

When expressions get more complicated, it is important to follow the order of operations. Therefore, start by simplifying the inside of any parentheses before distributing any exponents or using the power rule for exponents. Finally, do any remaining multiplication or division using the product and quotient rules for exponents.

EXAMPLE   4   COMBINING THE RULES FOR EXPONENTS

Simplify the following expressions.

a. $(7d^3g^2)^2(5dg^3)^3$

b. $\left(\dfrac{16x^5y^{12}z^9}{2xy^8z^6}\right)^3$

c. $\left(\dfrac{a^4b^3c^5}{a^3bc}\right)^4\left(\dfrac{5ab^3}{bc}\right)^2$

Solution

a.

$$(7d^3g^2)^2(5dg^3)^3$$ — Distribute the exponents to each base and the coefficients.

$$[7^2(d^3)^2(g^2)^2][5^3d^3(g^3)^3]$$ — Using the power rule, multiply the exponents and raise the coefficients to their powers.

$$(49d^6g^4)(125d^3g^9)$$

$$6125d^9g^{13}$$ — Multiply the coefficients and use the product rule to add exponents.

b.

$$\left(\dfrac{16x^5y^{12}z^9}{2xy^8z^6}\right)^3$$ — Simplify inside the parentheses by reducing the coefficients and using the quotient rule to subtract exponents.

$$(8x^4y^4z^3)^3$$ — Distribute the exponent to each base and coefficient.

$$8^3(x^4)^3(y^4)^3(z^3)^3$$

$$512x^{12}y^{12}z^9$$ — Raise the coefficient to the third power and multiply the exponents.

Concept Connection

Order of Operations:
Recall that the order of operations are an agreement by mathematicians about what order operations are to be done in.
1: Perform all operations within parentheses or other grouping symbols.
2: Evaluate all exponents.
3: Multiply or divide in order from left to right.
4: Add or subtract in order from left to right.

The order of operations is sometimes remembered using the acronym:
PEMDAS

c.

$$\left(\frac{a^4b^3c^5}{a^3bc}\right)^4\left(\frac{5ab^3}{bc}\right)^2$$

Simplify inside each parentheses using the quotient rule for exponents.

$$(ab^2c^4)^4\left(\frac{5ab^2}{c}\right)^2$$

Distribute the exponents to each base and coefficient.

$$[a^4(b^2)^4(c^4)^4]\left[\frac{5^2a^2(b^2)^2}{c^2}\right]$$

Using the power rule, multiply the exponents and raise the coefficient to the second power.

$$(a^4b^8c^{16})\left(\frac{25a^2b^4}{c^2}\right)$$

Multiply the two expressions adding the exponents of like bases.

$$\frac{25a^6b^{12}c^{16}}{c^2}$$

Use the quotient rule to simplify.

$$25a^6b^{12}c^{14}$$

EXAMPLE PRACTICE PROBLEM

Simplify the following expressions. Be sure to follow the order of operations.

a. $(3x^2y^5)(2x^3y)^3$ **b.** $\dfrac{(2m^2n^3)^4}{10m^5n^{10}}$ **c.** $\left(\dfrac{m^2n^3}{mn^2}\right)^4\left(\dfrac{m^5}{n^2}\right)$

NEGATIVE EXPONENTS AND ZERO AS AN EXPONENT

When using the quotient rule for exponents in some situations, you will get negative numbers when you subtract the denominator's exponent from the numerator's exponent. A result like this makes us think about ways to define how **negative exponents** work. Let's look at a division problem.

$$\frac{x^3}{x^5}=\frac{\cancel{xxx}}{\cancel{xxxxx}}=\frac{1}{xx}=\frac{1}{x^2}$$

Using the basic definition of exponents.

$$\frac{x^3}{x^5}=x^{3-5}=x^{-2}$$

Using the quotient rule for exponents

Because using the basic definition of exponents gives us $\dfrac{1}{x^2}$ and the quotient rule for exponents gives us x^{-2}, these two expressions need to be the same for both of these methods to agree and be reliable. Therefore, we have that

$$x^{-2}=\frac{1}{x^2}$$

The negative part of the exponent represents a reciprocal of the base. Notice that once you take the reciprocal of (flip) the base, the exponent becomes positive. The negative exponent only moves the base; it does not make that base negative. If a base with a negative exponent is in the denominator of a fraction, it will also be a reciprocal and will end up in the numerator of the fraction. Answers without negative exponents are easier to understand and work with, so we will write all our answers with only positive exponents.

NEGATIVE EXPONENTS

$$x^{-n} = \frac{1}{x^n}$$

When raising a base to a negative exponent, you get the reciprocal of that base raised to the absolute value of the exponent.

$$x^{-3} = \frac{1}{x^3} \qquad \left(\frac{1}{2}\right)^{-1} = 2 \qquad \frac{2}{x^{-4}} = 2x^4$$

EXAMPLE **5** **WORKING WITH NEGATIVE EXPONENTS**

Simplify the following expressions. Write all answers without negative exponents.

a. $x^{-5}y^2$ **b.** $\dfrac{a^5 b}{a^3 b^4}$ **c.** $-2x^{-4}$

d. $\dfrac{21 m^3 n^{-2}}{7 m^{-5} n^3}$ **e.** $\dfrac{25 a^3 b^{-7} c^{-2}}{15 a^8 b^3 c^5}$

Solution

a.

$$x^{-5}y^2 = \frac{y^2}{x^5}$$

The x has a negative exponent, so take the reciprocal of the base x.

b.

$$\frac{a^5 b}{a^3 b^4}$$

First, subtract the exponents.

$$a^2 b^{-3}$$

Notice that the result from subtracting exponents is always placed on top of the fraction and then moved if negative exponents remain.

$$\frac{a^2}{b^3}$$

Move any negative exponents to the bottom.

c.

$$-2x^{-4} = \frac{-2}{x^4}$$

The negative exponent moves the x to the bottom.
The negative 2 does not move because it is not an exponent.

d.

$$\frac{21 m^3 n^{-2}}{7 m^{-5} n^3}$$

First, subtract the exponents. Use caution with the negatives.

$$3 m^8 n^{-5}$$

Again, the results are in the numerator and move to the denominator if they have a negative exponent.

$$\frac{3 m^8}{n^5}$$

Move any negative exponents to the bottom.

e.

$$\frac{25a^3b^{-7}c^{-2}}{15a^8b^3c^5}$$

$$\frac{5a^{-5}b^{-10}c^{-7}}{3}$$ Subtract exponents and reduce the constants.

$$\frac{5}{3a^5b^{10}c^7}$$ Move any negative exponents to the bottom.

EXAMPLE PRACTICE PROBLEM

Simplify the following expressions. Write all answers without negative exponents.

a. $5c^{-3}d^4$

b. $\dfrac{12x^4y^3z^8}{10xy^5z^2}$

c. $\dfrac{-2g^{-2}h^3}{g^5h^{-4}}$

 **CONCEPT INVESTIGATION** WHAT DOES AN EXPO-
NENT OF ZERO DO? ••••••••••••••••••••••••••••••••••••

a.   Fill in the missing values in the following table.

x	2^x
5	32
4	16
3	
2	
1	
0	
−1	$\frac{1}{2}$
−2	
−3	

b.   When the exponent is reduced by 1, how does the value of the exponential expression change?

c.   What is the value of the exponential when the exponent is zero?

d.   On your own paper, create a table for the exponential expression 5^x. You should include positive, zero, and negative values for x.

e. What was the value for 5^0?

f. Pick any base b $(b \neq 0)$ that you want for an exponential and determine the value of b^0.

g. Make a statement about what you think any base to the power of zero should be.

h. Now make one more table for an exponential with a base zero, that is, 0^x.

i. Finally, make a statement about the value of any exponential with exponent zero. Check this statement with others in your class or your instructor.

··· ◆

This last statement about exponents equal to zero should lead you to the following property of exponents.

> ## ZERO AS AN EXPONENT
>
> $$x^0 = 1 \qquad x \neq 0$$
>
> When raising any exponential expression with a base other than zero to the power of zero, the expression will equal 1.
>
> $$25^0 = 1 \qquad (-17.4)^0 = 1$$
>
> 0^0 does not exist

The rule for zero as an exponent states that the base must not equal zero. To see why this is necessary, look at the following two patterns.

$$
\begin{array}{ll}
4^0 = 1 & 0^4 = 0 \\
3^0 = 1 & 0^3 = 0 \\
2^0 = 1 & 0^2 = 0 \\
1^0 = 1 & 0^1 = 0 \\
0^0 = 1? & 0^0 = 0?
\end{array}
$$

The last line of these two patterns says that 0^0 equals both 1 and 0, which, of course, it can't. This is a conflict and is one demonstration of why 0^0 is undefined. Any non zero expression that is raised to the zero power is 1.

$$\left(\frac{25x^5 y^{16} z^{-8}}{157a^4 b^{-2}} \right)^0 = 1$$

E X A M P L E   6   COMBINING THE RULES FOR EXPONENTS

Simplify the following expressions. Write all answers without negative exponents.

a. $(2x^3y^2)^{-2}(3x^{-3}y^6)$ **b.** $\left(\dfrac{5a^5}{a^3b^4}\right)^{-1}$ **c.** $\left(\dfrac{18m^2n^{-2}}{9m^{-4}n^5}\right)^{-3}$

d. $(123a^{15}b^{-20}c^{30})^0$

Solution

a.

$$(2x^3y^2)^{-2}(3x^{-3}y^6)$$

$(2^{-2}x^{-6}y^{-4})(3x^{-3}y^6)$ Distribute the outside exponents.

$\left(\dfrac{1}{4}x^{-6}y^{-4}\right)(3x^{-3}y^6)$ Take the reciprocal of any numbers with negative exponents.

$\dfrac{3}{4}x^{-9}y^2$ Add exponents when you multiply.

$\dfrac{3y^2}{4x^9}$ Take the reciprocal of any variables with negative exponents.

b.

$$\left(\dfrac{5a^5}{a^3b^4}\right)^{-1}$$

$\left(\dfrac{5a^2}{b^4}\right)^{-1}$ Simplify the inside of the parentheses. Subtract exponents when you divide.

$\dfrac{5^{-1}a^{-2}}{b^{-4}}$ Distribute the outside exponent.

$\dfrac{b^4}{5a^2}$ Take the reciprocal of any bases with negative exponents.

c.

$$\left(\dfrac{18m^2n^{-2}}{9m^{-4}n^5}\right)^{-3}$$

$(2m^6n^{-7})^{-3}$ Simplify the inside of the parentheses.

$2^{-3}m^{-18}n^{21}$ Distribute the outside exponent.

$\dfrac{n^{21}}{2^3m^{18}}$ Take the reciprocal of any bases with negative exponents.

$\dfrac{n^{21}}{8m^{18}}$ Multiply out the constant.

d.

$$(123a^{15}b^{-20}c^{30})^0 = 1$$ This expression to the zero power is 1.

EXAMPLE PRACTICE PROBLEM

Simplify the following expressions. Write all answers without negative exponents.

a. $(5x^2y^{-3})^{-3}(7x^5y^{-4})^2$ **b.** $\left(\dfrac{3g^4h^{-5}}{6g^{-2}h^5}\right)^3$ **c.** $\left(\dfrac{4}{3}a^3b^7c\right)^0$

RATIONAL EXPONENTS

All of the problems that we have worked on so far have had integer exponents. Fractions are often used as exponents in exponential problems, so we will want to know how to work with them. In this section, we will learn the basic meaning of a rational exponent and use them to simplify expressions. In chapter 5, we will re-visit rational exponents and use them to solve some problems.

   A rational exponent is another way of writing a radical such as a square root or cube root.

$$\sqrt{25} = 25^{1/2} = 5 \qquad \sqrt[3]{27} = 27^{1/3} = 3$$

The rational exponent $\dfrac{1}{2}$ represents a square root, the rational exponent $\dfrac{1}{3}$ represents a cube root, and so on.

RATIONAL EXPONENTS

$$x^{1/n} = \sqrt[n]{x}$$

Raising a base to a rational exponent with a denominator of n is the same as taking the nth root of the base.

$$8^{1/3} = \sqrt[3]{8} = 2$$

If x is negative, n must be odd. If x is positive, n can be any whole number greater than or equal to 2.

   We will use this definition to rewrite some rational exponents into radical form and some radical expressions into exponent form. The power rule for exponents helps us deal with rational exponents that have numerators other than 1.

$$x^{5/3} = \left(x^5\right)^{1/3} = \sqrt[3]{x^5}$$

EXAMPLE 7 REWRITING RATIONAL EXPONENTS IN RADICAL FORM

Rewrite the following exponents in radical form.

a. $x^{1/5}$ **b.** $w^{2/3}$ **c.** $t^{3/7}$

Solution

a. The denominator of the exponent becomes the radical's index. Therefore, we get

$$\sqrt[5]{x}$$

Using Your TI Graphing Calculator

To calculate square roots and other higher roots on the graphing calculator, you can use either of two methods.

If you are taking the square root of a number, the square root function is above the

button on the left-hand side of the calculator. First you must press the [2nd] button, then the [x^2] button, and then the number you want to take the square root of.

If you want to take a higher root, you can use the caret button

and a fraction exponent in parentheses. In this case, you will need to enter the number or variable you want to raise to a power, then press the [$\wedge$] button, followed by the power you wish to use.

For example, $\sqrt[3]{8} = 8^{1/3}$ would be entered as

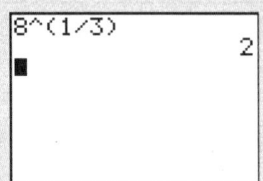

Remember all fraction exponents must be in parentheses.

b. The denominator of the exponent becomes the radical's index, and the numerator stays as the exponent of the variable. Therefore, we get

$$w^{2/3} = (w^2)^{1/3} = \sqrt[3]{w^2}$$

c. The denominator of the exponent is 7, so the index of the radical will be 7. The 3 will stay as the exponent of the variable. Therefore, we get

$$t^{3/7} = (t^3)^{1/7} = \sqrt[7]{t^3}$$

EXAMPLE ⑦ PRACTICE PROBLEM

Rewrite the following exponents in radical form.

a. $g^{4/9}$ **b.** $m^{7/10}$

EXAMPLE ⑧ REWRITING RADICALS USING RATIONAL EXPONENTS

Rewrite the following radicals using rational exponents.

a. $\sqrt{5x}$ **b.** $\sqrt[3]{w^2}$ **c.** $(\sqrt[5]{t})^3$

Solution

a. This is a square root so the index is 2. Therefore, the exponent is 1/2. We get $(5x)^{1/2}$.

b. The index of the radical is 3, so the denominator of the fraction exponent is 3. Therefore, we get $w^{2/3}$.

c. The index of the radical is 5, so the denominator of the fraction exponent is 5. Therefore, we get $t^{3/5}$.

EXAMPLE ⑧ PRACTICE PROBLEM

Rewrite the following radicals using rational exponents.

a. $\sqrt{7a}$ **b.** $\sqrt[4]{t^3}$

All of the rules that we have learned in this section apply to positive and negative numbers as well as to whole numbers or fractions. When simplifying expressions that contain fractional exponents, simply follow the rules for exponents as if the exponents were whole numbers. Then reduce the fractional exponents if possible.

EXAMPLE ⑨ SIMPLIFYING EXPRESSIONS WITH RATIONAL EXPONENTS

Simplify the following expressions. Write all answers without negative exponents.

a. $(25x^4 y^{10})^{1/2}$ **b.** $\left(\dfrac{24a^5 b^6 c^{-2}}{3a^2 b^{-3} c} \right)^{1/3}$

c. $(5g^{1/4} h^{3/4})^2 (3g^{1/2} h^{1/2})$

Solution

a.

$$(25x^4 y^{10})^{1/2}$$

$$25^{1/2}(x^4)^{1/2}(y^{10})^{1/2}$$

$$5x^2 y^5$$

There is no simplifying to do inside the parentheses, so use the power rule for exponents and distribute the fraction exponent.

b.

$$\left(\frac{24a^5 b^6 c^{-2}}{3a^2 b^{-3} c}\right)^{1/3}$$

Simplify the inside of the parentheses using the quotient rule.

$$(8a^3 b^9 c^{-3})^{1/3}$$

$$8^{1/3}(a^3)^{1/3}(b^9)^{1/3}(c^{-3})^{1/3}$$

Distribute the fraction exponent using the power rule for exponents.

$$2ab^3 c^{-1}$$

$$\frac{2ab^3}{c}$$

Use the reciprocal of variable with a negative exponent.

c.

$$(5g^{1/4} h^{3/4})^2 (3g^{1/2} h^{1/2})$$

Distribute the exponent using the power rule for exponents.

$$[5^2 (g^{1/4})^2 (h^{3/4})^2](3g^{1/2} h^{1/2})$$

$$(25g^{1/2} h^{3/2})(3g^{1/2} h^{1/2})$$

$$75gh^2$$

Multiply using the product rule for exponents.

EXAMPLE 9 PRACTICE PROBLEM

Simplify the following expressions. Write all answers without negative exponents.

a. $(32a^5 b^{10})^{1/5}$

b. $\left(\dfrac{32n^3 m^{-3}}{2nm^{-1}}\right)^{1/2}$

3.1 Exercises

For Exercises 1 through 50, simplify the given expression using the order of operations and exponent rules. Write each answer without negative exponents.

1. $2^5 + 3^4$

2. $4^3 + 7^4$

3. $2^4(2^5)$

4. $5^3(5^8)$

5. $w^2 w^5 w^3$

6. $15g^3 g^7$

7. $\dfrac{7^{23}}{7^{20}}$

8. $\dfrac{6^{30}}{6^{27}}$

9. $\dfrac{z^{12}}{z^8}$

10. $\dfrac{g^{17}}{g^9}$

11. $3^{15}(3^{-12})$

12. $4^{53}(4^{-49})$

13. $s^5(s^{-3})$

14. $k^7(k^{-2})$

15. $4r^5 t^{-4}$

16. $12p^{-2} q^6$

17. $\dfrac{3}{b^{-2}}$

18. $\dfrac{5c^{-3}}{d^{-2}}$

19. $\dfrac{x^{-3}y^2}{3x^5 y^{-7}}$

20. $\dfrac{-7c^5 d^{-8}}{c^{-9}d^3}$

21. $\dfrac{(7x^2 y)^{10}}{(7x^2 y)^8}$

22. $\dfrac{(5ab^3)^{14}}{(5ab^3)^{11}}$

23. $\dfrac{200a^5 bc^3}{25a^2 bc}$

24. $\dfrac{45x^7 y}{3x^5 y}$

25. $3x^2 y(5x^4 y^3)$

26. $5w^3 z(2w^2 z^3)$

27. $\dfrac{(x + 9y)^7}{(x + 9y)^5}$

28. $\dfrac{(2a - c)^{10}}{(2a - c)^8}$

29. $\left(\dfrac{2}{3}a^3 b^7 c\right)^4$

30. $\left(\dfrac{4}{5}x^5 y^3 z\right)^2$

31. $(2x^2 y^3)^{-2}$

32. $(5a^3 bc^7)^{-3}$

33. $\left(\dfrac{5w^3 v^7 x^{-4}}{17wx^3}\right)^0$

34. $\left(\dfrac{28a^6 b^{12} c}{148ab^9 c^4}\right)^0$

35. $\left(\dfrac{3}{5}\right)^{-2}$

36. $\left(\dfrac{2}{3}\right)^{-2}$

37. $\left(\dfrac{2x^3 y^{-4}}{5xy^5}\right)\left(\dfrac{15xy^2}{7x^5 y^{-3}}\right)$

38. $\left(\dfrac{3g^5h^{-7}}{4g^2h^3}\right)\left(\dfrac{10g^{-2}h^6}{11gh^{-2}}\right)$

39. $\left(\dfrac{2g^{-2}h^{-3}}{5gh^{-6}}\right)^2$

40. $\left(\dfrac{3c^5d^2}{5c^3d^2}\right)^{-2}$

41. $(2x^3y^{-4})^{-3}(3x^2y^{-6})^2$

42. $(4a^2b^{-4})^3(10a^5b^7)^{-2}$

43. $\left(\dfrac{1}{5}a^{-2}b^3c\right)^{-2}\left(\dfrac{2}{3}a^4b^{-6}c\right)^{-1}$

44. $\left(\dfrac{2}{3}xy^5z^{-4}\right)^{-1}\left(\dfrac{5}{7}x^{-3}y^2z^{-2}\right)^{-3}$

45. $(8x^3y^9)^{\frac{1}{3}}$

46. $(243t^{20}u^{15})^{\frac{1}{5}}$

47. $\left(\dfrac{16m^3n^6p}{mn^{-2}p^3}\right)^{\frac{1}{2}}$

48. $\left(\dfrac{9x^5y^4z^{-7}}{x^3y^{-2}z^{-1}}\right)^{\frac{1}{2}}$

49. $(100ab^3c^2)^{-\frac{1}{2}}(a^6b^{-2})^{\frac{1}{4}}$

50. $(5x^2y^7z)^{\frac{3}{2}}(5x^2y^7z)^{\frac{7}{2}}$

51. Rewrite the following radicals using exponents.

  a. $\sqrt{x}$

  b. $\sqrt[3]{k}$

  c. $\sqrt[5]{m}$

52. Rewrite the following radicals using exponents.

  a. $\sqrt{c^3}$

  b. $(\sqrt[3]{t})^2$

  c. $\sqrt[4]{m^3}$

53. Rewrite the following radicals using exponents.

  a. $\sqrt{5xy}$

  b. $\sqrt[3]{2x^2y}$

  c. $\sqrt[7]{4m^3n^6p^2}$

54. Rewrite the following exponents in radical form.

  a. $r^{\frac{1}{3}}$

  b. $x^{\frac{1}{2}}$

  c. $n^{\frac{2}{3}}$

55. Rewrite the following exponents in radical form.

  a. $x^{\frac{1}{3}}y^{\frac{1}{3}}$

  b. $(mn)^{\frac{1}{5}}$

56. Rewrite the following exponents in radical form.

  a. $r^{\frac{1}{5}}s^{\frac{2}{5}}$

  b. $(xy^3z)^{\frac{1}{2}}$

  c. $n^{\frac{2}{3}}m^{\frac{1}{3}}p^{\frac{2}{3}}$

Combining Functions

DEFINING POLYNOMIALS

In this section we will look at a family of functions called polynomials. To start, we will review some basics about polynomials.

The most fundamental component of a polynomial is a **term.** A term can be either a constant, a variable, or a combination of constants and variables multiplied together. The constant part of a term is called the coefficient. A **polynomial** is any combination of terms that are added together. The powers of all variables in a polynomial must be positive integers.

Say What?

Many specific names are given to different types of polynomials.

All of the following are different names for specific types of polynomials:

*   Monomial: polynomial with one term

*   Binomial: polynomial with two terms

*   Trinomial: polynomial with three terms

The prefixes used in these names are Greco-Roman and mean numbers.

  mono = one.
  bi = two.
  tri = three.

DEFINITIONS

Term: A constant, a variable or the product of any number of constants and variables. Terms can include constants and/or variables raised to exponents.

Examples: 12 $-3x$ $5xy^2$

Polynomial: Any combination of terms that are added together. The powers of all variables in a polynomial must be positive integers.

Polynomial	**Not a Polynomial**
$5x^2y$	$3x^2 + \dfrac{5}{x} - 4$
$3m^2 + 2m - 7$	$3\sqrt{x} + 5$

Linear functions are an example of polynomials because they are a combination of constants and variables multiplied and added together. Basic linear functions usually have two terms, mx and b.

EXAMPLE   **1**   RECOGNIZING TERMS IN A POLYNOMIAL

Determine the number of terms in the following polynomial expressions, and list the terms as either constant terms or variable terms.

a. $3x^2 + 5x - 9$ **b.** $5t + 6$ **c.** $5m^2 - 3$

d. 104 **e.** $-98xy^2$

Solution

a. This polynomial expression has three terms; -9 is a constant term, and both $3x^2$ and $5x$ are variable terms because they contain at least one variable.

b. This linear expression has two terms; 6 is a constant term, and $5t$ is a variable term.

c. This polynomial expression has two terms; -3 is a constant term, and $5m^2$ is a variable term.

d. This is a single constant term.

e. This is a single variable term. Although two variables are involved, they are multiplied together and thus are part of one term. Terms are always separated by addition.

DEGREE

One important feature of a polynomial is its **degree.** The degree of an individual term is the sum of all the exponents for the variables. The degree of a polynomial is the same as the highest-degree term.

> **DEFINITIONS**
>
> **Degree of a term:** The sum of all the exponents of the variables in the term.
>
> **Degree of a polynomial:** The degree of the highest-degree term.

EXAMPLE **2** **IDENTIFYING THE DEGREE OF TERMS AND POLYNOMIALS**

For the given expressions, list the degree of each term and the entire polynomial.

a. $3x^2 + 5x + 6$ **b.** $7t - 4$ **c.** $2x^2y + 5xy - 7y + 2$

d. $4a^3b^4c - 6ab^2c^2$ **e.** 12

Solution

a. $3x^2$ has degree 2, $5x$ has degree 1, 6 has degree 0, and the polynomial has degree 2.

b. $7t$ has degree 1, -4 has degree 0, and the polynomial has degree 1.

c. $2x^2y$ had degree 3 because x has an exponent of 2, and y has an exponent of 1. $5xy$ has degree 2, $-7y$ has degree 1, 2 has degree 0, and the polynomial has degree 3.

d. $4a^3b^4c$ has degree 8, $-6ab^2c^2$ has degree 5, and the polynomial has degree 8.

e. 12 has degree zero. This is because there are no variables in this term, so there are no variable exponents to sum. The polynomial has degree 0.

EXAMPLE **2** **PRACTICE PROBLEM**

For the given expressions, list the degrees of each term and the entire polynomial.

a. $-0.23x^2 + 4x - 3$ **b.** $8m^3n + 6mn^2 - 2n^3$**c.** $-9s^5t^4u^2 - 7s^3t^2u^2$

ADDING AND SUBTRACTING FUNCTIONS

Because functions can represent many things in our lives, it is often helpful to combine them using different operations such as adding, subtracting, multiplying, or

Say What?

When adding, subtracting, multiplying, or dividing functions, you can write these operations in function notation in a few ways.

Addition

$$f(x) + g(x)$$
$$= (f + g)(x)$$

Subtraction

$$f(x) - g(x)$$
$$= (f - g)(x)$$

Multiplication

$$f(x)g(x)$$
$$= (fg)(x)$$

Division

$$\frac{f(x)}{g(x)}$$
$$= f(x) \div g(x)$$
$$= \left(\frac{f}{g}\right)(x)$$

When combining functions without a context or meaning for the variables, you are free to combine any functions you are given as long as the input variables are the same. The only caution would be that you can divide as long as you do not divide by zero. This is a typical restriction that is assumed in all cases.

dividing. Many functions can be combined to make more complicated functions or to create a new function that gives you the information that you want without having to work with several functions at once. When working with polynomials and functions, you will need to pay close attention to a few details. To add or subtract two terms, they must be **like terms.** Like terms have the same variables with the same exponents. The next few examples and problems will show you some applications of addition and subtraction of polynomials.

E X A M P L E   COMBINING REVENUE AND COSTS TO FIND PROFIT

The revenue and costs for Southwest Airlines can be modeled by the following functions.

$$R(t) = 183t^2 - 581t + 5964$$

$$C(t) = 102t^2 - 100t + 4897$$

where $R(t)$ represents the revenue for Southwest Airlines in millions of dollars and $C(t)$ represents the costs for Southwest Airlines in millions of dollars t years since 2000.

Source: Models derived from data found at finance.google.com.

a. Find the revenue and costs for Southwest Airlines in 2006.

b. Find a new function that will give the profit of Southwest Airlines. (*Hint:* profit equals revenue minus costs.)

c. Use the new profit function to find the profit in 2006.

Solution

a. 2006 is represented by $t = 6$, so we get

$$R(6) = 9066$$
$$C(6) = 7969$$

Therefore, Southwest Airlines had a revenue of about $9,066 million and costs of about $7,969 million in 2006.

b. If we define $P(t)$ to be the profit in millions of dollars for Southwest Airlines, we know that profit is revenue minus the costs, so we get

$$P(t) = R(t) - C(t)$$ Substitute the Revenue and Cost functions.

$$P(t) = (183t^2 - 581t + 5964) - (102t^2 - 100t + 4897)$$

$$P(t) = 183t^2 - 581t + 5964 - 102t^2 + 100t - 4897$$ Distribute the negative and combine like terms.

$$P(t) = 81t^2 - 481t + 1067$$

c. Again, 2006 is represented by $t = 6$, so we get $P(6) = 1097$. So the profit for Southwest Airlines was about $1,097 million in 2006.

EXAMPLE PRACTICE PROBLEM

The revenue and profit for the California lottery can be modeled by the following functions.

$$R(t) = 59.6t^2 - 299.4t + 3022.8$$

$$P(t) = 23t^2 - 121t + 1157.2$$

where $R(t)$ represents the revenue in millions of dollars and $P(t)$ represents the profit in millions of dollars from the California lottery t years since 2000.

Source: Models derived from data in Survey of Government Finances 2000 - 2006.

a. Calculate the revenue and profit for the California lottery in 2010.

b. Find a new function for the cost associated with the California lottery. Hint: remember Profit = Revenue - Cost.

c. Using the new cost function, determine the cost for the California lottery in 2015.

It is important to note that the revenue, cost, and profit functions had the same input variable. The output variables were also measured with the same units. You can add or subtract functions for revenue, cost, and profit only when the units are the same. If revenue was measured in millions of dollars and cost was measured in thousands of dollars you could not simply subtract the two functions to find profit.

EXAMPLE A GEOMETRY EXAMPLE OF COMBINING FUNCTIONS

The area of a Norman window (a rectangle with a semicircle on top) can be determined by adding the area functions of the rectangle and semicircle pieces that it is made of. If the rectangular part has a height of 99 inches, find the following.

a. A function for the total area of the Norman window as a function of the width of the window.

b. The area of the Norman window if its width is 48 inches.

Solution

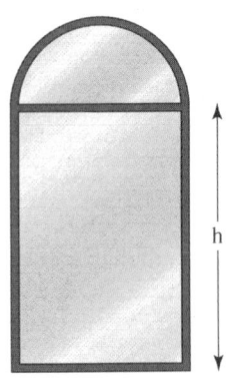

a. The area function will be the sum of the rectangle's area function and the semicircle's area function. We are missing the width of the window, so let's use the following definition:

w = the width of the window in inches.

The area of the rectangular piece of the window can be calculated as length times width. Since the height of the rectangular piece is 99 inches, we can represent the area with the function.

$$R(w) = 99w \qquad \text{Area of a rectangle is length times width.}$$

where $R(w)$ represents the area of the rectangle part in square inches.

The area of the circular piece can be calculated using the formula for the area of a circle, $A = \pi r^2$. In this window the radius is half of the width of the window. Only half of a circle is part of the window, so we multiply the area by 1/2. We can represent the area of the half-circle piece with the function

$$C(w) = \frac{1}{2}\pi\left(\frac{w}{2}\right)^2 \qquad \text{The area of a half circle is} \\ \text{half of pi times radius squared.}$$

where $C(w)$ represents the area of the semicircle part in square inches. To get the total area of the Norman window, we need to add these two functions together. This results in a total area function of

$$N(w) = 99w + \frac{1}{2}\pi\left(\frac{w}{2}\right)^2 \qquad \text{The total area is the sum of the} \\ \text{two functions.}$$

$$N(w) = 99w + \frac{1}{2}\pi\frac{w^2}{4} \qquad \text{Simplify.}$$

$$N(w) = \frac{\pi}{8}w^2 + 99w$$

where $N(w)$ represents the total area in square inches of a Norman window with a height of 99 inches and a width of w inches for the rectangle part.

b. Because the width is given as 48 inches, we can substitute 48 for w and find the total area $N(48) = 5656.8$. Therefore a Norman window with these dimensions will have a total area of 5656.8 square inches.

Units are the key to combining functions in any application. To determine whether you can combine two functions by addition or subtraction, be sure the input variables are the same and that the output variables have the same units.

COMBINING FUNCTIONS IN APPLICATIONS USING ADDITION OR SUBTRACTION

- The inputs for both functions must be measured in the same units.
- The outputs must be measured in the same units.

EXAMPLE 5 COMBINING FUNCTIONS USING FUNCTION NOTATION

Combine the following functions using addition or subtraction to write a new function that will give you the result requested. Represent the new function using function notation.

$S(t) = $ The total number of students at Clark State University t years since 2000.

$F(t) = $ The number of full-time students at Clark State University t years since 2000.

$B(u) = $ The average amount in dollars a student pays for textbooks when taking u units a semester.

$T(u) = $ The average amount in dollars a student pays for tuition and fees when taking u units a semester.

   a. The average total amount a student pays for textbooks, tuition, and fees when taking u units a semester.

   b. The number of part-time students at Clark State University t years since 2000.

Solution

   a. $B(u) + T(u) =$ The average total amount a student pays for textbooks, tuition, and fees when taking u units a semester.

   Since we want the average total amount spent on textbooks as well as tuition and fees, we need to add the functions $B(u)$ and $T(u)$ together. The input variables are the same and the output units are both dollars and make sense to add together.

   b. $S(t) - F(t) =$ The number of part-time students at Clark State University t years since 2000.

   We want the number of part-time students at Clark State University but only have a function for the total number of students and another function for the number of full-time students. To find the number of part-time students, we can subtract the number of full-time students from the total number of students. Again the input variables are the same and the output units are both types of students, so subtracting them makes sense in this situation.

EXAMPLE 6   ADD AND SUBTRACT FUNCTIONS ALGEBRAICALLY

Combine the following functions.

$$f(x) = 5x + 6 \qquad g(x) = 2x - 9 \qquad h(x) = 3x + 4$$

   a. $f(x) + g(x)$ **b.** $g(x) - f(x)$ **c.** $(h - g)(x)$

Solution

   a. To add these two functions, we combine like terms and simplify.

$$f(x) + g(x) = (5x + 6) + (2x - 9)$$
$$f(x) + g(x) = 7x - 3$$

   b. When subtracting functions, be careful to distribute the negative throughout the second function.

$$g(x) - f(x) = (2x - 9) - (5x + 6)$$
$$g(x) - f(x) = 2x - 9 - 5x - 6 \qquad \text{Distribute the negative through the parentheses.}$$
$$g(x) - f(x) = -3x - 15 \qquad \text{Combine like terms.}$$

   c. This function notation is another way to write subtraction of two functions.

$$(h - g)(x) = (3x + 4) - (2x - 9)$$
$$(h - g)(x) = 3x + 4 - 2x + 9 \qquad \text{Distribute the negative through the parentheses.}$$
$$(h - g)(x) = x + 13 \qquad \text{Combine like terms.}$$

EXAMPLE PRACTICE PROBLEM

Combine the following functions.

$$f(x) = 7x + 5 \qquad g(x) = 4x - 8$$

a. $f(x) + g(x)$ **b.** $g(x) - f(x)$

MULTIPLYING AND DIVIDING FUNCTIONS

Multiplication and division are two more ways that we can combine functions and polynomial expressions. When using multiplication or division, be sure that the input variables are the same and that when you combine the output variables, the units make sense. The output units can be the same if you are computing something like area. On the other hand, the output units can be different, but when they are combined they have to simplify to an appropriate unit for the application.

> ### COMBINING FUNCTIONS IN APPLICATIONS USING MULTIPLICATION OR DIVISION
> - The inputs for both functions must be the same.
> - The outputs must make sense together when combined.
>   - **i.** feet • feet = square feet.
>   - **ii.** pounds • $\dfrac{\text{dollars}}{\text{pound}}$ = pounds • $\dfrac{\text{dollars}}{\text{pound}}$ = dollars
>   - **iii.** hours • $\dfrac{\text{miles}}{\text{hour}}$ = hours • $\dfrac{\text{miles}}{\text{hour}}$ = miles
>   - **iv.** dollars ÷ people = $\dfrac{\text{dollars}}{\text{person}}$ = dollars per person

EXAMPLE **7** COMBINING FUNCTIONS USING FUNCTION NOTATION

Combine the following functions using addition, subtraction, multiplication, or division to write a new function that will give you the result requested. Represent the new function using function notation.

$U(t)$ = The population of the United States t years since 1900

$F(t)$ = The number of females in the United States t years since 1900

$D(t)$ = The United States National Debt in dollars t years since 1900

$A(t)$ = The average dollars spent per person on dining out t years since 1900

a. The total amount spent by Americans on dining out t years since 1900.

b. The number of males in the United States t years since 1900.

c. The average amount that each American would have to contribute to pay off the national debt t years since 1900.

Solution

a. $U(t) \bullet A(t) =$ The total amount spent by Americans on dining out, in dollars, t years since 1900

To get the total, we took the average amount spent and multiplied by the total number of people. The units of dollars per person times people simplifies to dollars, which makes sense in this situation.

b. $U(t) - F(t) =$ The number of males in the United States t years since 1900

Subtracting the number of females in the United States from the total population gives us the number of males.

c. $\dfrac{D(t)}{U(t)} =$ The average amount that each American would have to contribute to pay off the national debt t years since 1900

Dividing the national debt by the number of people in the United States gives an average amount for each person.

EXAMPLE (7) PRACTICE PROBLEM

Use the following functions to write a new function that will give you the result requested.

$F(t) =$ The number of people employed by Ford Motor Company in year t

$I(t) =$ The average cost, in dollars per employee, for health insurance at Ford Motor Company in year t

$V(t) =$ The total cost, in dollars, for vacations taken by Ford Motor Company non-management employees in year t

$M(t) =$ The number of employees of Ford Motor Company who are in management in year t

a. The total amount spent on health insurance for Ford Motor Company employees in year t.

b. The number of non-management employees at Ford Motor Company in year t.

c. The average cost per non-management employee for vacations at Ford Motor Company in year t.

Unlike addition and subtraction, in which you must have like terms, when multiplying, the terms do not have to be like. You will, however, have to use the distributive property and add exponents for any variables that are the same.

EXAMPLE 8 AN APPLICATION OF MULTIPLYING FUNCTIONS

The willingness of digital camera producers to produce cameras depends on the price they can sell the cameras for. This situation can be modeled by

$$S(p) = 0.015p^2 - 1.5p + 80$$

where $S(p)$ represents the supply of digital cameras in thousands when p is the price in dollars and the price is greater than \$50.

a.   Find the number of digital cameras that producers are willing to supply when the price is $100.

b.   Find a function for the projected revenue if all the cameras that are supplied sell at price p.

c.   Estimate the projected revenue if digital cameras are sold for $150 each.

Solution

a.   The given price of $100 can be substituted into p to get $S(100) = 80$. Therefore, the digital camera producers are willing to supply 80 thousand cameras if the cameras will sell for $100 each.

b.   Revenue is calculated by taking the number of items sold by the price. In this case, the function $S(p)$ represents the number of items sold and p represents the price. Using these variables, we get

$$R(p) = \text{price (quantity sold)}$$
$$R(p) = p \cdot S(p)$$
$$R(p) = p(0.015p^2 - 1.5p + 80)$$
$$R(p) = 0.015p^3 - 1.5p^2 + 80p$$

where $R(p)$ represents the revenue in thousands of dollars for selling digital cameras at a price of p dollars.

c.   The given price of $150 can be substituted into p to get $R(150) = 28875$. Therefore, the digital camera producers will earn $28,875 thousand if cameras sell for $150 each.

EXAMPLE ⑧ PRACTICE PROBLEM

The willingness of a car manufacturer to produce SUVs depends on the price they can sell the vehicles for. This can be modeled by

$$S(p) = 0.008p^2 - 0.5p + 30$$

where $S(p)$ represents the supply of SUVs in thousands when p is the price in thousands of dollars and the price is greater than $25,000.

a.   Find the number of SUVs this car manufacturer is willing to supply when the price is $45,000.

b.   Find a function for the projected revenue if all the SUVs that are supplied sell at price p.

c.   Estimate the projected revenue if SUVs are sold for $40,000 each.

EXAMPLE 9   COMBINING FUNCTIONS ALGEBRAICALLY

Combine the following functions.

$$f(x) = 5x + 6 \qquad g(x) = 2x - 9 \qquad h(x) = 3x + 4$$

a. $f(x)g(x)$

b. $\dfrac{g(x)}{h(x)}$

Solution

a. When multiplying two functions, be sure to use the distributive property.

$$f(x)g(x) = (5x + 6)(2x - 9)$$
$$f(x)g(x) = 10x^2 - 45x + 12x - 54$$
$$f(x)g(x) = 10x^2 - 33x - 54$$

b. For now, we will write the division of two functions as a fraction and we will learn how to simplify these in Chapter 7.

$$\frac{g(x)}{h(x)} = \frac{2x-9}{3x+4}$$

EXAMPLE ⑨ PRACTICE PROBLEM

Combine the following functions.

$$f(x) = 5x + 6 \qquad g(x) = 2x - 9 \qquad h(x) = 3x + 4$$

a. $f(x)g(x)$ b. $\frac{g(x)}{h(x)}$

At times, we will combine polynomial expressions the same way we combine functions. In these problems, the function notation is left out and the operations are done to simplify the expression.

EXAMPLE 10 MULTIPLYING POLYNOMIAL EXPRESSIONS

Perform the indicated operation and simplify.
a. $(2ab^2 + 5b) + (4a^2b + 3ab^2)$ b. $(x^4 - 3x^2 + 4x) - (3x^2 - 5x + 3)$

c. $(2x + 5)(4x - 8)$ d. $(x + 2)(x^2 + 3x - 7)$

Solution

a. Combine like terms.

$$(2ab^2 + 5b) + (4a^2b + 3ab^2)$$
$$4a^2b + 5ab^2 + 5b$$

b. First, we need to distribute the negative sign through the second parentheses and then combine like terms.

$$(x^4 - 3x^2 + 4x) - (3x^2 - 5x + 3)$$
$$x^4 - 3x^2 + 4x - 3x^2 + 5x - 3$$
$$x^4 - 6x^2 + 9x - 3$$

Skill Connection

The **distributive property** is the basis for multiplication of polynomials. To use the distributive property multiply each term of the first expression by each term of the second expression in the product.

SC-Example 1:
Perform the indicated operation and simplify.

a. $5x(3x + 8)$
b. $(x + y)(2x + 3y)$

Solution:

a. Use the distributive property to multiply the 5x through the parentheses.

$$5x(3x + 8)$$
$$15x^2 + 40x$$

b. Use the distributive property to multiply the x and then the y through the second parentheses.

$$(x + y)(2x + 3y)$$
$$2x^2 + 3xy + 2xy + 3y^2$$
$$2x^2 + 5xy + 3y^2$$

In multiplying two binomials, the distributive property can be remembered by using the acronym **FOIL**: Multiply the First terms, the Outer terms, the Inner terms, and finally the Last terms.

c. Using FOIL (distributive property), we multiply the $2x$ through and then the 5.

$$(2x + 5)(4x - 8)$$
$$8x^2 - 16x + 20x - 40$$
$$8x^2 + 4x - 40$$

d. Use the distributive property to multiply the x and then to multiply the 2 through and combine like terms.

$$(x + 2)(x^2 + 3x - 7)$$
$$x^3 + 3x^2 - 7x + 2x^2 + 6x - 14$$
$$x^3 + 5x^2 - x - 14$$

EXAMPLE ⑩ PRACTICE PROBLEM

Perform the indicated operation and simplify.

a. $(4x^2y + 3xy - 7y) + (3x^2y - 2y + 10)$

b. $(3m^4n^2 - 5m^2n + 2mn) - (8m^2n - 3mn + 4n^2)$

c. $3(7x + 3)(3x - 5)$

d. $(4x^2 + 5x)(2x^2 - 3x + 7)$

3.2 Exercises

For Exercises 1 through 6 determine the number of terms in the given polynomial expression and list the terms as either constant terms or variable terms.

1. $5x + 9$

2. $-2x + 4$

3. $-3x^2 + 2x - 8$

4. $0.5t^2 + 5t - 7$

5. $12a^3b^2c + 3abc^2 - 8bc$

6. $m^3n^2 - 3m^2n + 4mn^3 - 4n$

In Exercises 7 through 16, list the degrees of each term and the degree of the entire polynomial.

7. $2x^5 - 7$

8. $3x^2 + 5x - 7$

9. $5p^2 + 4p - 87$

10. $1.023t + 6.2$

11. $4m + 8$

12. $-14s^3t + 6t^4 - 9$

13. $2a^2b^3 + 3ab^2 - 8b$

14. $23a^4b^2 - 62a^2b + 9b^3$

15. $\frac{2}{3}gh + \frac{1}{4}g^3h^5 - \frac{2}{9}g + 7$

16. $\frac{5}{7}r^2s^3t^2 - \frac{3}{8}r^4st^2 + \frac{4}{9}rs - \frac{5}{11}$

In Exercises 17 through 34, perform the indicated operation and simplify.

17. $(5x + 6) + (2x + 8)$

18. $(4x + 5) + (8x + 3)$

19. $(7x + 6) - (3x + 2)$

20. $(10x - 9) - (4x + 3)$

21. $(5x^2 - 6x - 12) - (-3x^2 - 10x + 8)$

22. $(14x^2 - 4x - 6) - (-2x^2 - 3s + 12)$

23. $(5x^3z^2 - 4x^2z + 6z) - (3x^3z^2 - 17xz + 3z)$

24. $(a^2b + 5ab^2 - 9b) - (3ab^2 + 8b)$

25. $(5ab - 8b)(3a + 4b)$

26. $(3xy + 2x)(5x + 7y)$

27. $(2x + 5)(3x^2 - 5x + 7)$

28. $(4a + 7)(3a^2 - 4a + 2)$

29. $(3x + 7y)^2$

30. $(x + 4y)^2$

31. $(2x^2 + 3x - 9) + (7x + 2)^2$

32. $(5x^2 + 2x - 8) + (3x - 5)^2$

33. $7x(3y^2 + 5y - 6) + 8xy$

34. $(7x - 3)(2x + 8) - 9x^2 + 8$

35. The average number of pounds of fruits and vegetables per person that each American eats can be modeled by

$$T(y) = -4.5y + 712.3$$

pounds per person y years since 2000.

The average number of pounds of fruit per person each American eats can be modeled by

$$F(y) = -3.5y + 289.6$$

pounds per person y years since 2000.

Source: Based on data from the Statistical Abstract 2008.

a. How many pounds of fruits and vegetables did the average American eat in 2000?

b. How many pounds of fruit did the average American eat in 2000?

c. Using your previous results, how many pounds of vegetables did the average American eat in 2000?

d. Write a new function that will give the average number of pounds of vegetables each American eats per year.

e. Using your new function, how many pounds of vegetables did each American eat in 2000?

f. Estimate the number of pounds of vegetables each American will eat in 2010, 2015, and 2020.

36. In exercise 23 of Section 1.6 you found a model for the amount spent by individuals on health care expenses in the United States close to

$$I(t) = 17t + 147$$

billions of dollars t years since 2000.

The amount spent by insurance companies on health care expenses in the United States can be modeled by

$$H(t) = 58t + 370$$

billions of dollars t years since 2000.

Source: Based on data from the Statistical Abstract 2008.

a. How much was spent by individuals on health care in 2000?

b. How much was spent by insurance companies on health care in 2000?

c. What was the total amount spent by individuals and insurance companies on health care in 2000?

d. Find a new function that gives the total amount spent on health care by individuals and insurance companies.

e. Use your new function to determine the total amount spent on health care by individuals and insurance companies in 2010, 2015, and 2020.

37. The per capita consumption of milk products in the United States can be modeled by

$$M(t) = -0.29t + 22.42$$

gallon per person t years since 2000.

The per capita consumption of whole milk in the United States can be modeled by

$$W(t) = -0.21t + 8.09$$

gallons per person t years since 2000.

Source: Based on data from the Statistical Abstract 2008.

a. Use these models to find a new function that gives the per capita consumption of milk products other than whole milk.

b. How much whole milk was consumed per person in 2005?

c. How much milk other than whole milk was consumed per person in 2005 and in 2010?

d. What is the slope of $W(t)$? Explain its meaning in this context?

e. What is the M-intercept for $M(t)$? Explain its meaning in this context?

38. The number of United States residents who are black or of Hispanic origin can be modeled by

$$B(t) = 400.99t - 5240.19$$

$$H(t) = 11.51t^2 - 1181.14t + 35414.03$$

where $B(t)$ represents the number of black residents in thousands and $H(t)$ represents the number of residents of Hispanic origin in thousands t years since 1900.

Source: Models derived from data in Statistical Abstract 2001.

a. Find a function that will represent the total number of residents in the United States who are either black or of Hispanic origin.

b. Use this function to determine the total number of residents who are either black or of Hispanic origin in 2000.

c. Find when the number of residents of Hispanic origin was the same as the number of black residents.

In Exercises 39 through 46, combine the functions by adding, subtracting, multiplying, dividing, or composing the two functions. Write the function notation for the combination you use and list the units for the inputs and outputs of the new function.

39. Let $U(t)$ be the population of the United States, in millions, t years since 1900. Let $M(t)$ be the number of men in the United States, in millions, t years since 1900. Find a function for the number of women in the United States.

40. Let $S(t)$ be the number of students at Suncoast College, t years since 2010. Let $A(t)$ be the number of students who are involved in the athletics program at Suncoast College, t years since 2010. Find a function for the number of students at Suncoast College who are not involved in the schools athletics program.

© Thomson Learning/Heinle

41. Let $S(d)$ be the average speed, in miles per hour, traveled on a cross-country trip during day d of the trip. Let $T(d)$ be the time, in hours, traveled during a cross-country trip on day d of the trip. Find a function for the number of miles traveled on day d of the cross-country trip.

42. Let $P(w)$ be the price of gas in dollars per gallon during week w of the year. Let $G(w)$ be the number of gallons of gas your car needs during week w of the year. Find a function for the total cost of gas used by your car during week w of the year.

43. Let $U(t)$ be the population of the United States, in millions, t years since 1900. Let $D(t)$ be the national debt, in millions of dollars, t years since 1900. Find a function for the average amount of national debt per person.

44. Let $S(d)$ be the number of students in your math class on day d of the semester. Let $U(d)$ be the total number of units being taken by all students in your math class on day d of the semester. Find a function for the average number of units a student in your math class is taking on day d of the semester.

45. Let $U(t)$ be the population of the United States, t years since 1900. Let $D(t)$ be the average amount of personal debt in dollars per person in the United States t years since 1900. Find a function for the total amount of personal debt in the United States.

46. Let $P(b)$ be the profit, in thousands of dollars, that Pride Bike Co. makes if they sell b bikes per year. Let $R(b)$ be the revenue, in thousands of dollars, that Pride Bike Co. makes if they sell b bikes per year. Find a function that gives the cost that Pride Bike Co. has if they sell b bikes per year.

47. The percent of jail inmates who are male in U.S. federal and state prisons can be modeled by $M(t) = -0.24t + 88.64$ percent t years since 2000.

Source: Based on data from Statistical Abstract 2008.

   a. Let $F(t)$ be the percent of jail inmates who are female in U.S. federal and state prisons t years since 2000. Explain why $M(t) + F(t) = 100$.

   b. Substitute the function $M(t)$ into the equation in part a and solve for $F(t)$.

   c. Find $F(5)$ and interpret it in this context.

   d. What is the slope of $F(t)$? Explain its meaning in this context?

   e. What is the M-intercept for $M(t)$? Explain its meaning in this context?

48. The percent of mortgage loans in United States that are considered subprime loans can be modeled by $S(t) = 3.5t - 1$ percent t years since 2000.

Source: Based on data from 2007 Federal Reserve Bank of Chicago Chicago Fed Letter.

   a. Let $P(t)$ be the percent of mortgage loans in United States that are not considered subprime t years since 2000. Explain why $S(t) + P(t) = 100$.

   b. Substitute the function $S(t)$ into the equation in part a and solve for $P(t)$.

   c. Find $P(10)$ and interpret it in this context.

   d. What is the slope of $P(t)$? Explain its meaning in this context?

   e. What is the P-intercept for $P(t)$? Explain its meaning in this context?

49. The percent of occupied housing units that are owner occupied is given in the following chart.

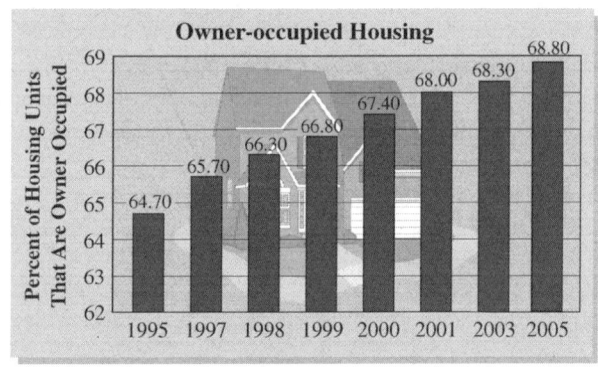

Source: Statistical Abstract 2008.

   a. Let $O(t)$ be the percent of occupied housing units that are owner occupied t years since 1990. Find a model for $O(t)$.

   b. Let $R(t)$ be the percent of occupied housing units that are renter occupied t years since 1990. Using your model from part a, find a model for $R(t)$.

   c. Find $R(20)$ and explain its meaning in this context.

   d. What is the slope of $R(t)$? Explain its meaning in this context?

50. The percent of all drug prescriptions that are written for brand name drugs is given in below.

Year	Percent for Brand Name Drugs
2002	57.9
2003	55.0
2004	51.9
2005	48.7
2006	45.7

Source: Statistical Abstract 2008.

a. Let $B(t)$ be the percent of all drug prescription that are written for brand name drugs t years since 2000. Find a model for $I(t)$.

b. Let $G(t)$ be the percent of all drug prescription that are written for generic drugs t years since 2000. Using your model from part a, find a model for $G(t)$.

c. Find $G(15)$ and explain its meaning in this context.

d. What is the slope of $G(t)$? Explain its meaning.

51. The revenue and costs for goods sold by International Business Machines (IBM) can be modeled by the functions

$$R(t) = 3.1t^2 - 33.7t + 180.9$$
$$C(t) = 2.5t^2 - 28.9t + 136.1$$

where $R(t)$ represents the revenue for goods sold in billions of dollars and $C(t)$ represents the costs of selling those goods in billions of dollars t years since 2000.

Source: Models derived from data found at finance.google.com.

a. Find the revenue and costs for IBM in 2010.

b. Find a new function that will give the profit of IBM. (*Hint:* Profit equals revenue minus costs.)

c. Use the new profit function to find the profit in 2010.

52. The revenue and costs for goods sold by Nike, Inc. can be modeled by the functions

$$R(t) = 1.3t + 7.0$$
$$C(t) = 0.7t + 4.5$$

where $R(t)$ represents the revenue for goods sold in billions of dollars and $C(t)$ represents the costs of selling those goods in billions of dollars t years since 2000.

Source: Models derived from data found at finance.google.com.

a. Find the revenue and costs for Nike in 2009.

b. Find a new function that will give the gross profit of Nike. (*Hint:* Gross profit equals revenue minus costs.)

c. Use the new profit function to find the profit in 2009

53. The revenue and profit for Pearson Publishing Company can be modeled by

$$R(t) = 164.1t^2 - 1353.9t + 6380.0$$
$$P(t) = 87.9t^2 - 708.9t + 3322.8$$

where $R(t)$ represents the revenue in millions of British pounds and $P(t)$ represents the profit in millions of British pounds for Pearson t years since 2000.

Source: Models derived from data found at finance.google.com.

a. Calculate the revenue and profit earned by Pearson in 2011.

b. Find a new function for the costs at Pearson Publishing.

c. Using the new cost function determine the costs at Pearson Publishing in 2011.

54. The cost and profit for Home Depot, Inc. can be modeled by

$$C(t) = -0.9t^2 + 10.56t + 21.7$$
$$P(t) = -0.6t^2 + 7.1t + 5.8$$

where $C(t)$ represents the cost in billions of dollars and $P(t)$ represents the profit in billions of dollars for Home Depot, Inc. t years since 2000.

Source: Models derived from data found at finance.google.com.

a. Calculate the costs and profit of Home Depot in 2010.

b. Find a new function for the revenue at Home Depot Inc.

c. Using the new revenue function estimate the revenue at Home Depot in 2010.

55. The willingness of car stereo producers to produce stereos depends on the price the stereos can be sold for. This relationship can be modeled by

$$S(p) = 0.009p^2 - 0.5p + 20$$

where $S(p)$ represents the supply of car stereos in thousands when p is the price in dollars and the price is greater than $30.

a. Find the number of car stereos that producers are willing to supply when the price is $100.

b. Find a function for the projected revenue if all the car stereos that are supplied sell at price p.

c. Estimate the projected revenue if car stereos are sold for $90 each.

56. The willingness of sports equipment producers to produce tennis rackets depends on the price that the tennis rackets can be sold for. This relationship can be modeled by

$$S(p) = 0.09p^2 - 4.2p + 51$$

where $S(p)$ represents the supply of tennis rackets in millions when p is the price in dollars and the price is greater than $25.

a. Find the number of tennis rackets that producers are willing to supply when the price is $40.

b. Find a function for the projected revenue if all the tennis rackets that are supplied sell at price p.

c. Estimate the projected revenue if tennis rackets are sold for $75 each.

57. The willingness of car manufacturers to produce minivans depends on the price the minivans can be sold for. This can be modeled by

$$S(p) = 0.11p^2 - 3.2p + 45$$

where $S(p)$ represents the supply of minivans in thousands when p is the price in thousands of dollars and the price is greater than $14,500.

a. Find the number of minivans manufacturers are willing to supply when the price is $25,000.

b. Find a function for the projected revenue if all the minivans that are supplied sell at price p.

c. Estimate the projected revenue if minivans are sold for $31,000 each.

58. The willingness of a computer manufacturer to produce laptop computers depends on the price that laptops can be sold for. This economic relationship can be modeled by

$$S(p) = 55.7p^2 - 728.6p + 4321.6$$

where $S(p)$ represents the supply of laptop computers when p is the price in hundreds of dollars and the price is greater than $700.

a. Find the number of laptop computers that manufacturers are willing to supply when the price is $1,500.

b. Find a function for the projected revenue if all the laptop computers that are supplied sell at price p.

c. Estimate the projected revenue if laptop computers are sold for $3,000 each.

59. The total number of immigrants admitted to the United States as permanent residents under refugee acts t years since 1990 can be represented by

$$I(t) = -17853.5t^2 + 212449.1t - 501783.9$$

Of those admitted under refugee acts, the number coming from Europe can be represented by

$$E(t) = -6431.5t^2 + 74006.3t - 161816.2$$

Source: Models derived from data in Statistical Abstract 2001.

a. Estimate the total number of immigrants admitted to the United States as permanent residents under refugee acts in 1995.

b. Estimate the number of immigrants admitted under refugee acts who were from Europe in 1995.

c. Find a new function to represent the number of non-European immigrants who were admitted to the United States as permanent residents under refugee acts.

d. Use your model to determine the number of non-European immigrants who were admitted under refugee acts in 1998.

60. The number of men and women in the United States since 1850 can be modeled by

$$M(t) = 2.86t^2 + 97.22t - 137.68$$

$$F(t) = 3.65t^2 - 48.46t + 4883.84$$

where $M(t)$ represents the number of men in thousands and $F(t)$ represents the number of females in thousands in the United States t years since 1800.

Source: Models derived from data in Statistical Abstract 2001.

   a. Estimate the number of males and females in the United States in 2000.

   b. Find a function for the total population in the United States.

   c. Use the new function to estimate the total population in the United States in 2002.

   d. Find when the number of men was equal to the number of women in the United States.

   e. Find the vertex of the total population function and explain its meaning in this context.

61. The number of U.S. residents who are white can be modeled by

$$W(t) = 1.693t + 209.107$$

where $W(t)$ represents the number of U.S. residents who are white in millions t years since 1990.

    The number of white residents who are 18 years old or older can be modeled by

$$O(t) = 0.014t^2 + 1.169t + 157.528$$

where $O(t)$ represents the number of white residents 18 years old or older in millions t years since 1990.

Source: Models derived from data in Statistical Abstract 2001.

   a. Find a new function for the number of white residents who are under 18 years old.

   b. Estimate the number of white residents under 18 years old in 2002.

62. The number of U.S. residents who are white and 65 years old or older can be modeled by

$$R(t) = -0.011t^2 + 0.409t + 28.098$$

where $R(t)$ represents the number of U.S. residents who are white and 65 years old or older in millions t years since 1990. Using the function you found in Exercise 11 for the number of white residents under 18 years old, determine when the number of white residents under 18 years old is the same as the number 65 years old or older.

Source: Model derived from data in Statistical Abstract 2001.

63. The area of a Norman window (a rectangle with a semicircle on top) can be determined by adding the area functions of the rectangle and semicircle pieces it is made of. If the rectangle part has a height of 75 inches, find the following.

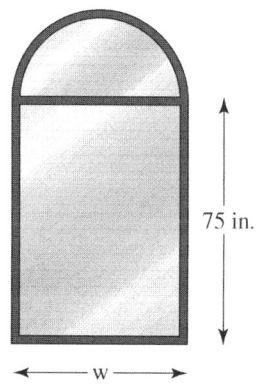

   a. A function for the total area of the Norman window.

   b. The area of the Norman window if its width is 36 inches.

64. If the rectangle part of a Norman window has a height of 80 inches, find the following.

   a. A function for the total area of the Norman window.

   b. The area of the Norman window if its width is 45 inches.

65. Let $f(x) = 3x + 8$ and $g(x) = -6x + 9$.

   a. Find $f(x) + g(x)$.

   b. Find $f(x) - g(x)$.

   c. Find $f(x)g(x)$.

   d. Find $\dfrac{f(x)}{g(x)}$

66. Let $f(x) = 19x + 28$ and $g(x) = -17x + 34$.

    a. Find $f(x) + g(x)$.

    b. Find $g(x) - f(x)$.

    c. Find $f(x)g(x)$.

    d. Find $\dfrac{f(x)}{g(x)}$

67. Let $f(x) = \dfrac{1}{2}x + \dfrac{1}{5}$ and $g(x) = 2x - \dfrac{3}{5}$.

    a. Find $f(x) + g(x)$.

    b. Find $g(x) - f(x)$.

    c. Find $f(x)g(x)$.

    d. Find $\dfrac{f(x)}{g(x)}$

68. Let $f(x) = \dfrac{2}{3}x + \dfrac{1}{3}$ and $g(x) = 5x - 3$.

    a. Find $f(x) + g(x)$.

    b. Find $f(x) - g(x)$.

    c. Find $f(x)g(x)$.

    d. Find $\dfrac{f(x)}{g(x)}$

69. Let $f(x) = 3x + 5$ and $g(x) = x^2 + 4x + 10$.

    a. Find $f(x) + g(x)$.

    b. Find $g(x) - f(x)$.

    c. Find $f(x)g(x)$.

    d. Find $\dfrac{f(x)}{g(x)}$

70. Let $f(x) = 4x - 7$ and $g(x) = x^2 - 3x + 12$.

    a. Find $f(x) + g(x)$.

    b. Find $f(x) - g(x)$.

    c. Find $f(x)g(x)$.

    d. Find $\dfrac{f(x)}{g(x)}$

71. Let $f(x) = x + 2$ and $g(x) = 4x^2 + x + 1$.

    a. Find $f(x) + g(x)$.

    b. Find $g(x) - f(x)$.

    c. Find $f(x)g(x)$.

    d. Find $\dfrac{f(x)}{g(x)}$

72. Let $f(x) = 5x - 2$ and $g(x) = 2x^2 - 7x + 6$.

    a. Find $f(x) + g(x)$.

    b. Find $f(x) - g(x)$.

    c. Find $f(x)g(x)$.

    d. Find $\dfrac{f(x)}{g(x)}$

73. Let $f(x) = 3x + 8$ and $g(x) = -6x + 9$.

    a. Find $f(5) + g(5)$.

    b. Find $f(5) - g(5)$.

    c. Find $f(5)g(5)$.

    d. Find $\dfrac{f(5)}{g(5)}$

74. Let $f(x) = 19x + 28$ and $g(x) = -17x + 34$.

    a. Find $f(2) + g(2)$.

    b. Find $g(2) - f(2)$.

    c. Find $f(2)g(2)$.

    d. Find $\dfrac{f(2)}{g(2)}$

75. Let $f(x) = \dfrac{1}{2}x + \dfrac{1}{5}$ and $g(x) = 2x - \dfrac{3}{5}$.

    a. Find $f(2) + g(2)$.

    b. Find $g(2) - f(2)$.

    c. Find $f(2)g(2)$.

    d. Find $\dfrac{f(2)}{g(2)}$

76. Let $f(x) = \dfrac{2}{3}x + \dfrac{1}{3}$ and $g(x) = 5x - 3$.

    a. Find $f(4) + g(4)$.

    b. Find $f(4) - g(4)$.

    c. Find $f(4)g(4)$.

    d. Find $\dfrac{f(4)}{g(4)}$

77. Let $f(x) = 3x + 5$ and $g(x) = x^2 + 4x + 10$.

    a. Find $f(5) + g(5)$.

    b. Find $g(5) - f(5)$.

    c. Find $f(5)g(5)$.

    d. Find $\dfrac{f(5)}{g(5)}$

78. Let $f(x) = 4x - 7$ and $g(x) = x^2 - 3x + 12$.

   a. Find $f(3) + g(3)$.

   b. Find $f(3) - g(3)$.

   c. Find $f(3)g(3)$.

   d. Find $\dfrac{f(3)}{g(3)}$

79. Let $f(x) = x + 2$ and $g(x) = 4x^2 + x + 1$.

   a. Find $f(2) + g(2)$.

   b. Find $g(2) - f(2)$.

   c. Find $f(2)g(2)$.

   d. Find $\dfrac{f(2)}{g(2)}$

80. Let $f(x) = 5x - 2$ and $g(x) = 2x^2 - 7x + 6$.

   a. Find $f(4) + g(4)$.

   b. Find $f(4) - g(4)$.

   c. Find $f(4)g(4)$.

   d. Find $\dfrac{f(4)}{g(4)}$

Composing Functions

COMPOSING FUNCTIONS

Many things in life require you to do a step-by-step process in which a subsequent step requires the result of the previous step(s). A simple example of this would be if you are trying to get a drink from a vending machine that does not take dollar bills. If all you have is dollar bills, you will need first to get change for that bill and then use that change to purchase the drink. The process of buying that drink is a combination of getting change and using the vending machine to get the drink. Today most vending machines have combined these two steps into one by combining a change machine and the vending machine together into one machine so that dollar bills can be used to make a purchase.

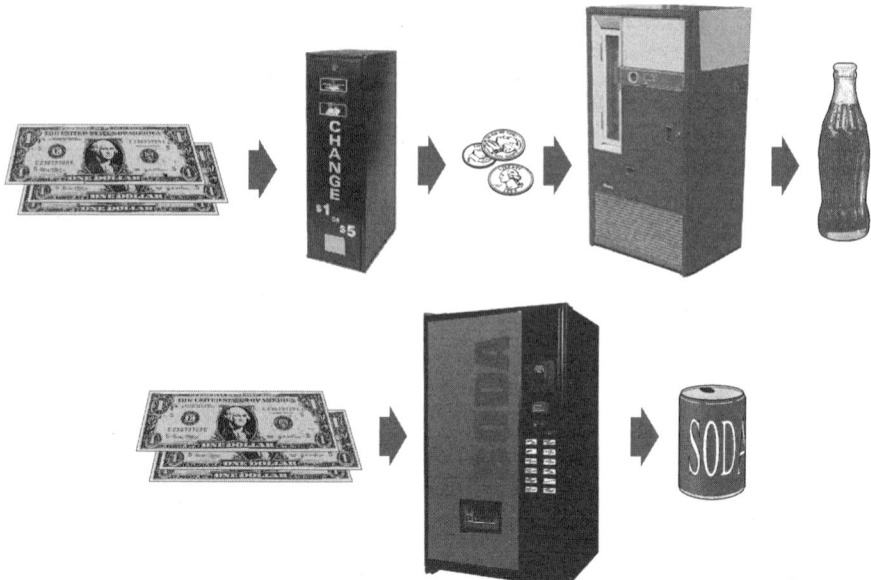

In mathematics, this way of combining functions is called a **composition** of functions. Whenever one function, or the output of the function, is substituted into another function, you are composing the two functions.

EXAMPLE   **1**   COMPOSING TWO FUNCTIONS

A company has analyzed data and come up with the following two functions.

　　$C(e)$ = The cost of health insurance in dollars for a company with e employees

　　$E(t)$ = The number of employees at the company in year t

Find a function that will give the company's cost for health insurance in year t.

Solution

If the company wants to know the cost for health insurance for a certain year, they cannot immediately get that information from either of these functions. To use $C(e)$, they need to know the number of employees. They can use $E(t)$ to find the number of employees in year t. Because both e and $E(t)$ represent the number of employees, we have

$$e = E(t)$$

We can substitute $E(t)$ for e and get

$$C(e) = C(E(t))$$

Doing so, in effect, is a composition of these two functions. $C(E(t))$ is the cost of health insurance for this company in year t.

This notation can be understood by starting from the inner parentheses and working out. $C(E(t))$ says that t is the input for the function and C is the final output of the function. Therefore, the input is a year t and the output is the cost C of health insurance for this company.

Say What?

Two notations are used for the composition of functions:

$$(f \circ g)(x)$$

or

$$f(g(x))$$

These are both read as the function *f* of *g* of *x*.

The function *g*(*x*) is being substituted into the input of the function *f*(*x*).

EXAMPLE 2 AN APPLICATION OF COMPOSING FUNCTIONS

NatureJuice, Inc. processes oranges for several Florida citrus growers. NatureJuice, Inc. is concerned with their annual cost projections and are working with the following two functions.

$$C(t) = 135.87t + 25000$$

where $C(t)$ is the annual cost, in dollars, to process t thousand tons of oranges per year, and

$$T(y) = 2.4y - 4500$$

where $T(y)$ is the annual production of oranges, in thousands of tons, in year y.

© AGStock USA, Inc./Alamy

a. Find a function that will calculate the annual cost to process all oranges in year y.

b. Find the annual cost to process oranges in 2015.

c. When was the annual cost to process oranges $72,000?

Solution

a. NatureJuice, Inc. wants a function that has an output of annual cost but an input of years. The function $T(y)$ depends on the year y and has an output of thousand of tons of oranges. The function $T(y)$ and the variable t both represent thousands of tons of oranges, so

$$T(y) = t$$

we can compose $T(y)$ with $C(t)$ to get the annual cost when given the year y. $C(T(y))$ will give the annual cost, in dollars, to process all oranges in year y.

$C(T(y)) = 135.87(2.4y - 4500) + 25000$

$C(T(y)) = 326.088y - 611415 + 25000$ Distribute 135.87.

$C(T(y)) = 326.088y - 586415$ Combine like terms.

b. We can do this problem in two ways. We will do it once using the two functions given separately and then a second time using the newly composed function we found in part a.

i. Using the two equations separately, we first must find the number of oranges produced in 2015.

$$T(y) = 2.4y - 4500$$
$$T(2015) = 2.4(2015) - 4500$$
$$T(2015) = 4836 - 4500$$
$$T(2015) = 336$$

Thus 336 thousand tons of oranges will be produced in 2015. Now, we must substitute this into the cost equation.

$$C(t) = 135.87t + 25000$$
$$C(336) = 135.87(336) + 25000$$
$$C(336) = 45652.32 + 25000$$
$$C(336) = 70652.32$$

Therefore, the annual cost to process oranges in 2015 will be about $70,652.32.

ii. Using the new composed function from part a, we have

$$C(T(y)) = 326.088y - 586415$$
$$C(T(2015)) = 326.088(2015) - 586415$$
$$C(T(2015)) = 657067.32 - 586415$$
$$C(T(2015)) = 70652.32$$

This is the same result as the first method. It just takes much less effort once the two functions have been composed. If you had to do this same calculation for several years, perhaps to show a trend in the costs from year to year, the composed function would be much more efficient.

c. Using the new composed function, we have

$$C(T(y)) = 326.088y - 586415$$

$$72000 = 326.088y - 586415$$

$$\underline{586415 \qquad\qquad\qquad 586415}$$

$$658415 = 326.088y$$

$$\frac{658415}{326.088} = \frac{326.088y}{326.088}$$

$$2019.13 \approx y$$

In the year 2019, the annual cost to process oranges was about \$72,000.

EXAMPLE ② PRACTICE PROBLEM

Consider the following functions for a metropolitan area in the United States.

$$P(t) = 1.2t + 3.4$$

where $P(t)$ is the population of the metropolitan area in millions of people t years since 2000 and

$$A(p) = 15.48p - 104$$

where $A(p)$ is the amount of air pollution in parts per million when p million people live in the metropolitan area.

a. Find the population of this metropolitan area in 2005.

b. Find the amount of air pollution for this metropolitan area in 2005.

c. Find a new function that will give the amount of air pollution for this metropolitan area depending on the number of years since 2000.

d. Use the function that you found in part c to find the air pollution in 2005, 2007, and 2010.

In composing two functions, it is very important that you check that the output of one function is the same as the input needed by the other function. This is the only way in which you can compose the two functions together.

COMBINING FUNCTIONS USING COMPOSITION

- The input for one function must be the same as the output of the other function.
- The inputs do **not** have to be the same.

EXAMPLE 3 AN APPLICATION OF COMPOSING FUNCTIONS

The amount of drywall damaged by a small leak in a pipe can be modeled by the function

$$D(h) = 0.75h + 1$$

where $D(h)$ is the drywall damaged in square feet h hours after the leak started. The cost to repair drywall damage from a water leak can be modeled by

$$C(a) = 20a + 150$$

where $C(a)$ is the cost in dollars to repair a square feet of water damaged drywall.

a. Find the cost to repair 3 square feet of water damage to drywall.

b. Find a new function that will give the cost to repair the drywall damage depending on the number of hours the leak occurred.

c. Use the function that you found to determine the cost to repair the drywall damage caused by a leak that has been going on for 2 days.

Solution

a. We are given the amount of damage as 3 square feet, so we can substitute that into the $C(a)$ function.

$$C(3) = 20(3) + 150$$
$$C(3) = 210$$

Therefore, the cost to repair 3 square feet of drywall damage is about $210.

b. We want to find a cost function that depends on the hours a water leak continued. The cost function that we have has an input of square feet of damaged drywall, not hours. The function $D(h)$ has an input of hours and output of drywall damage. Since $D(h)$ represents square feet of drywall damage and the variable a is also square feet of drywall damage, we also have $D(h) = a$, so we can compose the two functions.

$$C(a) = 20a + 150$$
$$C(D(h)) = 20(0.75h + 1) + 150$$
$$C(D(h)) = 15h + 20 + 150$$
$$C(D(h)) = 15h + 170$$

$C(D(h))$ is the cost to repair the drywall damage depending on the number of hours the leak occurred.

c. We are given a time of 2 days, which must be converted to hours so it can be substituted for h. Because 2 days is 48 hours, we will substitute 48 for h and find the cost.

$$C(D(48)) = 15(48) + 170$$
$$C(D(48)) = 890$$

The drywall damage caused by a leak that continues for 2 days will cost about $890 to repair.

EXAMPLE PRACTICE PROBLEM

The number of snowmobiles entering Yellowstone National Park on a winter weekend can be modeled by the function

$$S(h) = -0.04h^2 + 1.11h - 5.76$$

where $S(h)$ is the number of snowmobiles in hundreds h hours after midnight. The carbon monoxide (CO) concentration at Old Faithful can be modeled by

$$C(n) = 0.68n + 0.10$$

where $C(n)$ is the CO concentration at Old Faithful in parts per million (ppm) when n hundred snowmobiles are visiting the park.

a. Find a new function that will give the CO concentration at Old Faithful depending on the number of hours after midnight.

b. Find the CO concentration at Old Faithful at 3:00 p.m.

EXAMPLE 4 COMPOSING FUNCTIONS ALGEBRAICALLY

Combine the following functions.

$$f(x) = 3x + 9 \qquad g(x) = 2x - 7 \qquad h(x) = 3x^2 + 5x - 10$$

a. $f(g(x))$ 　　 b. $g(f(x))$ 　　 c. $h(g(x))$

Solution

a. We will take the entire function $g(x)$ and substitute it into the input variable for $f(x)$.

$$f(x) = 3x + 9$$
$$f(g(x)) = 3(2x - 7) + 9$$
$$f(g(x)) = 6x - 21 + 9$$
$$f(g(x)) = 6x - 12$$

Say What?

Composing two functions in the opposite order does not always result in the same new function.

$$f(g(x))$$

and

$$g(f(x))$$

Note that th answers to Example 4a and 4b are not the same.

b. Now we substitute $f(x)$ into the input variable for the function $g(x)$.

$$g(x) = 2x - 7$$
$$g(f(x)) = 2(3x + 9) - 7$$
$$g(f(x)) = 6x + 18 - 7$$
$$g(f(x)) = 6x + 11$$

c. We will substitute $g(x)$ into the input variable for the function $h(x)$. Be sure to substitute $g(x)$ into every x of the $h(x)$ function.

$$h(x) = 3x^2 + 5x - 10$$
$$h(g(x)) = 3(2x - 7)^2 + 5(2x - 7) - 10 \qquad \text{Use foil to square the } 2x\text{-}7.$$
$$h(g(x)) = 3(2x - 7)(2x - 7) + 5(2x - 7) - 10$$
$$h(g(x)) = 3(4x^2 - 14x - 14x + 49) + 10x - 35 - 10$$
$$h(g(x)) = 3(4x^2 - 28x + 49) + 10x - 45 \qquad \text{Distribute the 3.}$$
$$h(g(x)) = 12x^2 - 84x + 147 + 10x - 45$$
$$h(g(x)) = 10x^2 - 74x + 102 \qquad \text{Combine like terms.}$$

As you can see when the two functions in parts a and b are composed in different orders, they do not end with the same function. This implies that you should use caution when deciding which function goes into which.

EXAMPLE 4 PRACTICE PROBLEM

Combine the following functions.

$$f(x) = 4x + 5 \qquad g(x) = 5x - 8 \qquad h(x) = 2x^2 + 4x - 1$$

a. $f(g(x))$ 　　 b. $h(g(x))$

3.3   Exercises

1. Far North Manufacturing is starting to manufacture a new line of toys for this Christmas season. The number of toys they can manufacture each week can be modeled by

$$T(w) = 500w + 3000$$

where $T(w)$ represents the number of toys manufactured during week w of production.

   The total weekly cost for manufacturing these toys can be modeled by

$$C(t) = 1.75t + 5000$$

where $C(t)$ represents the total weekly cost in dollars from producing t toys a week.

   a. Use these functions to find a new function that will give the total weekly cost for week w of production.
   b. Find the number of toys produced during the fifth week of production.
   c. Find the total weekly cost to produce 5000 toys.
   d. Find the total weekly cost for the seventh week of production.
   e. In what week will the total weekly cost reach $18,500.00?

2. The number of custom computer systems sold to the biotech industry by a new computer manufacturer m months after starting the business can be modeled by
$$C(m) = 4m + 2$$
where $C(m)$ represents the number of computer systems sold during the month.

   The monthly profit made by the new computer manufacturer can be modeled by
$$P(c) = 300c - 6000$$
where $P(c)$ represents the monthly profit in dollars when c computers are sold during the month.

   a. Use these functions to find a new function that will give the monthly profit for m months after starting the business.
   b. Find the number of computer systems sold during the fifth month.

   c. Find the total monthly profit if 20 computer systems are sold in a month.
   d. Find the total monthly profit for the tenth month.
   e. In what month will the monthly profit reach $7800.00?

3. West Tech, Inc.'s vacation policy increases each employee's vacation time on the basis of the number of years he or she has been employed at the company. The number of weeks of vacation an employee gets in a year can be modeled by

$$v(y) = 0.25y + 1.5$$

weeks of vacation after working for the company for y years.

   West Tech, Inc.'s cost of an employee's vacation time in a year can be modeled by

$$C(w) = 1500w + 575$$

dollars when w weeks of vacation are taken.

   a. Find the number of weeks of vacation a West Tech employee will get per year after working with the company for 10 years.
   b. What is West Tech's cost for a 10-year employee's vacation?
   c. Use these functions to find a new function that determines the cost for vacation taken by an employee who has been with the company for y years. Assume that the employee will take all of his or her vacation time.
   d. What is West Tech's cost for a 20-year employee's vacation?
   e. What is West Tech's cost for a 30-year employee's vacation?

4. The number of employees working at a small magazine publisher can be modeled by
$$E(m) = 3m + 2$$
where $E(m)$ represents the number of employees m months after starting to publish the magazine.

   The total cost for health benefits at the publisher can be modeled by
$$C(e) = 1500e + 2000$$

where $C(e)$ represents the total monthly cost for health benefits in dollars when e employees work for the publisher.

a. Find the number of employees at the publisher during the sixth month.

b. Find the total cost for health benefits when the publisher has 10 employees.

c. Use these functions to find a new function that determines the total monthly cost for health benefits for the publisher m months after starting to publish the magazine.

d. What is total monthly cost of health benefits eight months after starting to publish the magazine?

e. Find the total cost for health benefits during the first year of publishing the magazine. (Hint: you will add the monthly cost for each month of the year.)

5. The number of people who visit a local Renaissance fair can be modeled by
$$A(d) = -400d^2 + 3500d - 1000$$
where $A(d)$ represents the attendance at the Renaissance fair d days after it opens.

The profit made by the fair promoters can be modeled by
$$P(a) = 2a - 2500$$
where $P(a)$ represents the profit in dollars for the Renaissance fair on a day when a people attend.

a. Find the number of people in attendance on the 4th day of the Renaissance fair.

b. Find the profit made on the fourth day of the fair.

c. Use these functions to find a new function that will give the profit made at the Renaissance fair d days after the fair opens.

d. How much profit does the fair make on the third day of the fair.

e. If the fair lasts a total of seven days find the total amount of profit for the Renaissance fair. (hint: you will need to add up the profit from each day of the fair.)

6. The depth of the water flowing in a local stream w weeks after the start of spring can be modeled by
$$D(w) = -0.75w^2 + 10w + 2$$
where $D(w)$ represents the depths in inches of the stream.

The amount of water per minute going over a waterfall at the end of the same stream can be modeled by
$$G(d) = 25d + 1$$
where $G(d)$ represents the amount of water going over the waterfall in gallons per minute when the steam is d inches deep.

a. Find the depth of the stream 2 weeks after the start of spring.

b. Find the amount of water going over the waterfall 2 weeks after the start of spring.

c. Use these functions to find a new function that will give the amount of water going over the waterfall w weeks after the start of spring.

d. Find the amount of water going over the waterfall 8 weeks after the start of spring.

For Exercises 7 through 10 use function notation to show how you would compose the two functions to get the requested new function. List the units for the inputs and outputs of the new function.

7. Let $P(b)$ be the profit, in thousands of dollars, that Pride Bike Co. makes if they sell b bikes per year. Let $B(t)$ be the number of bikes that Pride Bike Co. sells in year t. Find a function for the amount of profit Pride Bike Co. makes in year t.

8. Let $D(v)$ be the amount of environmental damage, in thousands of dollars, done to a national park when v visitors come to the park in a year. Let $V(t)$ be the number of visitors who come to a national park in year t. Find a function that gives the amount of environmental damage at a national park in year t.

9. Let $T(d)$ be the low night-time temperature in degrees Fahrenheit on day d of the week. Let $H(t)$ be the number of homeless who seek space in a local homeless shelter when the low night-time temperature is t degrees Fahrenheit. Find a function for the number of homeless who seek space at the homeless shelter on day d of the week.

10. Let $B(m)$ be the advertising budget for a new product m months after the product launch. Let $R(a)$ be the monthly revenue from selling this product when the advertising budget for the month is a dollars. Find a function that gives the monthly revenue from selling this product m months after the product launch.

For Exercises 11 through 15, assume that the following functions refer to a specific tour company offering European tours and that t represents the year.

- Let $T(t)$ be the total number of people on the tour.

- Let $K(t)$ be the number of children under 12 years old on the tour.

- Let $C(k)$ be the cost for k children under 12 years old to take the tour.

- Let $A(a)$ be the cost for a people over 12 years old to take the tour.

- Let $B(n)$ be the number of buses needed when n people travel on the tour.

Use function notation to show how you would combine the given functions by adding, subtracting, multiplying, dividing, or composing. List the units for the inputs and outputs of the new function.

11. Find a function that gives the number of people over 12 years old who go on the tour in year t.

12. Find a function that gives the cost for all children under 12 years old who traveled on the tour in year t.

13. Find a function that gives the cost for all people over 12 years old who traveled on the tour in year t.

14. Find a function that gives the total cost for all people who traveled on the tour in year t.

15. Find a function that gives the number of busses needed for the tour in year t.

16. Let $f(x) = 2x + 4$ and $g(x) = 6x + 9$.
    a. Find $f(g(x))$.
    b. Find $g(f(x))$.

17. Let $f(x) = 5x - 6$ and $g(x) = 7x + 3$.
    a. Find $f(g(x))$.
    b. Find $g(f(x))$.

18. Let $f(x) = 3x + 8$ and $g(x) = -6x + 9$.
    a. Find $f(g(x))$.
    b. Find $g(f(x))$.

19. Let $f(x) = 19x + 28$ and $g(x) = -17x + 34$.
    a. Find $f(g(x))$.
    b. Find $g(f(x))$.

20. Let $f(x) = \frac{1}{2}x + \frac{1}{5}$ and $g(x) = 2x - \frac{3}{5}$.
    a. Find $f(g(x))$.
    b. Find $g(f(x))$.

21. Let $f(x) = \frac{2}{3}x + \frac{1}{3}$ and $g(x) = 5x - 3$.
    a. Find $f(g(x))$.
    b. Find $g(f(x))$.

22. Let $f(x) = 0.68x + 2.36$ and $g(x) = 3.57x + 6.49$.
    a. Find $f(g(x))$.
    b. Find $g(f(x))$.

23. Let $f(x) = -0.6x - 3.2$ and $g(x) = -7x - 4.8$.
    a. Find $f(g(x))$.
    b. Find $g(f(x))$.

24. Let $f(x) = 3x + 5$ and $g(x) = x^2 + 4x + 10$.
    a. Find $f(g(x))$.
    b. Find $g(f(x))$.

25. Let $f(x) = 4x - 7$ and $g(x) = x^2 - 3x + 12$.
    a. Find $f(g(x))$.
    b. Find $g(f(x))$.

26. Let $f(x) = x + 2$ and $g(x) = 4x^2 + x + 1$.
    a. Find $f(g(x))$.
    b. Find $g(f(x))$.

27. Let $f(x) = 5x - 2$ and $g(x) = 2x^2 - 7x + 6$.

    a. Find $f(g(x))$.

    b. Find $g(f(x))$.

28. Let $f(x) = 3x + 8$ and $g(x) = -6x + 9$.

    a. Find $f(g(5))$.

    b. Find $g(f(5))$.

29. Let $f(x) = 19x + 28$ and $g(x) = -17x + 34$.

    a. Find $f(g(2))$.

    b. Find $g(f(2))$.

30. Let $f(x) = \dfrac{1}{2}x + \dfrac{1}{5}$ and $g(x) = 2x - \dfrac{3}{5}$.

    c. Find $f(g(2))$.

    d. Find $g(f(2))$.

31. Let $f(x) = \dfrac{2}{3}x + \dfrac{1}{3}$ and $g(x) = 5x - 3$.

    a. Find $f(g(4))$.

    b. Find $g(f(4))$.

32. Let $f(x) = 0.68x + 2.36$ and $g(x) = 3.57x + 6.49$.

    a. Find $f(g(-3))$.

    b. Find $g(f(-3))$.

33. Let $f(x) = -0.6x - 3.2$ and $g(x) = -7x - 4.8$.

    a. Find $f(g(7))$.

    b. Find $g(f(7))$.

34. Let $f(x) = 3x + 5$ and $g(x) = x^2 + 4x + 10$.

    a. Find $f(g(5))$.

    b. Find $g(f(5))$.

35. Let $f(x) = 4x - 7$ and $g(x) = x^2 - 3x + 12$.

    a. Find $f(g(3))$.

    b. Find $g(f(3))$.

36. Let $f(x) = x + 2$ and $g(x) = 4x^2 + x + 1$.

    a. Find $f(g(2))$.

    b. Find $g(f(2))$.

37. Let $f(x) = 5x - 2$ and $g(x) = 2x^2 - 7x + 6$.

    a. Find $f(g(4))$.

    b. Find $g(f(4))$.

Factoring Polynomials

FACTORING USING THE AC METHOD

In section 3.2, we multiplied two polynomials using the distributive property.

$$a(x + y) = ax + ay$$

When working with polynomials, we sometimes want to "undo" the distributive property by changing a polynomial into a product of two or more factors. The process of changing a polynomial from terms that are added together to factors that are multiplied together is called **factoring**. Factoring is one method that is used to solve some polynomial equations and has several other uses in algebra. The basis of factoring is to take a polynomial and rewrite it into simpler pieces that are multiplied together.

Standard Form	Factored Form
$f(x) = x^2 + 8x + 15$	$f(x) = (x + 3)(x + 5)$

$$(x + 3)(x + 5)$$
$$x^2 + 5x + 3x + 15$$
$$x^2 + 8x + 15$$

Multiply out the factored form and we get the standard form.

The most basic step in any factoring process is to factor out the common elements from the terms of an expression. When we factor out a common element, it does not go away but becomes a factor that is in front of the remaining expression. The common elements of an expression are called a common factor. The terms of the expression $12x + 10$ both are divisible by 2, so a 2 can be factored out of the expression and be put in front as a factor.

$$12x + 10$$
$$2(6x + 5)$$

If we multiply the 2 back into the expression $6x + 5$, we would be right back to the original expression $12x + 10$. The terms in some expressions will have constants, variables, or both variables and constants in common.

EXAMPLE   1   FACTORING OUT THE GREATEST COMMON FACTOR

Factor out the greatest common factor.

a. $5x^2 + 20x$ **b.** $15w^3z - 9w^2z^2$ **c.** $2x^2 + 6x - 4$

d. $5x(x - 8) + 2(x - 8)$ **e.** $3z(z - 5) - (z - 5)$

Solution

a. Both terms in this expression have at least one x and also are divisible by 5, so we can factor out $5x$ from both terms.

$$5x(x + 4)$$

The first term of this expression had x^2, but one is factored out, leaving a single x.

Say What?

Factor or factor.

The word factor is used in a couple of ways in mathematics.

When multiplying, the expressions being multiplied together are called factors.

$$24 = (4)(6)$$

4 and 6 are called factors of 24.

When factoring, the act of taking out a common element is sometimes called "to factor out".

In the phrase:

"Factor out the common factor"

The first word factor is an action saying to take out. The second word factor is referring to the expression that is going to be taken out.

$$10x^3 + 4x$$

$$2x(5x^2 + 2)$$

$2x$ is the common factor that is being factored out of the original experssion.

b. Both of these terms have a w^2, have a z, and are divisible by 3. Factoring $3w^2z$ out, we have

$$3w^2z(5w - 3z)$$

c. All of the terms in this polynomial are divisible by 2. Because the last term does not have an x in it, x is not a common factor,

$$2(x^2 + 3x - 2)$$

d. Both terms here have the expression $(x - 8)$.

$$(x - 8)(5x + 2)$$

e. Both terms in this problem have the expression $(z - 5)$. You may suspect that nothing would remain after taking $(z - 5)$ out of the second term. This is not true. A 1 will remain in its place, because $(z - 5)$ divided by $(z - 5)$ is 1.

$$(z - 5)(3z - 1)$$

EXAMPLE ① PRACTICE PROBLEM

Factor out the greatest common factor.

a. $12t^3 - 20t$

b. $6x^2 + 8x - 14$

c. $5x(x + 4) - 3(x + 4)$

Factoring out what is in common will always be the first step in any factoring process. In many cases, there will not be anything in common, and we will move on to other steps.

One of the most common polynomial functions that we will work with is called quadratics. Quadratic functions are second-degree polynomials of the form

$$f(x) = ax^2 + bx + c$$

where a, b, and c are real numbers and $a \neq 0$. In chapter 4, we will see how factoring a quadratic in this form will allow us to solve quadratic equations. Many methods are used to factor quadratics, but we will concentrate on one called the AC method. In working with a quadratic in standard form, this method will provide basic steps that will guide you through the factoring process.

AC METHOD OF FACTORING QUADRATICS

The standard form of a quadratic is $ax^2 + bx + c$.

1. Factor out the greatest common factor.
2. Multiply a and c together. (Do this step off to the side, in the margin.)
3. Find factors of ac that add up to b. (Do this off to the side also.)
4. Rewrite the middle (bx) term using the factors from step 3.
5. Group and factor out what is in common (the greatest common factor).

EXAMPLE   **2**   FACTOR USING THE AC METHOD

Factor the following.

a. $x^2 + 8x + 15$ **b.** $6x^2 + x - 35$ **c.** $4x^2 + 7xy + 3y^2$

Solution

a.   This quadratic is in standard form $ax^2 + bx + c$, so we will use the AC method.

Step 1   Factor out the greatest common factor.

The terms in this quadratic have no common factors.

Step 2   Multiply a and c together. (Do this step off to the side, in the margin.)

$$a = 1 \quad \text{and} \quad c = 15 \quad \text{so} \quad ac = 1(15) = 15$$

Step 3   Find factors of ac that sum to b. (Do this off to the side also.)

List both the positive and negative factors of 15.

$ac = 1(15) = 15$		$b = 8$	
1	15	$1 + 15 = 16$	Not equal to b.
3	5	$3 + 5 = 8$	Equal to b.
-1	-15	$-1 + (-15) = -16$	Not equal to b.
-3	-5	$-3 + (-5) = -8$	Not equal to b.

In this list, the factors 3 and 5 will add up to 8.

Step 4   Rewrite the middle (bx) term, using the factors from step 3.

The factors 3 and 5, found in the previous step, will be used to rewrite the bx term in the expression.

$$x^2 + 3x + 5x + 15$$

Step 5   Group and factor out what's in common.

If we group the first two terms together and the last two terms together, we can factor out some common elements.

$(x^2 + 3x) + (5x + 15)$	Group first and last two terms together.
$x(x + 3) + 5(x + 3)$	Factor out x from the first group and 5 from the second group.
$(x + 3)(x + 5)$	Factor out the $(x + 3)$.

To check the factorization, we can multiply out the factored form using the distributive property.

$$(x + 3)(x + 5)$$
$$x^2 + 5x + 3x + 15$$
$$x^2 + 8x + 15$$

We got the original expression back, so the factored form we found is correct.

b.   This quadratic is in standard form $ax^2 + bx + c$, so we will use the AC method.

Step 1   Take out anything in common.

The terms in this quadratic have nothing in common.

Step 2 Multiply a and c together. (Do this step off to the side, in the margin.)

$$a = 6 \quad \text{and} \quad c = -35 \quad \text{so} \quad ac = 6(-35) = -210$$

Step 3 Find factors of ac that sum to b. (Do this off to the side also.)

List the factors of -210.

	-210			
1	-210	-1	210	The product must be negative,
2	-105	-2	105	so one factor must be positive
3	-70	-3	70	and the other must be negative.
5	-42	-5	42	Be sure to give both lists of
6	-35	-6	35	positive and negative factors.
7	-30	-7	30	
10	-21	-10	21	
14	-15	-14	15	

In this list, the factors -14 and 15 will add up to 1.

Step 4 Rewrite the middle (bx) term, using the factors from step 3.

The factors -14 and 15 will replace the b value in the expression.

$$6x^2 - 14x + 15x - 35$$

Step 5 Group and factor out what is in common.

If we group the first two terms together and the last two terms together, we can factor out some common elements.

$$(6x^2 - 14x) + (15x - 35)$$
$$2x(3x - 7) + 5(3x - 7)$$
$$(3x - 7)(2x + 5)$$

We will check this factorization by multiplying the factored form out using the distributive property.

$$(3x - 7)(2x + 5)$$
$$6x^2 + 15x - 14x - 35$$
$$6x^2 + x - 35$$

This is the same expression that we started with, so the factored form is correct.

c. This polynomial has two variables, but each variable has a squared term and a first-degree term. We will factor it using the AC method focusing on the x variables and the y variables will follow along the process.

Step 1 Take out anything in common.

The terms in this polynomial have nothing in common.

Step 2 Multiply a and c together. (Do this step off to the side, in the margin.)

$$a = 4 \quad \text{and} \quad c = 3 \quad \text{so} \quad ac = 4(3) = 12$$

Step 3 Find factors of *ac* that sum to *b*. (Do this off to the side also.)

List the factors of 12.

12			
1	12	−1	−12
2	6	−2	−6
3	4	−3	−4

Group the first and last two terms together. Factor out $2x$ from the first group and 5 from the second group. Factor out the $(3x - 7)$.

In this list, the factors 3 and 4 will add up to 7.

Step 4 Rewrite the middle (*bx*) term, using the factors from step 3.

The factors 3 and 4 will replace the *b* value in the expression.

$$4x^2 + 4xy + 3xy + 3y^2$$

Step 5 Group and factor out what is in common.

If we group the first two terms together and the last two terms together, we can factor out some common factors.

$$(4x^2 + 4xy) + (3xy + 3y^2)$$
$$4x(x + y) + 3y(x + y)$$
$$(x + y)(4x + 3y)$$

Group the first two terms and the last two terms. Factor out $4x$ from the first group and $3y$ from the second group. Factor out the $(x + y)$.

We will check this factorization by multiplying the factored form out using the distributive property.

$$(x + y)(4x + 3y)$$
$$4x^2 + 3xy + 4xy + 3y^2$$
$$4x^2 + 7xy + 3y^2$$

This is the same expression that we started with, so the factored form is correct.

EXAMPLE ② PRACTICE PROBLEM

Factor

a. $x^2 - 2x - 48$ **b.** $10x^2 + 23x + 12$ **c.** $2x^2 + 11xy + 5y^2$

Some polynomials that are not quadratics may have something in common in every term of the expression. When this happens, you must remember to do the first step of the AC method and factor out the greatest common factor. After you have taken the common factor out, do not forget that it will be part of the factored form all the way to the end of the process.

EXAMPLE 3 FACTOR USING THE AC METHOD WITH A COMMON FACTOR

Factor the following.

a. $40x^2 + 46x + 12$ **b.** $3x^2y + 5xy - 28y$

Solution

a. This quadratic is in standard form $ax^2 + bx + c$, so we will use the AC method.

> **Step 1** Take out anything in common.
> The terms in this quadratic are all divisible by 2, so we will factor out a 2.
>
> $$2(20x^2 + 23x + 6)$$
>
> **Step 2** Multiply a and c together. (Do this step off to the side, in the margin.)
>
> $$a = 20 \quad \text{and} \quad c = 6 \quad \text{so} \quad ac = 20(6) = 120$$
>
> **Step 3** Find factors of ac that sum to b. (Do this off to the side also.)
> List the factors of 120.

120		120		
1	120	−1	−120	
2	60	−2	−60	**List both the positive**
3	40	−3	−40	**and negative factors.**
4	30	−4	−30	
5	24	−5	−24	
6	20	−6	−20	
8	15	−8	−15	
10	12	−10	−12	

> In this list, the factors 8 and 15 will add up to 23.
>
> **Step 4** Rewrite the middle (bx) term, using the factors from step 3.
> The factors 8 and 15, found in the previous step, will be used to rewrite the bx term in the expression.
>
> $$2(20x^2 + 8x + 15x + 6)$$
>
> **Step 5** Group and factor out what's in common.
> If we group the first two terms together and the last two terms together, we can factor out some common elements.

$$2[(20x^2 + 8x) + (15x + 6)]$$
$$2[4x(5x + 2) + 3(5x + 2)]$$
$$2(5x + 2)(4x + 3)$$

Group first and last two terms together.
Factor out $4x$ from the first group and 3 from the second group.
Finally factor out the $(5x + 2)$.
Do not forget the 2 that you factored out in step 1.

To check the factorization, we can multiply out the factored form using the distributive property.

$$2(5x + 2)(4x + 3)$$
$$2(20x^2 + 15x + 8x + 6)$$
$$2(20x^2 + 23x + 6)$$
$$40x^2 + 46x + 12$$

We got the original expression back, so the factored form we found is correct.

b. This expression is not a quadratic in standard form but has an extra variable in every term. Once we factor the common variable out, we will have a quadratic in standard form and will use the AC method to complete the factorization.

Skill Connection

When factoring out the greatest common factor in step 1 of the AC method, you must remember to keep the factor in the front of the expression at all times. One way to do this is to write it with brackets for several steps and then complete the factoring process inside those brackets.

$$x^2y + 2xy - 3y$$
$$y(x^2 + 2x - 3)$$
$$y[\qquad\qquad]$$
$$y[\qquad\qquad]$$
$$y[\qquad\qquad]$$
$$y(\quad)(\quad)$$

By writing the common factor in front of each bracket at the beginning, you will not forget it while you concentrate on the other steps of the AC method.

$$3x^2y + 5xy - 28y \qquad \text{Step 1 factor out the } y.$$
$$y(3x^2 + 5x - 28)$$

Step 2 and 3.

$$3(-28) = -84$$

-1	84	1	-84
-2	42	2	-42
-3	28	3	-28
-4	21	4	-21
-6	14	6	-14
-7	12	7	-12

$$y(3x^2 - 7x + 12x - 28) \qquad \text{Step 4.}$$

$$y[(3x^3 - 7x) + (12x - 28)] \qquad \text{Step 5.}$$
$$y[x(3x - 7) + 4(3x - 7)]$$
$$y(3x - 7)(x + 4)$$

We check the factorization using the distributive property.

$$y(3x - 7)(x + 4)$$
$$y(3x^2 + 12x - 7x - 28)$$
$$y(3x^2 + 5x - 28)$$
$$3x^2y + 5xy - 28y$$

The multiplication resulted in the original expression, so we have the correct factored form.

EXAMPLE ③ PRACTICE PROBLEM

Factor

a. $6x^2 + 15x - 36$ **b.** $16x^3 + 52x^2 + 12x$

When working with the AC method, we may ask if it matters in which order we rewrite the middle (bx) term during step 4. It does not make a difference mathematically, but many students find it best to put any negative values first so that the negative signs do not cause confusion. This is not always possible, since we may find that two negative numbers are needed to add up to b in step 3.

When this happens, we will have to work with the negative more carefully. One way to deal with this is to always put an addition between the groups and keep any negative signs with the number inside the parentheses.

$$x^2 - 5x + 6$$
$$x^2 - 3x - 2x + 6$$
$$(x^2 - 3x) + (-2x + 6)$$

Keeping the negative with the number will then cause us to have to factor out a negative number from the second parentheses so that the two remaining factors match.

$$(x^2 - 3x) + (-2x + 6)$$
$$x(x - 3) - 2(x - 3)$$

Now, we can see that the two pieces both have $(x - 3)$ in common, and it can be factored out to complete the factorization process.

$$x(x - 3) - 2(x - 3)$$
$$(x - 3)(x - 2)$$

Remember that you must use caution when working with negatives or your factorization may not multiply back together to give you the correct original expression.

EXAMPLE **4** FACTOR USING THE AC METHOD WITH 2 NEGATIVES

Factor the following.

a. $6x^2 - 13x + 5$ **b.** $10x^3 - 28x^2 + 16x$

Solution

a. Use the AC method. There is no common factor in every term so we continue on step 2.

$$6x^2 - 13x + 5$$

$$6x^2 - 3x - 10x + 5 \qquad \text{Step 4.}$$

$$(6x^2 - 3x) + (-10x + 5) \qquad \text{Step 5.}$$

$$3x(2x - 1) - 5(2x - 1)$$

$$(2x - 1)(3x - 5)$$

Steps 2 and 3.			
	$6(5) = 30$		
1	30	−1	−30
2	15	−2	−15
3	10	−3	−10
5	6	−5	−6

We check the factorization using the distributive property.

$$(2x - 1)(3x - 5)$$
$$6x^2 - 10x - 3x + 5$$
$$6x^2 - 13x + 5$$

When multiplied out, the factored form is the same as the original expression, so we have factored correctly.

b. All the terms in this expression are divisible by 2 and have an x in common, so we will factor $2x$ out and continue with the AC method.

$$10x^3 - 28x^2 + 16x \qquad \text{Step 1.}$$

$$2x(5x^2 - 14x + 8)$$

$$2x(5x^2 - 10x - 4x + 8) \qquad \text{Step 4.}$$

$$2x[(5x^2 - 10x) + (-4x + 8)] \qquad \text{Step 5.}$$

$$2x[5x(x - 2) - 4(x - 2)]$$

$$2x(x - 2)(5x - 4)$$

Steps 2 and 3.			
	$5(8) = 40$		
1	40	−1	−40
2	20	−2	−20
4	10	−4	−10
5	8	−5	−8

We check this factorization using the distributive property.

$$2x(x-2)(5x-4)$$
$$2x(5x^2 - 4x - 10x + 8)$$
$$2x(5x^2 - 14x + 8)$$
$$10x^3 - 28x^2 + 16x$$

The factored form multiplied out results in the original expression, so we have the correct factored form.

EXAMPLE ④ PRACTICE PROBLEM

Factor

a. $10x^2 - 43x + 28$ **b.** $90x^3 - 75x^2 + 10x$

FACTORING USING TRIAL AND ERROR

Another method used to factor quadratics and other polynomials is trial and error. This method uses our knowledge of multiplying and adding numbers to help us guess the factors of the quadratic. The first step of this method will also be to take out any common factors. After factoring out anything common, we will use factors of a and c to try and find the factored form of the expression. If $a = 1$ in the standard form we will only have to consider the factors of c.

$$a = 1$$
$$x^2 + 5x + 6$$
$$(x + \text{factor of } c)(x + \text{factor of } c)$$
$$(x + 2)(x + 3)$$

If a does not equal 1, we will have to consider both the factors of a and c.

$$a \neq 1$$
$$10x^2 + 33x + 20$$
$$(\text{factor of } a \cdot x + \text{factor of } c)(\text{factor of } a \cdot x + \text{factor of } c)$$
$$(2x + 5)(5x + 4)$$

The signs of the factors are another aspect of the factored form that we will have to be cautious about. If both of the terms in the original polynomial expression have plus signs, we will have two plus signs in the factored form. If the polynomial has one or more negative signs, we will have at least one negative sign in the factored form.

$x^2 + 9x + 20$	$x^2 + 2x - 8$	$x^2 - 3x + 2$	$x^2 - 4x - 5$
$(x + 5)(x + 4)$	$(x + 4)(x - 2)$	$(x - 2)(x - 1)$	$(x - 5)(x + 1)$

It is best if you do not attempt to memorize every situation. Learn to try something, check it, and try again if it does not work. Thus, the name trial and error.

EXAMPLE ⑤ FACTOR USING TRIAL AND ERROR

Factor the following.

a. $x^2 + 7x + 6$ **b.** $x^2 - 4x - 12$

c. $2x^2 + 9x + 4$ **d.** $24x^2 - 2x - 15$

Solution

a. Because a is 1, we only need to consider the factors of c. The factors of 6 are 1 and 6 or 2 and 3. Because 1 and 6 add up to 7, the factorization is

$$x^2 + 7x + 6$$
$$(x + 6)(x + 1)$$

We check this by multiplying the factored form out.

$$(x + 6)(x + 1)$$
$$x^2 + x + 6x + 6$$
$$x^2 + 7x + 6$$

b. Again, a is 1 so we consider the factors of c. The factors of 12 are 1 and 12, 2 and 6, or 3 and 4. Because we have negatives in the original expression, we will have one or more negatives in the factored form. The factors 2 and 6 can make - 4 if the 2 is positive and the 6 is negative. This gives us the factored form.

$$x^2 - 4x - 12$$
$$(x - 6)(x + 2)$$

We check this factored form by multiplying it out.

$$(x - 6)(x + 2)$$
$$x^2 + 2x - 6x - 12$$
$$x^2 - 4x - 12$$

c. Because a is not 1, we need to consider both the factors of a and c. The factors of 2 are 1 and 2, and the factors of 4 are 1 and 4 or 2 and 2. Using these, we find the factored form.

$$2x^2 + 9x + 4$$
$$(2x + 1)(x + 4)$$

We check this factored form by multiplying it out.

$$(2x + 1)(x + 4)$$
$$2x^2 + 8x + x + 4$$
$$2x^2 + 9x + 4$$

d. Again, a is not 1, so we need to consider the factors of a and c. The factors of 24 are 1 and 24, 2 and 12, 3 and 8, and 4 and 6. The factors of 15 are 1 and 15 or 3 and 5. The negatives in the original expression mean that we will have one or more negatives in the factored form.

$$24x^2 - 2x - 15$$

Try 1	Try 2
$(2x + 5)(12x - 3)$	$(3x + 5)(8x - 3)$
$24x^2 - 6x + 60x - 15$	$24x^2 - 9x + 40x - 15$
$24x^2 + 54x - 15$	$24x^2 + 31x - 15$
Wrong middle term.	**Wrong middle term.**

Try 3	Try 4
$(4x + 1)(6x - 5)$	$(4x + 3)(6x - 5)$
$24x^2 - 20x + 6x - 15$	$24x^2 - 20x + 18x - 15$
$24x^2 - 14x - 15$	$24x^2 - 2x - 15$
Wrong middle term.	This one works!

EXAMPLE ⑤ PRACTICE PROBLEM

Factor

a. $x^2 + 7x + 12$ **b.** $x^2 - 11x + 28$ **c.** $15x^2 - 14x - 8$

PRIME POLYNOMIALS

Say What?

Prime numbers
or
Prime polynomials?

Prime numbers and prime polynomials are similar in that neither can be broken up into smaller factors.

The number 13 is prime since the only factors of 13 are 1 and 13.

The polynomial
$$x^2 + 5x + 2$$
is prime because it is not factorable over the rationals.

Some polynomials will not be able to be factored using rational numbers. When a polynomial is not factorable it is called prime. When using the AC method, we know a polynomial is not factorable if in step 3 there is no combination of factors for ac that add up to b. If this happens, we will stop and say that the polynomial is not factorable.

When factoring by trial and error, we will need to try every combination of factors of a and c to see if any of the combinations work. If no combination of factors and positive and negative signs work, we will say the polynomial is not factorable over the rationals.

$6x^2 + 10x + 5$ $6(5) = 30$

1	30	−1	−30
2	15	−2	−15
3	10	−3	−10
5	6	−5	−6

None of the factors of ac add up to b, so this quadratic is not factorable over the rationals.

3.4 Exercises

In Exercises 1 through 16, factor out the greatest common factor.

1. $6x + 8$

2. $12x + 9$

3. $4h^2 + 7h$

4. $5r^2 - 3r$

5. $25x^3 + x^2 - 2x$

6. $10x^3 + 3x^2 - 7x$

7. $4a^2b + 6ab$

8. $18m^3n^2 - 15mn$

9. $4x^2 + 6x - 2$

10. $12x^2 - 9x + 3$

11. $15x^2yz + 6xyz^2 - 3xy^2z$

12. $24a^3b^2c - 18a^2b^2c^2 + 30a^2bc$

13. $3x(x + 5) + 2(x + 5)$

14. $7x(2x - 3) - 5(2x - 3)$

15. $7(5w + 4) - 2w(5x + 4)$

16. $9(6x - 7) - 2x(6x - 7)$

In Exercises 17 through 62, factor using the AC method or Trial and Error. Write prime if a polynomial is not factorable using the rational numbers.

17. $x^2 - 4x - 21$

18. $x^2 - 3x - 10$

19. $x^2 + 6x + 5$

20. $x^2 + 9x + 20$

21. $w^2 - 7w - 18$

22. $k^2 - 4k - 32$

23. $t^2 - 11t + 28$

24. $x^2 - 11x + 30$

25. $2x^2 + 13x + 15$

26. $3x^2 + 19x + 28$

27. $7m^2 - 25m + 12$

28. $4t^2 - 43t + 30$

29. $4x^2 - 31x - 45$

30. $8x^2 - 93x - 36$

31. $x^2 + 5x + 3$

32. $p^2 + 2p + 5$

33. $2x^2 + 16x + 2$

34. $10p^2 - 45p - 40$

35. $6w^2 + 29w + 28$

36. $20k^2 + 43k + 14$

37. $10t^2 - 41t - 18$

38. $77m^2 - 23n - 12$

39. $6x^2 + 2x + 5$

40. $8x^2 + 3x + 4$

41. $12p^2 - 4p - 9$

42. $16w^2 - 40w - 13$

43. $12x^2 - 43x + 56$

44. $20t^2 - 18t + 15$

45. $2m^2 + 14m + 24$

46. $3k^2 + 27k + 42$

47. $6x^2 - 66x + 168$

48. $10w^2 - 80w + 150$

49. $40x^2 + 30x - 45$

50. $24p^2 - 52p - 20$

51. $x^3 - 4x^2 - 21x$

52. $w^3 - 7w^2 + 10w$

53. $10x^3 + 35x^2 + 30x$

54. $6p^3 + 8p^2 + 2p$

55. $x^2y + 13xy + 36y$

56. $3x^2y - 15xy - 42y$

57. $2x^2 + 11xy + 15y^2$

58. $4a^2 + 13ab + 3b^2$

59. $6a^2 - 13ab - 5b^2$

60. $5m^2 - 22mn + 8n^2$

61. $-6mn^2 - 20mn - 16m$

62. $-6ab^2 + 75ab - 225$

Special Factoring Techniques

In this section, we will look at several other techniques of factoring, along with some special situations where you can use a pattern to factor. The key to most of these techniques is to recognize the form of the polynomial and follow the pattern to factor it. Many of these problems can be done by using the factoring techniques found in Section 3.4 but are often solved much more quickly by using these patterns.

PERFECT SQUARE TRINOMIALS

A perfect square trinomial comes as a result of squaring a binomial. Using the distributive property and combining like terms, we get the following.

$$(a + b)^2 = (a + b)(a + b) = a^2 + ab + ab + b^2 = a^2 + 2ab + b^2$$

$$(a - b)^2 = (a - b)(a - b) = a^2 - ab - ab + b^2 = a^2 - 2ab + b^2$$

If we look at the resulting trinomials, we can see a pattern for a perfect square trinomial. The first and last terms of the trinomial will be perfect squares, and the middle term will be either plus or minus twice the product of the first and last terms that are being squared. For factoring purposes, the relationship is better seen as follows.

$$a^2 + 2ab + b^2 = (a + b)^2$$

$$a^2 - 2ab + b^2 = (a - b)^2$$

EXAMPLE **1** FACTOR PERFECT SQUARE TRINOMIALS

Factor the following:

a. $x^2 + 6x + 9$ **b.** $m^2 - 14m + 49$

c. $4a^2 + 20a + 25$ **d.** $9x^2 - 24xy + 16y^2$

Solution

a. First, we must confirm that this is a perfect square trinomial. The first term is x squared, the last term is 3 squared, and the middle term is twice the product of these two terms ($2 \cdot 3 \cdot x$). So this trinomial can be factored into a binomial squared.

$$x^2 + 6x + 9$$

$$(x + 3)^2$$

Check the factorization.

$$(x + 3)^2$$

$$(x + 3)(x + 3)$$

$$x^2 + 3x + 3x + 9$$

$$x^2 + 6x + 9$$

b. The first term is m squared, the last term is 7 squared, and the middle term is twice the product of these two terms.

$$m^2 - 14m + 49$$
$$(m - 7)^2$$

Check the factorization.

$$(m - 7)^2$$
$$(m - 7)(m - 7)$$
$$m^2 - 7m - 7m + 49$$
$$m^2 - 14m + 49$$

c. The first term is $2a$ squared, the last term is 5 squared, and the middle term is twice the product of these two terms.

$$4a^2 + 20a + 25$$
$$(2a + 5)^2$$

Check the factorization.

$$(2a + 5)^2$$
$$(2a + 5)(2a + 5)$$
$$4a^2 + 10a + 10a + 25$$
$$4a^2 + 20a + 25$$

d. The first term is $3x$ squared, the last term is $4y$ squared, and the middle term is twice the product of these two terms.

$$9x^2 - 24xy + 16y^2$$
$$(3x - 4y)^2$$

Check the factorization.

$$(3x - 4y)^2$$
$$(3x - 4y)(3x - 4y)$$
$$9x^2 - 12xy - 12xy + 16y^2$$
$$9x^2 - 24xy + 16y^2$$

EXAMPLE ① PRACTICE PROBLEM

Factor the following:

a. $x^2 + 8x + 16$ **b.** $36b^2 - 24b + 4$ **c.** $r^2 - 10rt + 25t^2$

DIFFERENCE OF SQUARES

Skill Connection

A similar pattern that we can use is the difference of two squares. When multiplying two binomials, one that is the sum of two terms and the other that is the difference of those same two terms, we get the difference of two squares.

$$(a + b)(a - b) = a^2 - ab + ab - b^2 = a^2 - b^2$$

Perfect square trinomials and the difference of two squares can be factored nicely with the patterns shown in this section. They can also be factored using the AC method.
When factoring the difference of two squares the value of b will be zero because the middle term is not there.

From this we can see that a binomial made up of the difference of two perfect squares can be factored by following the pattern

$$a^2 - b^2 = (a + b)(a - b)$$

As you can see, the factorization is the sum of the two terms multiplied by the difference of the two terms. Although the difference of two squares is factorable, the sum of two squares is not factorable. Many students want to factor the sum of two square as

$$(a + b)(a + b)$$

but this is not a correct factorization. We can check this by multiplying the factored form out.

$$(a + b)(a + b)$$
$$a^2 + ab + ab + b^2$$
$$a^2 + 2ab + b^2$$

$$x^2 - 100$$
$$x^2 - 10x + 10x - 100$$
$$(x^2 - 10x) + (10x - 100)$$
$$x(x - 10) + 10(x - 10)$$
$$(x - 10)(x + 10)$$
You can see that it is possible to use the AC method here but it is much longer than using the pattern.

We can see that the middle term of the final polynomial does not cancel out, so the product does not result in the sum of two squares.

EXAMPLE 2 FACTORING THE DIFFERENCE OF TWO SQUARES

Factor the following:

a. $x^2 - 9$ **b.** $4a^2 - 25$

c. $x^2 + 49$ **d.** $9m^2 - 16n^2$

Solution

a. This expression is the difference of x squared and 3 squared, so we can factor it using the difference of two squares pattern.

$$x^2 - 9$$
$$x^2 - 3^2$$
$$(x + 3)(x - 3)$$

Check the factorization.
$$(x + 3)(x - 3)$$
$$x^2 - 3x + 3x - 9$$
$$x^2 - 9$$

b. This expression is the difference of $2a$ squared and 5 squared so we can factor it using the difference of two squares pattern.

$$4a^2 - 25$$
$$(2a)^2 - 5^2$$
$$(2a + 5)(2a - 5)$$

Check the factorization.

$$(2a + 5)(2a - 5)$$
$$4a^2 - 10a + 10a - 25$$
$$4a^2 - 25$$

c. $x^2 + 49$ is the sum of two squares, not the difference of two squares, and is not factorable.

d. This expression is the difference of $3m$ squared and $4n$ squared, so we can factor it using the difference of two squares pattern.

$$9m^2 - 16n^2$$
$$(3m)^2 - (4n)^2$$
$$(3m + 4n)(3m - 4n)$$

Check the factorization.

$$(3m + 4n)(3m - 4n)$$
$$9m^2 - 12mn + 12mn - 16n^2$$
$$9m^2 - 16n^2$$

EXAMPLE ② PRACTICE PROBLEM

Factor the following:

a. $x^2 - 36$ **b.** $16a^2 - 81b^2$ **c.** $m^2 + 25$

DIFFERENCE AND SUM OF CUBES

Remember that the *sum* of two squares is not factorable; only the *difference* of two squares is. Two other patterns that we can use are the sum and difference of two cubes. The sign changes in these patterns are very important, and you should be very careful when using these patterns to get the signs correct.

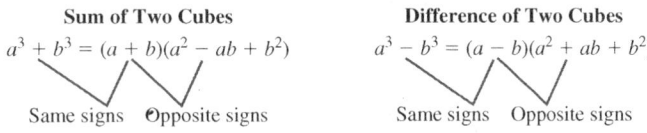

Sum of Two Cubes
$$a^3 + b^3 = (a + b)(a^2 - ab + b^2)$$
Same signs   Opposite signs

Difference of Two Cubes
$$a^3 - b^3 = (a - b)(a^2 + ab + b^2)$$
Same signs   Opposite signs

EXAMPLE ③ FACTOR SUM AND DIFFERENCES OF CUBES

Factor the following:

a. $x^3 + 27$ **b.** $m^3 - 8$ **c.** $8a^3 + 125b^3$ **d.** $2p^3 - 54r^3$

Solution

a. The first term is x cubed, and the second term is 3 cubed, so we can use the pattern for the sum of two cubes.

$$x^3 + 27$$

$$x^3 + 3^3$$

$$(x + 3)(x^2 - 3x + 9)$$

Check the factorization.

$$(x + 3)(x^2 - 3x + 9)$$

$$x^3 - 3x^2 + 9x + 3x^2 - 9x + 27$$

$$x^3 + 27$$

b. The first term is m cubed, and the second term is 2 cubed, so we can use the pattern for the difference of two cubes.

$$m^3 - 8$$

$$m^3 - 2^3$$

$$(m - 2)(m^2 + 2m + 4)$$

Check the factorization.

$$(m - 2)(m^2 + 2m + 4)$$

$$m^3 + 2m^2 + 4m - 2m^2 - 4m - 8$$

$$m^3 - 8$$

c. The first term is $2a$ cubed, and the second term is $5b$ cubed, so we can use the pattern for the sum of two cubes.

$$8a^3 + 125b^3$$

$$(2a)^3 + (5b)^3$$

$$(2a + 5b)(4a^2 - 10ab + 25b^2)$$

Check the factorization.

$$(2a + 5b)(4a^2 - 10ab + 25b^2)$$

$$8a^3 - 20a^2b + 50ab^2 + 20a^2b - 50ab^2 + 125b^3$$

$$8a^3 + 125b^3$$

d. We will first factor out the 2 that is in common and then use the pattern for the difference of two cubes.

$$2p^3 - 54r^3$$

$$2[p^3 - (3r)^3]$$

$$2(p - 3r)(p^2 + 3pr + 9r^2)$$

Check the factorization.

$$2(p - 3r)(p^2 + 3pr + 9r^2)$$
$$2(p^3 + 3p^2r + 9pr^2 - 3p^2r - 9pr^2 - 27r^3)$$
$$2(p^3 - 27r^3)$$
$$2p^3 - 54r^3$$

EXAMPLE ③ PRACTICE PROBLEM

Factor the following:

a. $x^3 + 8$ **b.** $x^3 - 27y^3$ **c.** $250a^3 - 16b^3$

MULTI-STEP FACTORIZATIONS

Some polynomials take several steps to factor completely or require slightly different thinking to find the key to factoring them. Always look for the greatest common factor and factor that out first. Then look for a pattern that you recognize and begin factoring. Several of the patterns that we have discussed can be found in more complicated expressions. The sum or difference of two cubes or the difference of two squares can be found in expressions with exponents that are multiples of 2 or 3.

$$x^6 - y^6 = (x^3)^2 - (y^3)^2$$
$$x^6 - y^6 = (x^2)^3 - (y^2)^3$$

This expression can be looked at as either a difference of two squares or a difference of two cubes. This leads to two different paths to factoring but the same result in the end. Using the difference of two squares first may make it easier to factor completely.

$x^6 - y^6$ $x^6 - y^6$

$(x^2)^3 - (y^2)^3$ $(x^3)^2 - (y^3)^2$

$(x^2 - y^2)(x^4 + x^2y^2 + y^4)$ $(x^3 + y^3)(x^3 - y^3)$

$(x + y)(x - y)(x^2 - xy + y^2)(x^2 + xy + y^2)$ $(x + y)(x^2 - xy + y^2)(x - y)(x^2 + xy + y^2)$

Trinomials of one variable can be of **quadratic form** by having the degree of the highest term be twice that of the next term and the final term be a constant.

$$a(\text{expression})^2 + b(\text{expression}) + c$$

These types of situations take practice to see, so consider the following example and pay close attention to the thinking behind each factorization.

EXAMPLE 4 MULTI-STEP FACTORING

Factor the following:

a. $x^4 - 81$ **b.** $3a^6 - a^3 - 10$

c. $12w^4 + 52w^2 + 35$ **d.** $3x^6 - 192y^6$

Solution

a. This is the difference of two squares, since the first term is x^2 squared and the second term is 9 squared.

$$x^4 - 81$$
$$(x^2)^2 - 9^2$$
$$(x^2 + 9)(x^2 - 9) \quad \text{The second factor is still the difference of two squares, so we can factor again.}$$
$$(x^2 + 9)(x + 3)(x - 3)$$

Check the factorization.

$$(x^2 + 9)(x + 3)(x - 3)$$
$$(x^2 + 9)(x^2 - 3x + 3x - 9)$$
$$(x^2 + 9)(x^2 - 9)$$
$$x^4 - 9x^2 + 9x^2 - 81$$
$$x^4 - 81$$

b. This trinomial is quadratic in form because the first term has degree twice that of the second term and the third term is a constant. We can use a substitution for a^3 to make the trinomial appear to be quadratic and then factor and replace the a^3 back into the expression. We can use any variable for this substitution, so for this problem, we will use u. Letting $u = a^3$, we get the following.

$$3a^6 - a^3 - 10$$
$$3(a^3)^2 - a^3 - 10$$
$$3u^2 - u - 10 \quad \text{Substitute in } u \text{ and factor.}$$
$$(3u + 5)(u - 2)$$
$$(3a^3 + 5)(a^3 - 2) \quad \text{Replace the } u \text{ with } a^3$$

Check the factorization.

$$(3a^3 + 5)(a^3 - 2)$$
$$3a^6 - 6a^3 + 5a^3 - 10$$
$$3a^6 - a^3 - 10$$

c. This trinomial is quadratic in form, and we can use substitution and then factor the remaining quadratic. Using $u = w^2$, we get the following.

$$12w^4 + 52w^2 + 35$$
$$12(w^2)^2 + 52w^2 + 35$$
$$12u^2 + 52u + 35 \quad \text{Substitute in } u \text{ and factor.}$$
$$(2u + 7)(6u + 5)$$
$$(2w^2 + 7)(6w^2 + 5) \quad \text{Replace } u \text{ with } w^2$$

Check the factorization.

$$(2w^2 + 7)(6w^2 + 5)$$

$$12w^4 + 10w^2 + 42w^2 + 35$$

$$12w^4 + 52w^2 + 35$$

d. After factoring out the 3 that is in common, this expression can be viewed as either the difference of two squares or the difference of two cubes. If we consider it to be the difference of two cubes, we end up needing to factor again, using the difference of two squares to completely finish the factoring.

$$3x^6 - 192y^6$$ Factor using the difference of two cubes.

$$3(x^6 - 64y^6)$$

$$3[(x^2)^3 - (4y^2)^3]$$ Finish by using the difference of two squares.

$$3(x^2 - 4y^2)(x^4 + 4x^2y^2 + 16y^4)$$

$$3(x - 2y)(x + 2y)(x^4 + 4x^2y^2 + 16y^4)$$

$$3(x - 2y)(x + 2y)(x^2 - 2xy + 4y^2)(x^2 + 2xy + 4y^2)$$

Check the factorization.

$$3(x - 2y)(x + 2y)(x^2 - 2xy + 4y^2)(x^2 + 2xy + 4y^2)$$

$$3(x^2 - 4y^2)(x^4 + 4x^2y^2 + 16y^4)$$

$$3(x^6 + 4x^4y^2 + 16a^2y^4 - 4x^4y^2 - 16a^2y^4 - 64y^6)$$

$$3(x^6 - 64y^6)$$

$$3x^6 - 192y^6$$

EXAMPLE ④ PRACTICE PROBLEM

Factor the following:

a. $t^6 - 64$ **b.** $3a^{10} - 5a^5 - 28$

3.5 Exercises

For Exercises 1 through 8, factor the following perfect square trinomials.

1. $x^2 + 8x + 16$

2. $w^2 + 10w + 25$

3. $9g^2 + 12g + 4$

4. $x^2 - 12x + 36$

5. $4t^2 - 28t + 49$

6. $100d^2 + 20d + 1$

7. $25x^2 + 30xy + 9y^2$

8. $49m^2 - 84mn + 36n^2$

For Exercises 9 through 22, factor the following difference of two squares and difference or sum of two cubes.

9. $x^2 - 36$

10. $m^2 - 49$

11. $9k^2 - 16$

12. $25b^2 - 4$

13. $m^3 - 64$

14. $x^3 - 27$

15. $x^3 + 125$

16. $p^3 + 8$

17. $8x^3 + 27$

18. $27d^3 - 64$

19. $3g^3 - 24$

20. $40x^3 + 625$

21. $50x^2 - 18$

22. $28h^2 - 63$

For Exercises 23 through 50, completely factor the following polynomials. Write prime if a polynomial is not factorable using the rational numbers.

23. $x^2 + 9$

24. $25t^2 + 81$

25. $r^6 - 64$

26. $x^6 - 729$

27. $16b^4 - 625c^4$

28. $81m^4 - 16p^4$

29. $h^4 + 3h^2 - 10$

30. $x^4 - 6x^2 + 5$

31. $g^8 + 6g^4 + 9$

32. $v^{10} - 10v^5 + 25$

33. $2t^{12} + 13t^6 + 15$

34. $3k^6 + 19k^3 + 28$

35. $8x^8 + 12x^4 + 18$

36. $6w^{10} - 27w^5 - 105$

37. $H + 6\sqrt{H} + 9$

38. $x + 2\sqrt{x} - 3$

39. $4t - 20t^{\frac{1}{2}} + 25$

40. $3m - 23m^{\frac{1}{2}} + 14$

41. $7g^4 - 567h^4$

42. $6a^4 - 96b^4$

43. $5w^5x^3z + 25w^3x^2z^2 - 120wxz^3$

44. $9a^5b^3c + 18a^3b^2c^2 - 27abc^3$

45. $24a^3b^2 + 11a^2b^3 - 35ab^4$

46. $-7g^7h^2 - 7g^5h^4 + 140g^3h^6$

47. $a^{12} - b^{36}$

48. $x^{12} - y^{36}$

49. $x^{16} - 1$

50. $5w^{16} - 5$

Chapter 3 Summary

CHAPTER
3

Section 3.1 Rules for Exponents

• Exponents follow the following rules and properties

Product rule for exponents: $x^m \cdot x^n = x^{m+n}$

Quotient rule for exponents: $\dfrac{x^m}{x^n} = x^{m-n}$

Power rule for exponents: $(x^m)^n = x^{mn}$

Distributing exponents over multiplication and division: $(xy)^m = x^m y^m$ and $\left(\dfrac{x}{y}\right)^m = \dfrac{x^m}{y^m}$

Negative exponents: $x^{-n} = \dfrac{1}{x^n}$ and $\dfrac{1}{x^{-n}} = x^n$

Zero as an exponent: $x^0 = 1$

Rational exponents: $x^{1/n} = \sqrt[n]{x}$

EXAMPLE

Simplify the following expressions.

a. $\dfrac{30x^3y^2z}{5xy^5z}$

b. $(4x^3y^4z)^2(3xy^2z^{-3})^3$

Solution

a. $\dfrac{30x^3y^2z}{5xy^5z}$

$6x^2y^{-3}$

$\dfrac{6x^2}{y^3}$

b. $(4x^3y^4z)^2(3xy^2z^{-3})^3$

$(16x^6y^8z^2)(27x^3y^6z^{-9})$

$432x^9y^{14}z^{-7}$

$\dfrac{432x^9y^{14}}{z^7}$

EXAMPLE 2

Rewrite the expression $x^{1/3}$ in radical form.

Solution

The denominator of the fraction exponent is three so the index of the radical will be three.

$$\sqrt[3]{x}$$

EXAMPLE 3

Rewrite the expression $\sqrt[5]{2ab^3}$ using exponents.

Solution

The index of the radical is five so the fraction exponent will have five in the denominator.

$$2^{1/5}a^{1/5}b^{3/5}$$

303

Section 3.2 Combining Functions

- A term is a constant, a variable, or the product of any combination of constants and variables.

- A polynomial is any combination of terms added together.

- The degree of a term equals the exponents of its variables added together.

- The degree of a polynomial the same as the degree of its highest-degree term.

- Like terms are terms that have exactly the same variable part.

- When adding or subtracting polynomials, you can combine only like terms.

- Use the distributive property to multiply polynomials.

- When multiplying two terms together, multiply the coefficients and add the exponents of the variables that are the same.

- When working in the context of an application, be sure your functions make sense when combined.

- To add or subtract two functions in a context, the input units must be the same, and the output units must be the same as well.

- To multiply or divide two functions in context, the input units must be the same, and the output units must make sense once they are combined.

- When you are not working in a context, you can combine two functions in whatever way you wish (as long as you do not perform an illegal mathematical operation, such as division by zero).

EXAMPLE 4

Give the degree of each term and the polynomial.

$$5x^3y^2z + 14x^2yz^2 - 3yz + 4z - 9$$

Solution

The degree of the terms are 6, 5, 2, 1, and 0, respectively. The degree of the polynomial is 6.

EXAMPLE 5

Perform the indicated operation and simplify.

a. $(4x^2 + 8x - 9) + (2x + 6)$

b. $(3x + 7) - (4x - 8)$

c. $3x(4x^2 + 2x + 8)$

d. $(2x + 5)(6x - 4)$

Solution

a. $(4x^2 + 8x - 9) + (2x + 6)$

$4x^2 + 10x - 3$

b. $(3x + 7) - (4x - 8)$

$3x + 7 - 4x + 8$

$-x + 15$

c. $3x(4x^2 + 2x + 8)$

$12x^3 + 6x^2 + 24x$

d. $(2x + 5)(6x - 4)$

$12x^2 - 8x + 30x - 20$

$12x^2 + 22x - 20$

EXAMPLE 6

Let $f(x) = 4x + 7$ and $g(x) = -5x + 12$.

a. Find $f(x) + g(x)$.   **b.** Find $(fg)(x)$.

c. Find $f(4) - g(4)$.

Solution

a. $f(x) + g(x) = (4x + 7) + (-5x + 12)$

$f(x) + g(x) = -x + 19$

b. $(fg)(x) = (4x + 7)(-5x + 12)$

$(fg)(x) = -20x^2 + 48x - 35x + 84$

$(fg)(x) = -20x^2 + 13x + 84$

c. $f(4) - g(4) = [4(4) + 7] - [-5(4) + 12]$

$f(4) - g(4) = 23 - (-8)$

$f(4) - g(4) = 31$

Section 3.3 Composing Functions

- When composing two functions one function will become the input for the other function. The notation for composing functions is $f(g(x))$.

- To compose two functions, the output unit of one function must be the same as the input unit of the other function.

EXAMPLE 7

Let $C(m)$ be the cost in thousands of dollars to improve m miles of city streets. Let $M(t)$ be the number of miles of city streets improved in year t. Use function notation to combine these two functions to create a new function that gives the cost to improve city streets in year t.

Solution

$C(M(t))$ is the cost in thousands of dollars to improve city streets in year t.

EXAMPLE 8

Let $f(x) = 4x + 7$ and $g(x) = -5x + 12$.
   Find $g(f(x))$

Solution

$$g(f(x)) = -5(4x + 7) + 12$$
$$g(f(x)) = -20x - 35 + 12$$
$$g(f(x)) = -20x - 23$$

Section 3.4 Factoring Polynomials

- To **factor** a quadratic using the AC method, you can use these steps.

1. Factor out the greatest common factor.
2. Multiply a and c together.
3. Find factors of ac that add up to b.
4. Rewrite the middle (bx) term using the factors from step 3.
5. Group and factor out what is in common (the greatest common factor).

- Factoring by trial and error:

   Consider the factors of a and c and find a combination of factors that will multiply together to give you the correct factorization.

- A prime polynomial is one that cannot be factored using rational numbers.

EXAMPLE 9

Factor the expression $6x^2 - 26x - 20$.

Solution

Every term in the expression is divisible by 2 so we will factor out a 2 and then continue with steps of the AC method.
$$6x^2 - 26x - 20$$
$$2(3x^2 - 13x - 10)$$
$$2(3x^2 - 15x + 2x - 10)$$
$$2[(3x^2 - 15x) + (2x - 10)]$$
$$2[3x(x - 5) + 2(x - 5)]$$
$$2(x - 5)(3x + 2)$$

Check the factorization.
$$2(x - 5)(3x + 2)$$
$$2(3x^2 + 2x - 15x - 10)$$
$$2(3x^2 - 13x - 10)$$
$$6x^2 - 26x - 20$$

EXAMPLE 10

Factor the expression $x^2 - 13x + 40$.

Solution

Using trial and error the factors of c are 1 and 40, 2 and 20, 4 and 10, or 5 and 8. 5 and 8 sum to 13 and if we use both negatives we will get the correct signs. Therefore this expression factors to

$$(x - 5)(x - 8)$$

Check the factorization.

$$(x - 5)(x - 8)$$
$$x^2 - 8x - 5x + 40$$
$$x^2 - 13x + 40$$

Section 3.5 Special Factoring Techniques

- A perfect square trinomial is of the form $a^2 + 2ab + b^2$ or $a^2 - 2ab + b^2$.
- Perfect square trinomials can be factored using the patterns
$$a^2 + 2ab + b^2 = (a + b)^2$$
or
$$a^2 - 2ab + b^2 = (a - b)^2$$
- The difference of two squares can be factored using the pattern
$$a^2 - b^2 = (a + b)(a - b)$$
- The sum of two squares cannot be factored using the rational numbers.
- The sum and difference of two cubes can be factored using the patterns
$$a^3 + b^3 = (a + b)(a^2 - ab + b^2)$$
$$a^3 - b^3 = (a - b)(a^2 + ab + b^2)$$
- An expression that is a quadratic in form may be easier to factor using substitution.

EXAMPLE 11

Factor the following expressions.

a. $x^2 + 16x + 64$

b. $36a^2 - 25b^2$

c. $2x^3 - 2$

d. $4x^{12} - 17x^6 - 15$

Solution

a. This expression is a perfect square trinomial and factors to

$$(x + 8)^2$$

Check the factorization.

$$(x + 8)^2$$
$$(x + 8)(x + 8)$$
$$x^2 + 8x + 8x + 64$$
$$x^2 + 16x + 64$$

b. This expression is the difference of two squares so it factors to

$$36a^2 - 25b^2$$
$$(6a)^2 - (5b)^2$$
$$(6a + 5b)(6a - 5b)$$

Check the factorization.

$$(6a + 5b)(6a - 5b)$$
$$36a^2 - 30ab + 30ab - 25b^2$$
$$36a^2 - 25b^2$$

c. Each term in this expression is divisible by 2 so we will factor out a 2 and then use the difference of two cubes patter to factor.

$$2x^3 - 2$$
$$2(x^3 - 1)$$
$$2(x - 1)(x^2 + x + 1)$$

Check the factorization.

$$2(x-1)(x^2 + x + 1)$$

$$2(x^3 + x^2 + x - x^2 - x - 1)$$

$$2(x^3 - 1)$$

$$2x^3 - 2$$

Check the factorization.

$$(4x^6 + 3)(x^6 - 5)$$

$$4x^{12} - 20x^6 + 3x^6 - 15$$

$$4x^{12} - 17x^6 - 15$$

d. This expression is quadratic in form since the degree of the first term is twice that of the second term. We will let $w = x^6$ and substitute w into the expression factor it and then replace w with x^6.

$$4x^{12} - 17x^6 - 15$$

$$4w^2 - 17w - 15$$

$$(4w + 3)(w - 5)$$

$$(4x^6 + 3)(x^6 - 5)$$

Chapter 3 Review Exercises

For Exercises 1 through 10, simplify the given expression using the order of operations and exponent rules. Write each answer without negative exponents.

1. $(2x^3y^4z)^3(5x^{-3}y^2z)^{-2}$ [3.1]

2. $(3x^2y^5)(5x^2y^2)$ [3.1]

3. $\left(\dfrac{225a^5b^3c}{15ab^6c^{-4}}\right)$ [3.1]

4. $\left(\dfrac{31m^5n^4}{a^5c^7}\right)^0$ [3.1]

5. $(32x^{10}y^{20})^{\frac{1}{5}}$ [3.1]

6. $(a^3b^{-5}c^{-2})^{-2}$ [3.1]

7. $(3m^{-1}n^3p^{-2})^2(5m^2n^{-6}p^4)$ [3.1]

8. $\left(\dfrac{a^2b^{-3}c^5}{a^3b^5c^7}\right)$ [3.1]

9. $\left(\dfrac{3a^{-1}b^3c^5}{ab^{-2}c^3}\right)^{-2}$ [3.1]

10. $\left(\dfrac{3x^2y^5z}{5xy^3z}\right)\left(\dfrac{15x^3yz^4}{10xz}\right)$ [3.1]

11. Write the following exponents in radical form.

   a. $x^{1/4}y^{3/4}$

   b. $3^{1/2}a^{1/2}$ [3.1]

12. Write the following radicals using rational exponents.

   a. $\sqrt[5]{4x^2}$

   b. $\sqrt{7ab}$ [3.1]

13. Hope's Pottery has the following functions to model her monthly revenue and costs.

$$R(v) = 155v$$
$$C(v) = 5000 + 65v$$

where $R(v)$ represents the monthly revenue in dollars for selling v vases and $C(v)$ represents the monthly costs in dollars for producing and selling v vases. Find a monthly profit function for Hope's Pottery.   [3.2]

14. The retail prescription drug sales in the United States is a multi-billion-dollar industry. Data for the total retail prescription drug sales and the mail-order sales is given in the following chart.

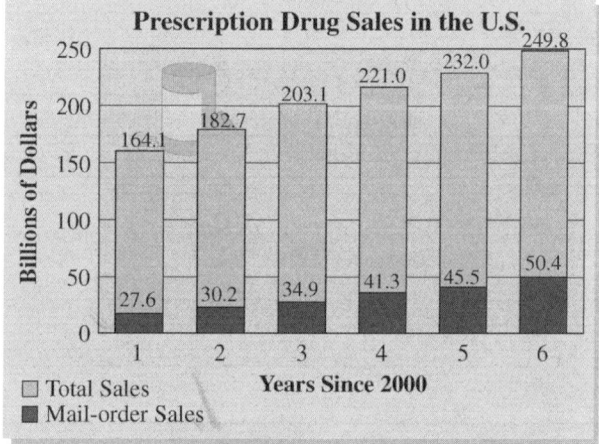

Source: National Association of Chain Drug Stores.

   a. Find a model for the total prescription drug sales.

   b. Find a model for the mail-order prescription drug sales.

   c. What was the total amount of prescription drug sales in 2010?

   d. What was the amount of prescription drugs sold through mail order in 2010?

   e. Using your two models, find a new function that gives the total amount of prescription drugs sold by non-mail-order retail stores.

   f. What was the amount of non-mail-order sales of prescription drugs in 2009 and 2015?   [3.2]

15. Johns Hopkins University and the University of Washington are the number one and two ranked universities in federal obligations for research and development funds. Their funding can be modeled as follows.

$$J(t) = 16.27t^3 - 321.0t^2 + 2089.32t - 3885.40$$

$$W(t) = 2.89t^3 - 53.94t^2 + 339.66t - 411.82$$

where $J(t)$ represents the federal funding for research and development at Johns Hopkins University in millions of dollars and $W(t)$ represents the federal funding for research and development at the University of Washington in millions of dollars t years since 1990.

Source: Models derived from data in Statistical Abstract, 2001.

 a. Find the federal funding for research and development at both universities in 1999.

 b. Find a new function for the difference in funding at these top-ranked schools.

 c. Use your new function to find the difference in federal funding at these universities in 2000. [3.2]

13. Perform the indicated operation and simplify.

 a. $(2x^2y + 5xy - 4y^2) - (7xy + 9y^2)$

 b. $(3x + 7)(2x - 9)$

 c. $(3x + 4y)(5x - 7y)$ [3.2]

In Exercises 8 through 12, combine the functions by adding, subtracting, multiplying, dividing, or composing the two functions. Write the function notation for the combination you use and list the units for the inputs and outputs of the new function.

16. Let $U(t)$ be the population of the United States, in millions, t years since 1900. Let $C(t)$ be the number of children under 20 years old in the United States, in millions, t years since 1900. Find a function for the number of adults 20 years old or older in the United States. [3.2]

17. Let $C(l)$ be the cost, in thousands of dollars, for Luxury Limousines, Inc. to produce l limousines per year. Let $R(l)$ be the revenue, in thousands of dollars, for Luxury Limousines, Inc. selling l limousines per year. Find a function for the profit of Luxury Limousines, Inc. from producing and selling l limousines per year. [3.2]

18. Let $E(t)$ be the number of employees of Disneyland in year t. Let $S(t)$ be the average number of sick days taken by Disneyland employees in year t. Find a function for the total number of sick days taken by Disneyland employees in year t. [3.2]

19. Let $E(t)$ be the number of employees of Disneyland in year t. Let $W(e)$ be the annual worker's compensation insurance cost in dollars when e employees work at Disneyland during the year. Find a function for the annual worker's compensation insurance cost at Disneyland in year t. [3.3]

20. Let $M(t)$ be the amount spent on treating cancer patients throughout the United States in year t. Let $R(t)$ be the amount spent on cancer research in the United States in year t. Find a function that represents the total amount spent on cancer treatments and research in the United States. [3.2]

21. The number of people in California's labor force and the number of these people who are unemployed are given below.

Year	Labor Force	Unemployment
1996	15,398,520	1,167,559
1997	15,747,435	1,061,231
1998	16,224,621	984,728
1999	16,503,004	927,430
2000	16,857,688	838,083
2001	17,246,969	807,290

Source: U.S. Bureau of Labor Statistics.

 a. Find a model for the labor force data.

 b. Find a model for the number of people who are unemployed.

 c. Estimate the number of people in the labor force in 2005.

 d. Estimate the number of people who are unemployed in 2005.

 e. Using your two models, find a new model for the percent of the labor force that is unemployed.

 f. What percent of the labor force is unemployed in 1998, 1999, 2000, and 2005? [3.2]

22. Let $f(x) = 6x + 3$ and $g(x) = -4x + 8$.

   a. Find $f(x) + g(x)$.

   b. Find $f(x) - g(x)$.

   c. Find $f(x)g(x)$.   [3.2]

   d. Find $f(g(x))$.

   e. Find $g(f(x))$.   [3.3]

23. Let $f(x) = 15x + 34$ and $g(x) = -17x + 34$.

   a. Find $f(x) + g(x)$.

   b. Find $g(x) - f(x)$.

   c. Find $f(x)g(x)$.   [3.2]

   d. Find $f(g(x))$.

   e. Find $g(f(x))$.   [3.3]

24. Let $f(x) = \dfrac{2}{5}x + \dfrac{3}{5}$ and $g(x) = 4x - 7$.

   a. Find $f(x) + g(x)$.

   b. Find $f(x) - g(x)$.

   c. Find $f(x)g(x)$.   [3.2]

   d. Find $f(g(x))$.

   e. Find $g(f(x))$.   [3.3]

25. Let $f(x) = \dfrac{1}{7}x + \dfrac{5}{7}$ and $g(x) = 2x - 7$.

   a. Find $f(4) + g(4)$.

   b. Find $f(4) - g(4)$.

   c. Find $f(4)g(4)$.   [3.2]

   d. Find $f(g(4))$.

   e. Find $g(f(4))$.   [3.3]

26. Let $f(x) = 0.6x + 2.5$ and $g(x) = 3.5x + 3.7$.

   a. Find $f(-3) + g(-3)$.

   b. Find $g(-3) - f(-3)$.

   c. Find $f(-3)g(-3)$.   [3.2]

   d. Find $f(g(-3))$.

   e. Find $g(f(-3))$.   [3.3]

27. Let $f(x) = 0.35x - 2.78$ and $g(x) = 2.4x - 6.3$.

   a. Find $f(2) + g(2)$.

   b. Find $g(2) - f(2)$.

   c. Find $f(2)g(2)$.   [3.2]

   d. Find $f(g(2))$.   [3.3]

For Exercises 28 through 40 factor the given expression using any method you choose.

28. $x^2 + 2x - 35$ [3.4]

29. $x^2 + 6x + 9$ [3.4]

30. $5b^2 - 14b - 3$ [3.4]

31. $6p^2 + 23p + 20$ [3.4]

32. $14k^2 - 21k + 7$ [3.4]

33. $20x^3 - 52x^2 - 24x$ [3.4]

34. $9m^2 - 100$ [3.5]

35. $25x^2 + 81$ [3.5]

36. $9x^2 + 30x + 25$ [3.5]

37. $27x^3 - 1$ [3.5]

38. $125w^3 + 8x^3$ [3.5]

39. $t^6 - 64$ [3.5]

40. $10m^6 - 29m^3 + 21$ [3.5]

Chapter 3 Test

For Exercises 1 through 3, simplify the expressions.
Write each answer without negative exponents.

1. $(2b^4 c^{-2})^5 (3b^{-3} c^{-4})^{-2}$

2. $\left(\dfrac{16b^{12} c^2}{2b^{-3}\ c^{-4}} \right)^{-1/3}$

3. $\dfrac{25x^{-9} y^{-8}}{35x^{-10}\ y^{-3}}$

4. The Pharmaceutical Management Agency of New Zealand (PHARMAC), manages the amount spent on various medical conditions and drugs throughout New Zealand. The amount spent on treating diabetes in New Zealand can be modeled by

$$T(t) = 1.1t + 6.84$$

where $T(t)$ represents the amount in millions of New Zealand dollars spent treating diabetes t years since 1990. Also the amount spent on diabetes research in New Zealand can be modeled by

$$R(t) = 1.3t + 0.52$$

where $R(t)$ represents the amount in millions of New Zealand dollars spent on diabetes research in New Zealand t years since 1990.

Source: PHARMAC.

   a. Find $T(9)$ and interpret its meaning in this context.

   b. Find $R(t) = 10$ and interpret its meaning in this context.

   c. Find a new function that represents the amount spent on both research and treatment of diabetes in New Zealand.

   d. Use the function from part c to determine the amount spent on research and treatment of diabetes in 2000.

5. Let $f(x) = 4x + 17$ and $g(x) = 2x - 7$.

   a. Find $f(x) - g(x)$.

   b. Find $f(x)g(x)$.

   c. Find $f(g(x))$.

6. Let $f(x) = 2.35x + 1.45$ and $g(x) = 2.4x - 6.3$.

   a. Find $(f + g)(4)$.

   b. Find $fg(-2)$.

   c. Find $f(g(6))$.

   d. Find $g(f(0))$.

7. The total number of graduate science/engineering students in doctoral programs can be modeled by

$$S(t) = 0.504t^3 - 8.596t^2 + 41.766t + 385.022$$

where $S(t)$ represents the total number of science/engineering graduate students in thousands t years since 1990.

   The number of female science/engineering graduate students in thousands t years since 1990 can be represented by

$$F(t) = 0.157t^3 - 2.953t^2 + 19.619t + 142.663$$

Source: Statistical Abstract 2001.

   a. Estimate the total number of graduate science/engineering students in doctoral programs in 2000.

   b. Find a new function for the number of male graduate science/engineering students in doctoral programs.

   c. Using this new model, estimate the number of male graduate science/engineering students in doctoral programs in 2000.

8. The number of murders, M, in thousands in the United States t years since 1990 can be represented by

$$M(t) = -0.42t^2 + 2.1t + 20.375$$

The percentage, P, of these murders that were committed by using a handgun can be represented by

$$P(t) = 0.048t^3 - 0.952t^2 + 4.898t + 49.454$$

Source: Models derived from data in Statistical Abstract 2001.

a. Estimate the percentage of murders in 1997 that were committed by using a handgun.

b. Find a new function that gives the number of murders committed by using a handgun.

c. Using your new function, estimate the number of murders committed by using a handgun in 1996.

For Exercises 9 through 12, assume that the following functions refer to a specific summer tennis camp and that t represents the year.

- Let $T(t)$ be the total number of children at the camp.
- Let $B(t)$ be the number of boys at the camp.
- Let $C(k)$ be the total cost for k children to go to the camp.
- Let $M(k)$ be the number of matches played at a camp with k children.
- Let $A(t)$ be the average length of time, in minutes, a match takes to play.

Combine the functions by adding, subtracting, multiplying, dividing, or composing the two functions. Write the function notation for the combination you use and list the units for the inputs and outputs of the new function.

9. Find a function that gives the number of girls attending the tennis camp.

10. Find a function for the number of boys' matches at the camp in year t.

11. Find a function for the total cost for the camp in year t.

12. Find the amount of time it takes for all the matches played in a certain year t.

13. Perform the indicated operation and simplify.

a. $(4x^2 + 2x - 7) - (3x^2 + 9)$

b. $(4x^2y + 3xy - 2y^2) + (5x^2y - 7xy + 8)$

c. $(2x + 8)(3x - 7)$

d. $(3a - 4b)(2a - 5b)$

For Exercises 14 through 20 factor the given expression using any method.

14. $x^2 - 14x + 45$

15. $2p^3 + 8p - 42$

16. $t^2 - 20t + 100$

17. $7x^3 - 28x$

18. $8m^3 + 125n^3$

19. $3x^2 + x + 10$

20. $6m^8 - 29m^4 + 28$

Chapter 3 Projects

Research Project

One or more people.

HOW DO I BREAK THAT CODE?

Factoring plays a critical role in many encryption codes used today and during history. Research how factoring is used in encryption and find an interesting example of the use of factoring in this field.

Write up

Write a one to two page paper describing some interesting historical use of factoring in the field of encryption.

Quadratic Functions

- Recognize a quadratic function from its graph and equation.

- Identify the vertex of a parabola and explain its meaning in a context.

- Graph quadratic functions by hand from the vertex form and the standard form.

- Find and use a quadratic model when appropriate.

- Find the domain and range of a quadratic function.

- Solve a quadratic equation by factoring.

- Solve quadratic equations using the square root property and completing the square.

- Solve quadratic equations using the quadratic formula.

- Solve systems of equations that contain quadratics.

- Find the equation of a quadratic from its graph.

© Royalty-Free/CORBIS

he U.S. apple industry's profits depend on the number of apples that an orchard can get out of each tree planted. The competition in the international apple market has increased. This increased competition has made it more important today for the apple industry to maximize its production of apples. In this chapter, we will discuss quadratic models that will help us to investigate some situations not represented by linear data or models. One of the chapter projects , on page 451, will ask you to use the skills from this chapter to investigate how fruit growers can maximize their fruit production.

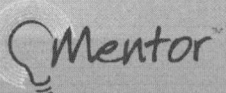

Introduction to Quadratics

INTRODUCTION TO QUADRATICS AND IDENTIFYING THE VERTEX

In this chapter, we will add another function to our list of possible models. Thus far we have only examined linear models and systems of linear equations. We will now examine data sets that do not follow a linear pattern and then introduce a new function to fit this pattern.

 **CONCEPT INVESTIGATION I** IS THAT DATA LINEAR?

a. The average monthly temperature for several months in San Diego, California, is given in the bar chart.

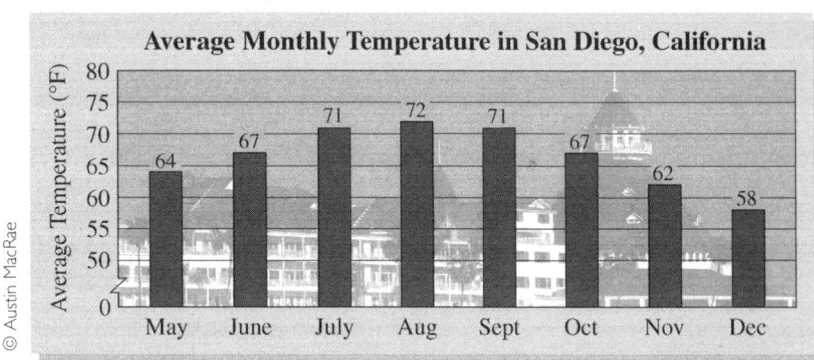

Source: Weatherbase.com.

Create a scatterplot of the data on your calculator and answer the following questions.

Define the variables as

T = The average monthly temperature, in degrees Fahrenheit, in San Diego, California

m = The month of the year; e.g., $m = 5$ represents May

i. Do these data follow a linear pattern? If not, describe the shape of the pattern.

ii. Graph the function $T(m) = -1(m - 8)^2 + 72$.

Describe how well this function fits the data.

Using Your TI Graphing Calculator

To enter exponents on the graphing calculator, you can use either of two methods.

If you are squaring a term, you can use the button on the left-hand side of the calculator. Enter the number or variable that you wish to square and press the button.

If you want to raise a variable or number to any other power, you should use the button in the far right column of the calculator. You will have to enter the number or variable you want to raise to a power, then press the button, followed by the power you wish to use.

For example, x^5 would be entered as

$\boxed{\text{X,T,θ,n}}$ $\boxed{5}$

When raising a negative number to a power you will need parentheses around the number in order for the negative to be raised to the exponent. For example to square -3 you would enter

$$(-3)^2$$

iii. A meeting planner wants to schedule a conference in San Diego during the hottest month of the year. Using the function as a model for the temperatures in San Diego, find the month for which they should plan the conference.

iv. According to the function, what is the average temperature in San Diego during the month of March? Do you think this is reasonable?

b. The number of public branch libraries in the United States for several years is given below:

Year	Public Library Branches
1994	6,223
1995	6,172
1996	6,205
1997	6,332
1998	6,435

Source: Statistical Abstract 2001.

Create a scatterplot of the data on your calculator and answer the following questions.

Define the variables as

B = The number of public branch libraries in the United States

t = Time in years since 1990

i. Describe the shape of this data set.

How is it similar to the pattern of the data in part a?

ii. Graph the function $B(t) = 28.7t^2 - 286.2t + 6899.3$.

Describe how well this function fits the data.

iii. At what point does this graph change directions? What does this point represent in this application?

iv. According to your model, how many branch libraries were there in 1990?

As you can see from the data given in Concept Investigation 1, not all data have a linear pattern. Both of the given data sets are curved and form a portion of what is called a **parabola**. Parabolas can be described as a "U" shape or an "upside-down U" shape. In chapter 1, we learned that a line is always increasing or always decreasing. A parabola will change from increasing to decreasing or from decreasing to increasing.

The point where the parabola turns around and goes in the opposite direction is called the **vertex.** The vertex represents the point where the lowest or minimum value occurs on the graph if the graph opens upward and represents the highest or maximum value on the graph if the graph opens downward. Examples of each of these types of graphs are shown below. Data that show this pattern can be modeled by using a **quadratic function.**

Say What?

When we discuss a quadratic function, the vertex is one of the most important points on the graph. The following phrases can all be used to refer to the vertex of a quadratic function.

- vertex
- maximum point
- minimum point
- highest point
- lowest point
- greatest value
- least value
- smallest value

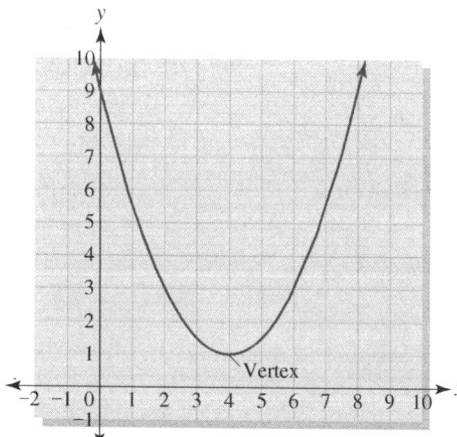

Parabola opens upward
Vertex: (4, 1)
    Lowest point
    Minimum point
Decreasing when $x < 4$
Increasing when $x > 4$

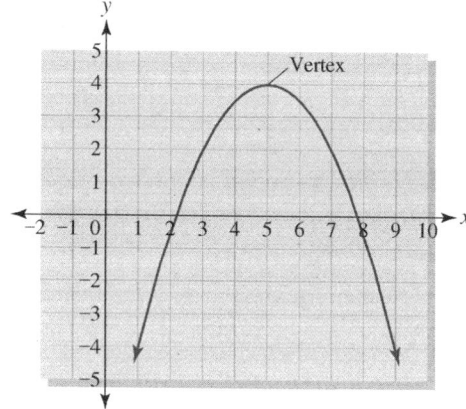

Parabola opens downward
Vertex: (5, 4)
    Highest point
    Maximum point
Increasing when $x < 5$
Decreasing when $x > 5$

DEFINITIONS

Parabola: The graph of a quadratic function.
A parabola has a "**U**" shape and can open either upward or downward.
Vertex: The point on a parabola where the graph changes direction.
The maximum or minimum function value occurs at the vertex of a parabola.

IDENTIFYING A QUADRATIC FUNCTION

 **CONCEPT INVESTIGATION** IS THAT A QUADRATIC FUNCTION? ···

Examine each of the following functions and state if you think it is linear, quadratic, or other. Then graph the function on your calculator to confirm your answer.

a. $f(t) = 3.5t - 7$

b. $R(t) = 5t^2$

c. $P(u) = 4u^2 + 5u - 8$

d. $G(d) = 7(d - 5)^2 + 5$

e. $Q(m) = 3.5(m - 4) - 9.8$

f. $F(x) = (x - 7)(x + 9)$

g. $T(c) = 0.5(c + 4)^3 - 5$

h. $E(w) = 3.5(2)^w$

i. $F(x) = (x + 4)^2$

j. $H(n) = 2n^3 - 5n^2 + n - 15$

k. $K(v) = 5v^2 - 9$

What characteristics do equations of quadratic functions have?

·· ◆

Quadratic functions are most commonly represented by one of the following two forms.

DEFINITIONS
Standard form:

$$f(x) = ax^2 + bx + c$$

where a, b, and c are real numbers and a cannot be zero.

Vertex form:

$$f(x) = a(x - h)^2 + k$$

where a, h, and k are real numbers and a cannot be zero.

You should note that in both of these definitions the only constant that cannot be zero is a. If a were zero, there would not be a squared term, and the function would no longer be quadratic. All of the other constants, b, c, h, and k, in these forms can be zero. The most basic quadratic function is $f(x) = x^2$, which has a vertex of $(0, 0)$ and an a value of 1.

Note that the two functions given in the definition box are simply two different forms of the same function. If we multiply out and simplify the vertex form the quadratic will be in standard form. For example if we take the function

$f(x) = 2(x-3)^2 + 5$ that is in vertex form and multiply it out we can put it into standard form.

$$f(x) = 2(x-3)^2 + 5$$ Start in vertex form.

$$f(x) = 2(x-3)(x-3) + 5$$ Multiply out the squared part.

$$f(x) = 2(x^2 - 3x - 3x + 9) + 5$$

$$f(x) = 2(x^2 - 6x + 9) + 5$$ Distribute the 2.

$$f(x) = 2x^2 - 12x + 18 + 5$$ Simplify

$$f(x) = 2x^2 - 12x + 23$$ End in standard form.

Concept Connection

Remember when reading a graph from left to right that a curve is increasing when it is going up, and decreasing when it is going down.

A graph like the ones in Example 1 continue on indefinitely from left to right and up or down.

RECOGNIZING GRAPHS OF QUADRATIC FUNCTIONS AND IDENTIFYING THE VERTEX

EXAMPLE **1** READING QUADRATIC GRAPHS

a. Use the graph of $f(x)$ to estimate the following.

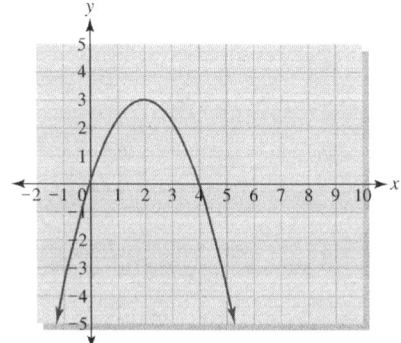

i. For what x values is this curve increasing? Decreasing?

ii. Vertex

iii. x-intercept(s)

iv. y-intercept

v. $f(5) = $?

vi. What x value(s) will make $f(x) = -2$?

b. Use the graph of $f(x)$ to estimate the following.

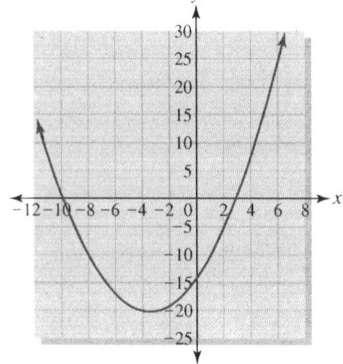

i.   For what x values is this curve increasing? Decreasing?

ii.   Vertex

iii. x-intercept(s)

iv. y-intercept

v. $f(6) = $?

vi.   What x value(s) will make $f(x) = -10$?

Solution

a.

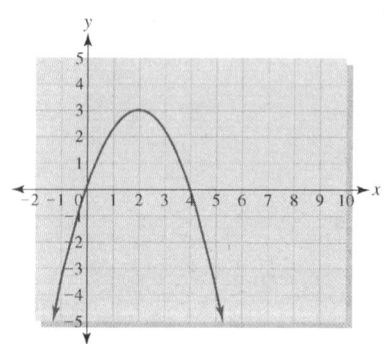

i.   Reading the graph from left to right, the curve is increasing for $x < 2$ and decreasing for $x > 2$.

ii.   This curve changes from increasing to decreasing when $x = 2$, so the vertex is (2, 3).

iii.   The curve crosses the x-axis at $x = 0$ and $x = 4$, so (0, 0) and (4, 0) are the x-intercept(s).

iv.   The curve crosses the y-axis at $y = 0$, so the y-intercept is (0, 0).

v.   When $x = 5$ the curve has an output of about $y = -3.5$, so $f(5) = -3.5$.

vi.   The output of the function is $y = -2$ when the input is about $x = -0.5$ and $x = -4.5$.

b.

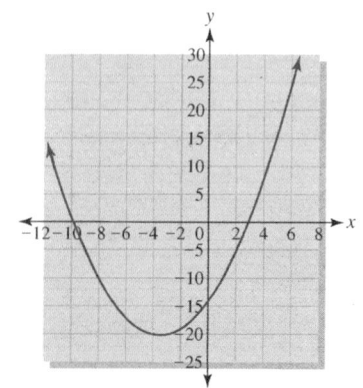

i.   This curve is increasing for $x > -3$ and decreasing for $x < -3$.

ii.   This curve changes from decreasing to increasing when $x = -3$, so the vertex is $(-3, -20)$.

iii.   The curve crosses the x-axis at $x = -10$ and $x = 3$, so $(-10, 0)$ and (3, 0) are the x-intercept(s).

iv.   The curve crosses the y-axis at $y = -15$, so the y-intercept is $(0, -15)$.

v.   When $x = 6$ the curve has an output of about $y = 25$, so $f(6) = 25$.

vi.   The output of the function is $y = -10$ when the input is about $x = -8$ and $x = 1$.

EXAMPLE ① PRACTICE PROBLEM

Use the graph of $f(x)$ to estimate the following.

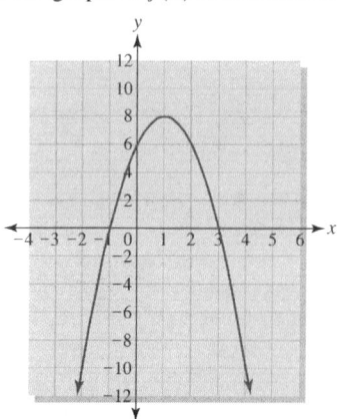

a. For what *x* values is this curve increasing? Decreasing?

b. Vertex

c. *x*-intercept(s)

d. *y*-intercept

e. $f(2) = $?

f. What *x* value(s) will make $f(x) = -10$?

From example 1, you can see that a parabola can have two horizontal (*x*-intercepts) intercepts. The following graphs show how a parabola can actually have two, one, or no horizontal intercepts.

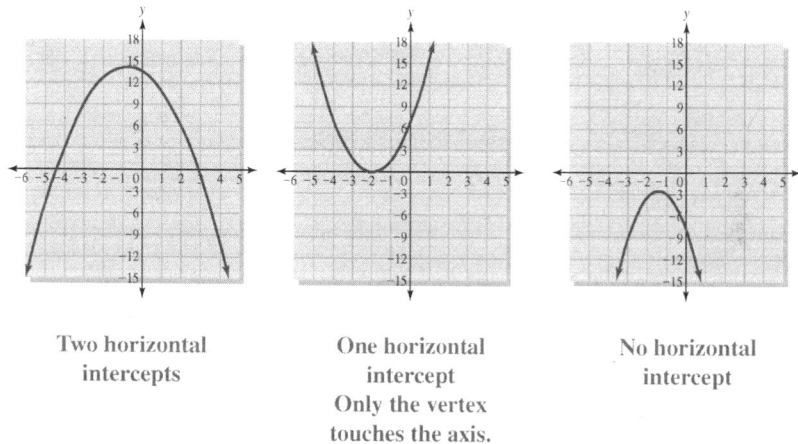

Two horizontal	One horizontal	No horizontal
intercepts	intercept	intercept
	Only the vertex	
	touches the axis.	

In Chapter 1 we also used a table of data to estimate the vertical and horizontal intercepts by looking for when the input and output variables were zero. We can use the same idea with quadratic data if we look carefully for the possibility of more than one horizontal intercept. Remember that the horizontal intercepts may not be whole numbers and therefore would be harder to find on a table.

EXAMPLE 2 USING A TABLE TO FIND INTERCEPTS

Use the table to estimate the vertical and horizontal intercept(s).

a.

Input	Output
−10	114
−7	0
−4	−78
0	−126
3	−120
9	0
12	114

b.

x	y
−8	−7.2
−6	0
−4	4.8
−2	7.2
0	7.2
2	4.8
4	0

Solution

a. The vertical intercept occurs when the input variable is zero. In this table, the vertical intercept is $(0, -126)$. The horizontal intercepts will occur when the output variable is zero. This table shows two horizontal intercepts, at $(-7, 0)$ and $(9, 0)$.

b. The vertical intercept is $(0, 7.2)$ and the horizontal intercepts are at $(-6, 0)$ and $(4, 0)$.

EXAMPLE 3 USING SCATTERPLOTS TO INVESTIGATE DATA

Create scatterplots for the following sets of data. Determine what type of model would best fit the data. If a quadratic function would be a good fit, give an estimate for the vertex and determine whether it is a maximum or minimum point.

a.

Input	Output
0	19
1	9
2	3
3	1
4	3
5	9
6	19

b.

x	y
−2.3	12.92
−1	12.4
1.2	11.52
2	11.2
3.2	10.72
4.2	10.32
5	10

c.

x	y
−7	−12.5
−6.5	−11.13
−4.3	−8.045
−4	−8
−3	−8.5
−2.6	−8.98

Solution

a.

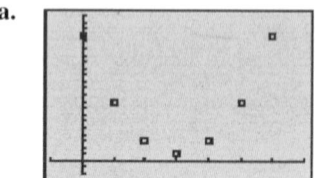

```
WINDOW
Xmin=-1
Xmax=7
Xscl=1
Ymin=-2
Ymax=22
Yscl=1
Xres=1
```

These data decrease and then increase and form the shape of a parabola. Therefore, a quadratic function would be used to model these data. The vertex is the minimum point on the graph and appears to be at about (3, 1).

b.

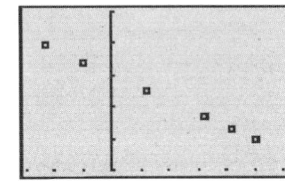 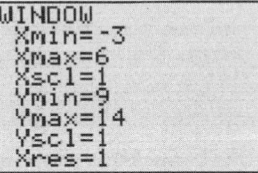

These data are decreasing at a constant rate so they have a linear pattern. Since these data have a linear pattern, the graph is not a parabola and therefore has no vertex.

c.

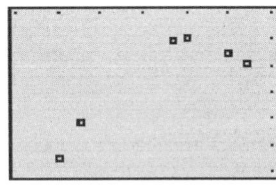 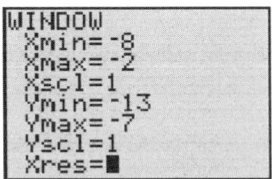

These data increase and then decrease and have the shape of a parabola. Therefore, a quadratic function would be used to model these data. The vertex is the maximum point on the graph and appears to be at about $(-4, -8)$. With this graph you, might want to trace the plotted points to determine an estimate for the vertex.

EXAMPLE PRACTICE PROBLEM

Create scatterplots for the following sets of data. If a quadratic function would be a good fit, give an estimate for the vertex and determine if it is a maximum or minimum point.

a.

Input	Output
2	10
3	0
4	−6
5	−8
6	−6
7	0
8	10

b.

x	y
−4	−2
−3	2.5
−2	4
−1	2.5
0	−2
1	−9.5

In the exercises for this section you will be given data that may follow a linear or quadratic pattern. If the data are quadratic, you will be given any needed models to use in your calculator. You can use the trace feature to estimate the vertex of any given model. In the next few sections you will learn more about quadratic functions and learn a method to model data that follow a quadratic pattern.

For Exercises 1 through 10, determine if the given function is linear, quadratic or other. check your answer by graphing the function on your calculator.

1. $f(x) = 5x^2 + 6x - 9$

2. $h(x) = 2(x + 5)^2 + 3$

3. $g(x) = 4x + 7$

4. $m(t) = 3(t + 4) - 7$

5. $w(b) = b^2 - 7$

6. $f(x) = 4^x + 7$

7. $c(p) = (p + 3)(p - 4)$

8. $h(x) = 5x^2$

9. $g(x) = 2x^3 + 4x^2 - 2x - 8$

10. $d(t) = 3(t + 5)^2$

11. Use the graph of $f(x)$ to estimate the following.

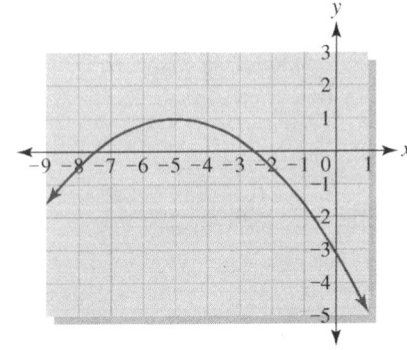

   a. Vertex

   b. For what x values is the graph increasing?

   c. For what x values is the graph decreasing?

   d. Horizontal intercept(s)

   e. Vertical intercept

12. Use the graph of $f(x)$ to estimate the following.

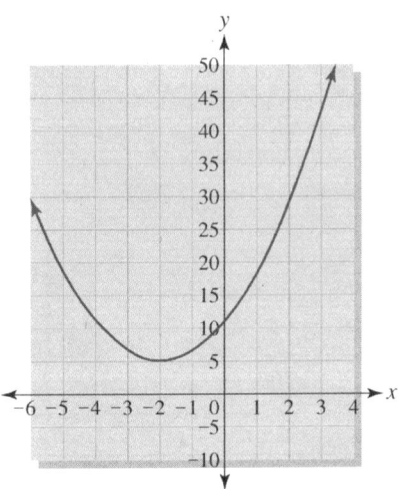

   a. Vertex

   b. For what x values is the graph increasing?

   c. For what x values is the graph decreasing?

   d. Horizontal intercept(s)

   e. Vertical intercept

13. Use the graph of $f(x)$ to estimate the following.

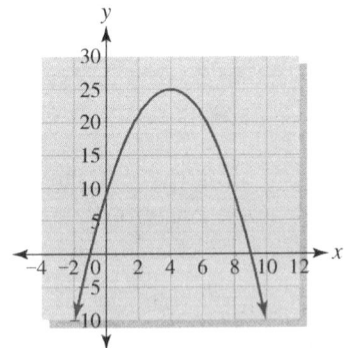

   a. Vertex

   b. For what x values is the graph increasing?

   c. For what x values is the graph decreasing?

   d. x-intercept(s)

   e. y-intercept

   f. $f(2) = $?

14. Use the graph of $f(x)$ to estimate the following.

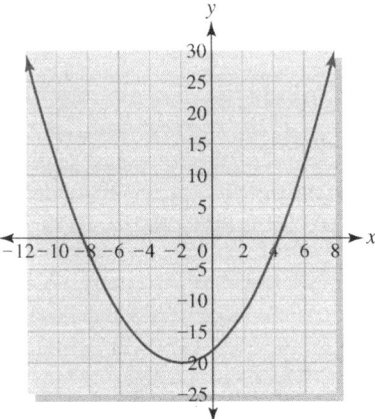

   a. Vertex

   b. For what x values is the graph increasing?

   c. For what x values is the graph decreasing?

   d. x-intercept(s)

   e. y-intercept

   f. $f(6) = $?

15. Use the graph of $f(x)$ to estimate the following.

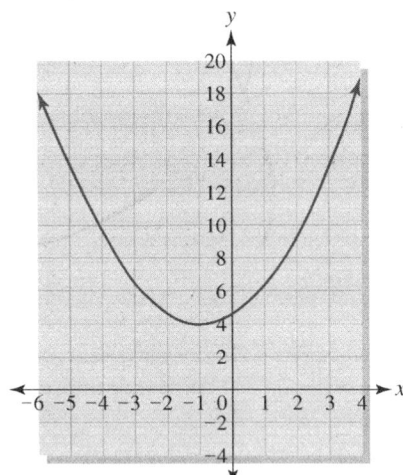

   a. Vertex

   b. For what x values is the graph increasing?

   c. For what x values is the graph decreasing?

   d. Horizontal intercept(s)

   e. Vertical intercept

   f. $f(-4) = $?

   g. What x value(s) will make $f(x) = 14$?

16. Use the graph of $f(x)$ to estimate the following.

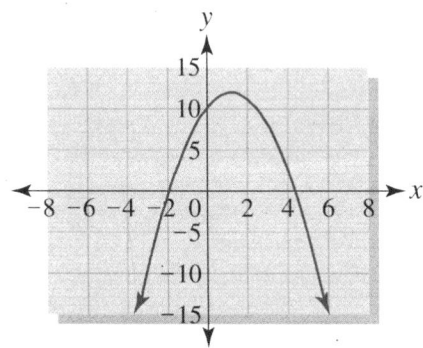

   a. Vertex

   b. For what x values is the graph increasing?

   c. For what x values is the graph decreasing?

   d. x-intercept(s)

   e. y-intercept

   f. $f(-2) = $?

   g. What x value(s) will make $f(x) = -10$?

For Exercises 17 through 20, use the table to find the horizontal and vertical intercepts.

17.

Input	Output
−6	−9
−4	0
−2	5
0	6
1	5
3	0

18.

Input	Output
−2	20
0	9
2	2
3	0
4	−1
6	0
7	2

19.

Input	Output
−15	15
−10	0
−5	−10
0	−15
5	−15
10	−10
15	0

20.

Input	Output
−2	180
0	80
2	20
4	0
6	20
8	80

21. The average number of days each month with a high temperature above 70°F in San Diego, California is given below.

Month	Average Number of Days Above 70°F
May	12
June	21
July	30
August	31
September	29
October	26
November	14
December	8

Source: Weatherbase.com.

a. Define variables and create a scatterplot.

b. Would a linear or quadratic function better model these data?

c. If a quadratic model would be better, estimate the vertex and determine if it is a maximum or minimum point.

d. Using the shape of the distribution estimate the number of days above 70°F in San Diego, California, during the month of April.

22. The number of cassette singles shipped by major recording media manufacturers is given below.

Year	Cassette Singles (in millions)	Year	Cassette Singles (in millions)
1987	5.1	1995	70.7
1988	22.5	1996	59.9
1991	69	1997	42.2
1992	84.6	1998	26.4
1993	85.6	1999	14.2
1994	81.1	2000	1.3

Source: Statistical Abstract 2001.

a. Let C be the number of cassette singles in millions shipped t years since 1980. Create a scatterplot for these data on your calculator.

b. Graph the function $C(t)$ on the same window as your scatterplot.

$$C(t) = -2.336(t - 13)^2 + 85.6$$

c. How well does this model fit the data?

d. Find $C(25)$ and explain its meaning in this context.

e. Estimate a vertex for this model.

f. Explain what the vertex means in this context.

g. Use the graph to estimate the year(s) in which no cassette singles were shipped.

23. The number of cable television systems in the United States during the 1990s are given in the following chart.

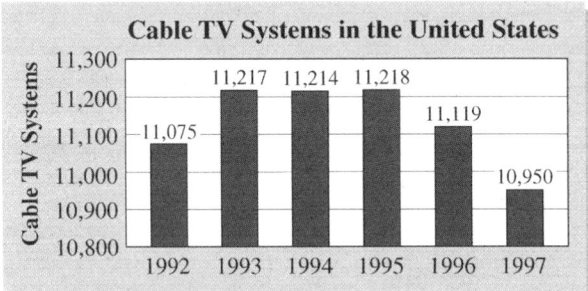

Source: Statistical Abstract 2001.

  a. Let C be the number of cable systems in the United States t years since 1990. Create a scatterplot on your calculator for these data.

  b. Graph the function $C(t)$ on the same window as your scatterplot.

  $$C(t) = -34.625t^2 + 285.482t + 10649.643$$

  c. How well does this model fit the given data?

  d. Estimate a vertex for this model. Is the vertex a maximum or minimum point for this function?

  e. Explain what the vertex means in this context.

24. The United States Golf Association (USGA) requires that golf balls be tested to see if they fall within the official rules of golf. Under one such test the height of a ball was measured after it was struck by the testing machine. These heights in feet are given in the table below.

Time (in seconds)	Height of Ball (in feet)
0	0.083
1	111.083
2	190.083
3	237.083
4	252.083
5	235.083
6	186.083
7	105.083

  a. Let H be the height of the golf ball in feet t seconds after being hit by the testing machine. Create a scatterplot for these data on your calculator.

  b. Graph the function $H(t)$ on the same window as your scatterplot.
  $$H(t) = -16t^2 + 127t + 0.083$$

  c. Find $H(6.5)$ and explain its meaning in this context.

  d. Estimate a vertex for this model. Is the vertex a maximum or minimum point for this function?

  e. Explain what the vertex means in this context.

  f. Use the model to estimate when the ball will hit the ground.

25. Out of frustration, a tennis player hits a tennis ball straight up in the air. The height of the ball can be modeled by the function
  $$H(t) = -4.9t^2 + 50t + 1$$

  where H is the height of the ball in meters t seconds after being hit.

  a. Graph the function $H(t)$ on your calculator using the window Xmin = -1, Xmax = 12, Ymin = -40, and Ymax = 150.

  b. Use the graph to estimate the vertex of this parabola, and explain its meaning in this context.

  c. Find $H(2)$ and explain its meaning in this context.

  d. Use the graph to estimate when the tennis ball will hit the ground.

26. Will launched a toy air pressure rocket in his front yard. The height of the rocket can be modeled by the function
  $$H(t) = -16(t-2)^2 + 64$$

  where H is the height of the rocket in feet t seconds after being launched.

  a. Graph the function $H(t)$ on your calculator using the window Xmin = -1, Xmax = 5, Ymin = -5, and Ymax = 70.

  b. Use the graph to estimate the vertex of this parabola, and explain its meaning in this context.

c. Find $H(1)$ and explain its meaning in this context.

d. Use the graph to estimate when the rocket will hit the ground.

27. The college bookstore is trying to find the price for school T-shirts that will result in the maximum monthly revenue for the store. By adjusting the price of the T-shirts over several weeks, the bookstore manager has created the following scatterplot for the monthly revenue in dollars from selling T-shirts for d dollars.

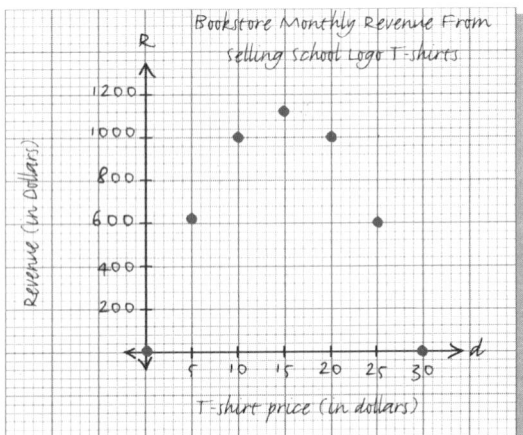

a. Use the scatterplot to estimate the revenue from selling T-shirts for $10 each.

b. Use the scatterplot to estimate the maximum monthly revenue.

c. What price should the bookstore charge to maximize the monthly revenue?

d. Explain why the revenue may go down after the vertex in this situation.

28. A baseball team has 30,000 cheap seats in the stadium and wants to maximize revenue from selling these seats. After some research, marketers have

created the following graph for the revenue per game from selling tickets for these cheap seats

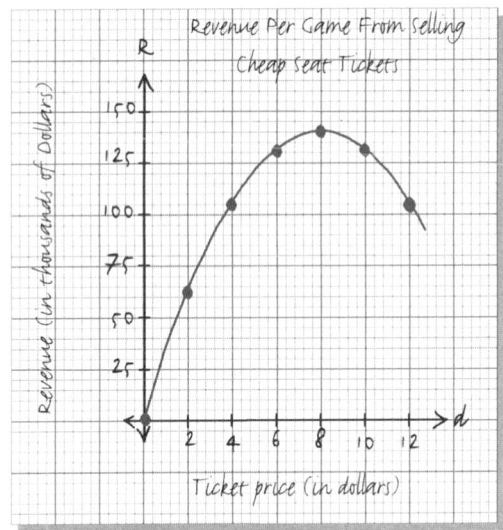

a. Use the graph to estimate the revenue per game from selling the cheap seats for $4 each.

b. Use the graph to estimate what price the baseball team should charge for these cheap seats to maximize revenue per game.

c. Use the graph to estimate what the maximum revenue from selling the cheap seat tickets will be.

d. For what ticket prices will the revenue be increasing?

For Exercises 29 through 36, create a scatterplot for the given set of data, determine what type of model (linear, quadratic, or other) would best fit the data. If quadratic, give an estimate for the vertex and determine if it is a maximum or minimum point.

29.

Input	Output
−7	−8
−5	8
−3	−8
−1	−56
1	−136
2	−188

30.

Input	Output
1	−57
2	−20
4	6
6	8
8	34
9	71

31.

Input	Output
1	12
2	5
4	−3
6	−3
8	5
9	12

32.

Input	Output
−2	−12
−1	−9.5
0	−7
3	0.5
4	3
6	8

33.

Input	Output
−7	42.8
−2	25.8
1	15.6
5	2
9	−11.6
18	−42.2

34.

Input	Output
−5	−48
−3	8
−2	15
−1	8
1	−48
2	−97

35.

Input	Output
−8	17
−5	19
0	22
7	27
16	32
27	37

36.

Input	Output
−8	−213.6
−5	−48.75
−2	−14.4
0	−20
3	−5.15
5	58.75
8	301.6

For Exercises 37 through 42 sketch a rough graph of a parabola that satisfies the given conditions and use your graph to answer the question.

37. Sketch a parabola with a vertex of (2, 4) that faces downward.

   a. How many horizontal intercepts does the graph have?

   b. How many vertical intercepts does the graph have?

38. Sketch a parabola with a vertex of (-3, 7) that faces upward.

    a. How many horizontal intercepts does the graph have?

    b. How many vertical intercepts does the graph have?

39. Sketch a parabola with a vertex of (4, -6) that faces upward.

    a. How many horizontal intercepts does the graph have?

    b. How many vertical intercepts does the graph have?

40. Sketch a parabola with a vertex of (0, 5) that faces up.

    a. How many horizontal intercepts does the graph have?

    b. How many vertical intercepts does the graph have?

41. Sketch a parabola that has two horizontal intercepts and faces downward.

    a. Is the vertex above or below the horizontal axis?

    b. Where does the vertex have to be for the parabola to face down and have no horizontal intercepts?

42. Sketch a parabola that has only one horizontal intercept and faces upward. Where does the vertex have to be for this graph?

SECTION 4.2 Graphing Quadratics in Vertex Form

In Section 4.1 we looked at some of the graphical characteristics of quadratic functions such as the "U" shape and vertex and used several quadratic functions as models for data. In this section we are going to investigate the vertex form and graphing quadratics.

VERTEX FORM

Because understanding the graph of a function is an important part of the modeling process, we are going to take a close look at the vertex form and notice how the constants a, h, and k affect the graph of the function.

CONCEPT INVESTIGATION WHAT DO h AND k DO TO THE GRAPH? ••

Consider the vertex form of a quadratic.

$$f(x) = a(x - h)^2 + k$$

We are going to study each component of this function one at a time. For each part of this investigation, consider how the graph is modified as you change one of the constants in the function. First, let's focus on k.

a. Graph the following group of functions on the same calculator window. (Use the standard window.)

   i. $f(x) = x^2$ This is the basic quadratic function.

   ii. $f(x) = x^2 + 2$

   iii. $f(x) = x^2 + 5$

   iv. $f(x) = x^2 + 6$

   v. $f(x) = x^2 + 8$

In these functions we, are considering how a positive k value changes the graph of a basic quadratic function. In your own words, what does a positive k value do to the graph?

b. Graph the following group of functions on the standard window.

   i. $f(x) = x^2$

   ii. $f(x) = x^2 - 2$

   iii. $f(x) = x^2 - 5$

   iv. $f(x) = x^2 - 6$

   v. $f(x) = x^2 - 8$

In these functions we are considering how a negative k value changes the graph of a basic quadratic function. In your own words, what does a negative k value do to the graph?

Concept Connection

When you use the vertex form

$$f(x) = a(x - h)^2 + k$$

notice that the constant h is being subtracted from the variable x. This subtraction can make the sign of h confusing.

Consider the following situations and how the value of h appears once the expression has been simplified.

CC-Example 1:
Consider the expression

$$x - h$$

Now substitute the following values for h and simplify.

$$h = 1, 3, -5, -3$$

$h = 1$

$$x - (1)$$
$$x - 1$$

The positive h value now looks negative.

$h = 3$

$$x - (3)$$
$$x - 3$$

Again the positive h value of 3 now looks like a negative 3.

$h = -5$

$$x - (-5)$$
$$x + 5$$

The negative h value looks like a positive 5.

$h = -3$

$$x - (-3)$$
$$x + 3$$

Now the negative h value of -3 looks like a positive 3.

Now Let's focus on h and how it affects the graph.

c. Graph the following functions on the same calculator window.

   i. $f(x) = x^2$

   ii. $f(x) = (x - 2)^2$

   iii. $f(x) = (x - 5)^2$

   iv. $f(x) = (x - 6)^2$

   v. $f(x) = (x - 8)^2$

In these functions we are considering how a **positive** h value changes the graph of a basic quadratic function. In your own words what does a **positive** h value do to the graph? (See the Concept Connection.)

d. Graph the following functions on the same calculator window.

   i. $f(x) = x^2$

   ii. $f(x) = (x + 2)^2$

   iii. $f(x) = (x + 5)^2$

   iv. $f(x) = (x + 6)^2$

   v. $f(x) = (x + 8)^2$

In these functions we are considering how a **negative** h value changes the graph of a basic quadratic function. In your own words, what does a **negative** h value do to the graph? (Remember that h is the only constant in the vertex form that is being subtracted and therefore its sign can be confusing.)

e. Graph the following functions and find the vertex of the parabola.

   i. $f(x) = x^2$ $h = 0$ $k = 0$ Vertex $(0, 0)$

   ii. $f(x) = (x - 2)^2$ $h = \_\_$ $k = \_\_$ Vertex $(\_\_ , \_\_)$

   iii. $f(x) = (x - 4)^2 + 3$ $h = \_\_$ $k = \_\_$ Vertex $(\_\_ , \_\_)$

   iv. $f(x) = (x - 6)^2 + 6$ $h = \_\_$ $k = \_\_$ Vertex $(\_\_ , \_\_)$

   v. $f(x) = (x - 8)^2 - 5$ $h = \_\_$ $k = \_\_$ Vertex $(\_\_ , \_\_)$

   vi. $f(x) = (x + 2)^2 - 4$ $h = \_\_$ $k = \_\_$ Vertex $(\_\_ , \_\_)$

   vii. $f(x) = (x + 5)^2 + 2$ $h = \_\_$ $k = \_\_$ Vertex $(\_\_ , \_\_)$

   viii. $f(x) = (x + 7)^2 + 4.5$ $h = \_\_$ $k = \_\_$ Vertex $(\_\_ , \_\_)$

   ix. $f(x) = (x - 2.5)^2 + 3.5$ $h = \_\_$ $k = \_\_$ Vertex $(\_\_ , \_\_)$

What is the relationship between the vertex form of a quadratic

$$f(x) = a(x - h)^2 + k$$

and the vertex of the parabola?

Be sure to use caution when interpreting the value of h and k. Remember that h will appear to have the opposite sign. These two constants control the location of the vertex of the parabola and, therefore, will also give you the values for the maximum or minimum point on the graph.

DEFINITIONS

Vertex in the vertex form: The vertex of a quadratic equation in vertex form can be read directly from the equation.

$$f(x) = a(x - h)^2 + k$$

$$\text{Vertex} = (h, k)$$

To complete our information about the vertex form, we still have to investigate what the value of a does to the graph of a parabola.

CONCEPT INVESTIGATION WHAT DOES a DO TO THE GRAPH? ••

a.   Graph the following functions on the same calculator window. Find the vertex.

i. $f(x) = (x + 2)^2 - 5$ (Note: $a = 1$.)   Vertex (__ , __)

ii. $f(x) = 2(x + 2)^2 - 5$ Vertex (__ , __)

iii. $f(x) = 5(x + 2)^2 - 5$ Vertex (__ , __)

iv. $f(x) = 18.7(x + 2)^2 - 5$ Vertex (__ , __)

In these functions, we are considering how a positive a value greater than 1 changes the graph of a quadratic function. Does the value of a affect the vertex of the graph?

In your own words, what does a positive a value greater than 1 do to the graph?

b.   Graph the following group of functions on the same calculator window.

i. $f(x) = (x - 4)^2 + 3$

ii. $f(x) = -2(x - 4)^2 + 3$

iii. $f(x) = -5(x - 4)^2 + 3$

iv. $f(x) = -8.7(x - 4)^2 + 3$

In these functions, we are considering how an a value less than -1 changes the graph of a quadratic function. In your own words, what does a negative a value do to the graph?

The value of a controls more than just whether the parabola faces upward or downward. Let's look at some other values of a and see how they can affect the graph.

c. Graph the following group of functions on the same calculator window.

i. $f(x) = x^2$

ii. $f(x) = 0.5x^2$

iii. $f(x) = -0.3x^2$

iv. $f(x) = -\dfrac{2}{3}x^2$ In your calculator use parentheses around the fractions.

v. $f(x) = \dfrac{1}{10}x^2$

In these functions, we are considering how an a value between -1 and 1 changes the graph of a quadratic function. In your own words, how do these values of a change the graph?

GRAPHING QUADRATICS IN VERTEX FORM

If we put all of the information we know about a, h, and k together, then we can sketch the graph of a quadratic function by hand. We now know that the vertex of a parabola can be read directly from the vertex form of the quadratic.

$$f(x) = a(x-h)^2 + k$$
$$\text{vertex} = (h, k)$$

Say What?

Symmetry is describing the shape of the curve, in particular the fact that the two sides of the curve will be the same.

Axis of Symmetry is an imaginary line that separates the two sides of a symmetric curve.

Unlike the x-axis or y-axis the axis of symmetry will move to wherever the curve is and divide the two symmetric sides.

Remember to always use caution when reading the h value because its sign will look the opposite of what it is. The value of a will affect the width of the graph as well as determine if the parabola faces upward (positive a) or faces downward (negative a).

One other characteristic of a parabola that is helpful is its **symmetry** about the vertical line going through the vertex. This vertical line is called the **axis of symmetry** and has the equation $x = h$, where the vertex is (h, k). Symmetry basically means that the curve looks the same on the right side of the axis of symmetry as it does on the left side. If you were to fold the parabola on the axis of symmetry, the sides of the curve would match up.

The symmetry of a parabola makes graphing even easier, since you need to have points on only one side of the vertex to copy them for the other side. In the graph of $f(x) = (x-4)^2 - 5$ below, you can see that if the input value changes one unit to the right and left of the vertex, the function gives the points $(3, -4)$ and $(5, -4)$ that are at the same height. Also changing the input value four units less than or greater than the vertex results in two points with the same height. Every point on the parabola to the left of the vertex will have a matching point on the right side of the vertex. These points are called *symmetric points*, since they are equal distance from the vertex both horizontally and vertically.

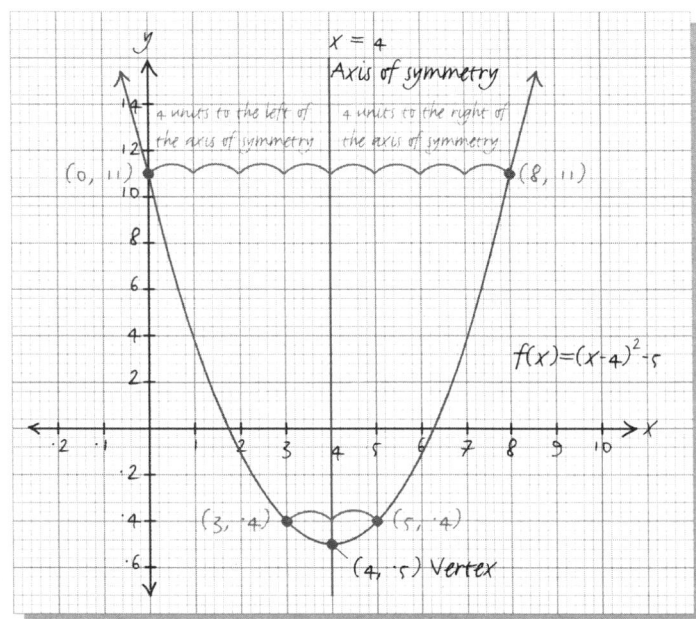

STEPS TO GRAPHING A QUADRATIC FROM THE VERTEX FORM

1. Determine whether the graph opens up or down.
2. Find the vertex and axis of symmetry.
3. Find the vertical intercept.
4. Find an extra point by picking an input value on one side of the axis of symmetry and calculating the output value.
5. Plot the points, plot their symmetric pairs and sketch the graph. (Find an additional symmetric pair if needed.)

EXAMPLE 1 GRAPHING QUADRATIC FUNCTIONS IN VERTEX FORM

Sketch the graph of the following quadratic function.

$$f(x) = 2.5(x - 6)^2 + 2$$

Solution

$$f(x) = 2.5(x - 6)^2 + 2$$

Step 1 Determine whether the graph opens up or down.

The value of a is 2.5. Since a is positive, the graph will open upward, and because it is greater than 1, the graph will be narrow.

Step 2 Find the vertex and axis of symmetry.

Since the quadratic is given in vertex form, the vertex is (6, 2). The axis of symmetry is the vertical line through the vertex $x = 6$.

Step 3 Find the vertical intercept.

To find the vertical intercept, we need to make the input variable zero.

$$f(x) = 2.5(x - 6)^2 + 2$$
$$f(0) = 2.5(0 - 6)^2 + 2 \quad \text{Be sure to follow the order of}$$
$$f(0) = 2.5(-6)^2 + 2 \quad\quad\text{operations.}$$
$$f(0) = 2.5(36) + 2$$
$$f(0) = 90 + 2$$
$$f(0) = 92$$

Therefore the vertical intercept is $(0, 92)$.

Step 4 Find an extra point by picking an input value on one side of the axis of symmetry and calculating the output value.

The axis of symmetry is $x = 6$, so we can choose any x values less than or greater than 6. We will choose $x = 4$. Substitute 4 into the input of the function to find the output.

$$f(4) = 2.5(4 - 6)^2 + 2$$
$$f(4) = 2.5(-2)^2 + 2$$
$$f(4) = 2.5(4) + 2$$
$$f(4) = 10 + 2$$
$$f(4) = 12$$

Remember that another way to evaluate a function for input values is to use the table feature on the calculator.

X	Y1	
4	12	

X=

Therefore we have the point $(4, 12)$.

Step 5 Plot the points, plot their symmetric pairs and sketch the graph. (Find an additional symmetric pair if needed.)

The vertex is $(6, 2)$, the vertical intercept is $(0, 92)$, and the other point we found in step 4 was $(4, 12)$. The x-values are all small but the y-values get up to 92 and above so we will need to scale the x-axis and y-axis differently.

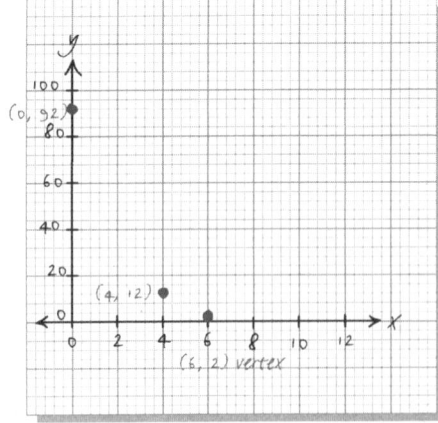

Because $(4, 12)$ is two units to the left of the axis of symmetry $x = 6$, the point two units to the right of the axis of symmetry $(8, 12)$ will be its symmetric point. The vertical intercept $(0, 92)$ is 6 units to the left of the axis of

Concept Connection

When graphing a parabola by hand or using a graphing calculator the scale used can distort the graph. In Example 1 we chose a scale of 20 for the y-axis to accommodate the large y-values. The scale on the x-axis was only 2 because the x-values were much smaller.

If we use the same scale on both axes we get a more realistic version of the curve. The same scale will allow us to see the narrow width and rapid decrease and increase of the curve.

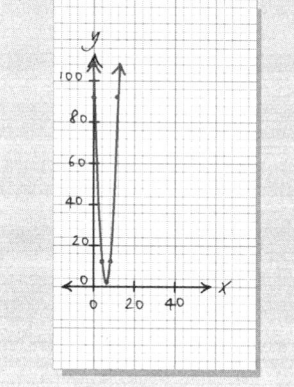

symmetry so its symmetric point will be 6 units to the right of the axis of symmetry at (12, 92).

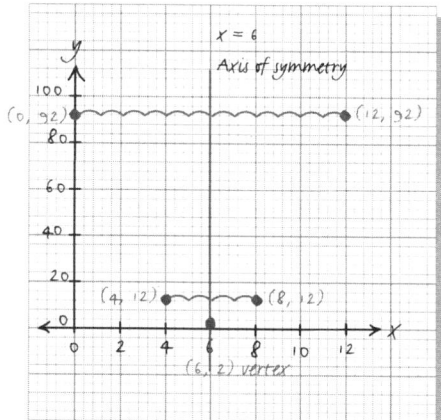

Now that all the points and their symmetric pairs are plotted we can complete the graph by connecting the points with a smooth curve.

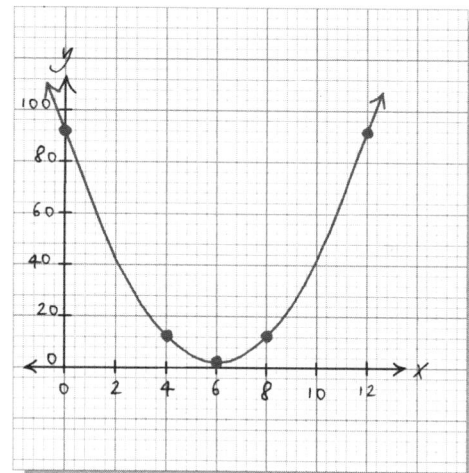

EXAMPLE 2 GRAPHING QUADRATIC FUNCTIONS IN VERTEX FORM

Sketch the graph of the following quadratic function.

$$f(x) = -0.25(x + 8)^2 + 10$$

Solution

Step 1 Determine whether the graph opens up or down.

The value of a is -0.25. Because a is negative, the graph will open downward, and because it is less than 1, the graph will be wide.

Step 2 Find the vertex and axis of symmetry.

Since the quadratic is given in vertex form, the vertex is $(-8, 10)$. The axis of symmetry is the vertical line through the vertex $x = -8$.

Step 3 Find the vertical intercept.

To find the vertical intercept, we make the input variable zero.

$$f(x) = -0.25(x+8)^2 + 10$$
$$f(0) = -2.5(0+8)^2 + 10$$
$$f(0) = -0.25(8)^2 + 10$$
$$f(0) = -0.25(64) + 10$$
$$f(0) = -16 + 10$$
$$f(0) = -6$$

Therefore, the vertical intercept is $(0, -6)$.

Step 4 Find an extra point by picking an input value on one side of the axis of symmetry and calculating the output value.

The axis of symmetry is $x = -8$, so we can choose any x value less than or greater than -8. We will choose $x = -3$. Substitute -3 into the input for the function to find the output value.

$$f(x) = -0.25(x+8)^2 + 10$$
$$f(-3) = -0.25(-3+8)^2 + 10$$
$$f(-3) = -0.25(5)^2 + 10$$
$$f(-3) = -0.25(25) + 10$$
$$f(-3) = -6.25 + 10$$
$$f(-3) = 3.75$$

Therefore, we have the point $(-3, 3.75)$.

Step 5 Plot the points, plot their symmetric pairs and sketch the graph. (Find an additional symmetric pair if needed.)

The point $(-3, 3.75)$ is 5 units to the right of the axis of symmetry so its symmetric point $(-13, 3.75)$ is 5 units to the left of the axis of symmetry. The vertical intercept $(0, -6)$ is 8 units to the right of the axis of symmetry so its symmetric point will be 8 units to the left of the axis of symmetry at $(-16, -6)$.

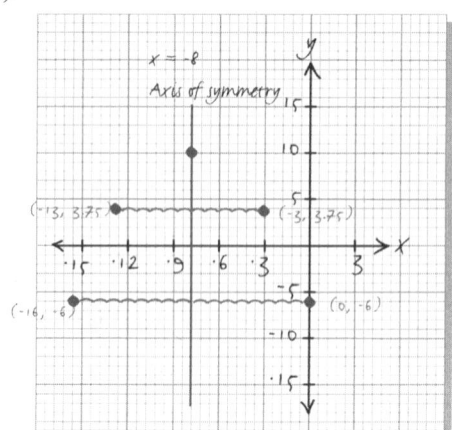

Now connect the points with a smooth curve.

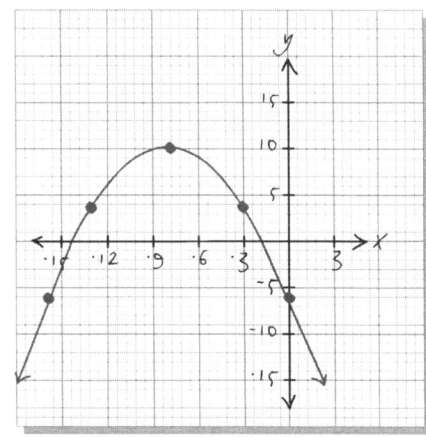

EXAMPLE 3 GRAPHING QUADRATIC FUNCTIONS IN VERTEX FORM

Sketch the graph of the following quadratic function.

$$g(x) = 2x^2 - 5$$

Solution

In this case, the quadratic is given in vertex form where $h = 0$, so the function could be written as $g(x) = 2(x - 0)^2 - 5$.

Step 1 Determine whether the graph opens up or down.

The value of a is 2. Because a is positive, the graph will open upward, and because it is more than 1, the graph will be narrow.

Step 2 Find the vertex and axis of symmetry.

The vertex for this quadratic is $(0, -5)$. The axis of symmetry is the vertical line through the vertex $x = 0$.

Step 3 Find the vertical intercept.

To find the vertical intercept, we make the input variable zero.

$$f(x) = 2x^2 - 5$$
$$f(0) = 2(0)^2 - 5$$
$$f(0) = -5$$

Therefore the vertical intercept is $(0, -5)$. Note that the vertex and the vertical intercept are the same point. Therefore we have only found one point on the parabola.

Step 4 Find an extra point by picking an input value on one side of the axis of symmetry and calculating the output value.

The axis of symmetry is $x = 0$, so we can choose any x value less than or greater than 0. We will choose $x = 3$. Substitute 3 into the input for the function to find the output value.

$$f(x) = 2x^2 - 5$$
$$f(3) = 2(3)^2 - 5$$
$$f(3) = 13$$

Therefore, we have the point (3, 13).

Step 5 Plot the points, plot their symmetric pairs and sketch the graph. (Find an additional symmetric pair if needed.)

Because we have only two points so far, we may want to find an additional symmetric pair to make the graph more accurate. To find another point we pick another input value and calculate the output value. Using $x = 1$ we get

$$f(1) = 2(1)^2 - 5$$
$$f(1) = -3$$
$$(1, -3)$$

The point (3, 13) is 3 units to the right of the axis of symmetry, so its symmetric point (−3, 13) is 3 units to the left of the axis of symmetry. The point (1, −3) is 1 unit to the right of the axis of symmetry, so its symmetric point (−1, −3) is 1 unit to the left of the axis of symmetry.

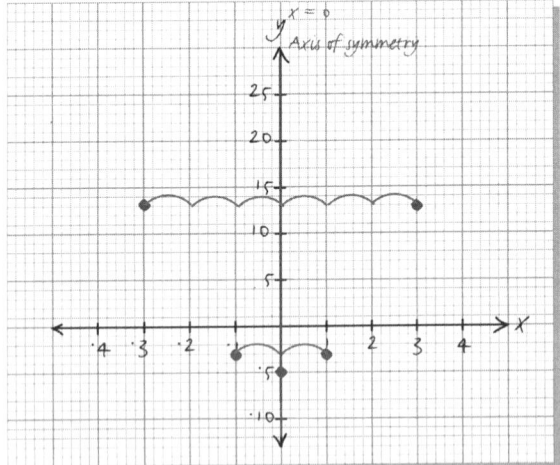

Now connect the points with a smooth curve.

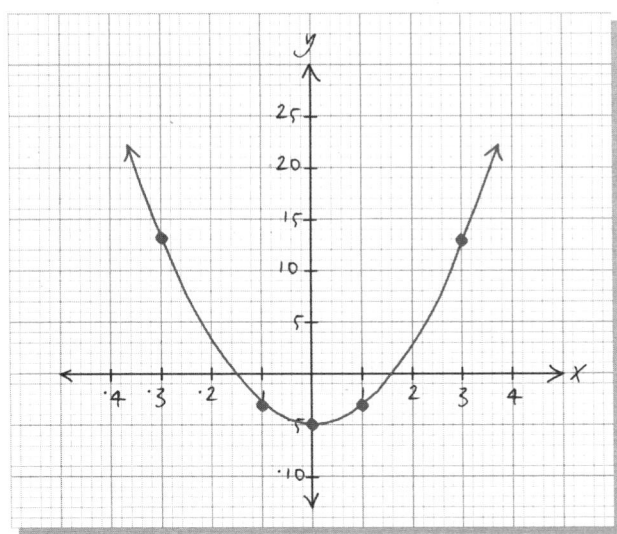

EXAMPLE ③ PRACTICE PROBLEM

Sketch the graph of the following quadratic function.

$$f(x) = -0.5(x + 2)^2 + 4$$

EXAMPLE 4 SOLVING APPLICATIONS USING THE GRAPH

The average number of hours per person per year spent watching television during the late 1990s can be modeled by

$$H(t) = 9.5(t - 7)^2 + 1544$$

where $H(t)$ represents the average number of hours per person per year spent watching television t years since 1990.

© Bill Aron/PhotoEdit

a. How many hours did the average person spend watching television in 1996?

b. Sketch a graph of this model.

c. In what year was the average number of hours spent watching television the least?

d. Use your graph to estimate in what year(s) the average number of hours spent watching television was 1800 hours per person per year.

e. Give a reasonable domain and range for this model.

Solution

a. $t = 6$ would represent 1996, so we substitute in 6 and solve for H.

$$H(t) = 9.5(t - 7)^2 + 1544$$
$$H(6) = 9.5(6 - 7)^2 + 1544$$
$$H(6) = 9.5(-1)^2 + 1544$$
$$H(6) = 9.5 + 1544$$
$$H(6) = 1553.5$$

The model predicts that in 1996 the average person spent 1553.5 hours per year watching television.

b. **Step 1** Determine whether the graph opens up or down.

The value of a is 9.5. Because a is positive, the graph will open upward, and because it is greater than 1, the graph will be narrower than $y = x^2$.

Step 2 Find the vertex and axis of symmetry.

Because the quadratic is given in vertex form, the vertex is (7, 1544). The axis of symmetry is the vertical line through the vertex $t = 7$.

Step 3 Find the vertical intercept.

To find the vertical intercept, we let the input variable equal zero.

$$H(t) = 9.5(t - 7)^2 + 1544$$
$$H(0) = 9.5(0 - 7)^2 + 1544$$
$$H(0) = 2009.5$$

Therefore, the vertical intercept is (0, 2009.5).

Step 4 Find an extra point by picking an input value on one side of the axis of symmetry and calculating the output value.

The axis of symmetry is $t = 7$, so we will choose $t = 3$. Substitute 3 for the input of the function, and calculate the output value.

$$H(t) = 9.5(t - 7)^2 + 1544$$
$$H(3) = 9.5(3 - 7)^2 + 1544$$
$$H(3) = 1696$$

Therefore, we have the point (3, 1696).

Step 5 Plot the points, plot their symmetric pairs, and sketch the graph. (Find an additional symmetric pair if needed.)

Because (3, 1696) is four units to the left of the axis of symmetry, the point (11, 1696) will be its symmetric point on the right side of the axis of symmetry. The vertical intercept (0, 2009.5) is 7 units to the left of the axis of symmetry, so its symmetric pair (14, 2009.5) is 7 units to the right.

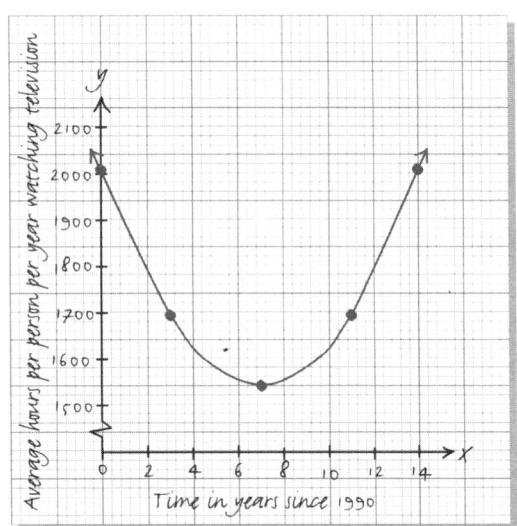

c. The lowest point on the graph is the vertex at (7, 1544). The model says that in 1997 people watched an average of 1544 hours of television per year. This was the minimum amount of television watched during the late 1990s.

d. According to the graph, the average hours per person per year spent watching television reached 1800 in year 2 and again in year 12. Since time is measured in years since 1990, 2 and 12 would be the years 1992 and 2002, respectively.

e. The problem states that the function models the late 1990s and television viewing seems to have continued to grow since the 1990s so we can extend the domain a couple of years and choose a domain of $5 \leq t \leq 12$. Within this domain, the graph has a lowest output of 1544 at the vertex and grows from there to the highest output when $t = 12$. Substituting $t = 12$ into the function, we will get the highest output.

$$H(t) = 9.5(t - 7)^2 + 1544$$
$$H(12) = 9.5(12 - 7)^2 + 1544$$
$$H(12) = 1781.5$$

Therefore, the range for this model is $1544 \leq H \leq 1781.5$.

EXAMPLE (4) PRACTICE PROBLEM

The number of cellular telephone subscribers, in thousands, can be modeled by

$$C(t) = 833.53(t + 1)^2 + 3200$$

where $C(t)$ represents the number of cellular telephone subscribers, in thousands, t years since 1990.

a. How many cellular telephone subscribers were there in 2000?

b. Sketch a graph of this model.

c. According to this model, in what year was the number of cellular telephone subscribers the least?

© BananaStock/Alamy

d. How many cellular telephone subscribers were there in 1985? Does your answer make sense?

e. Give a reasonable domain and range for this model.

When you work with larger numbers, it is usually best to gather as much information about the graph as possible before you start to create the graph. Doing so will allow you to get an idea of how to scale your axes.

The domain and range for quadratic models in a context are again restricted to what will make sense in the situation you are modeling. Remember that you may not want to spread out very far in the domain because model breakdown can occur more rapidly with a quadratic model than with a linear one. When you have no data to help guide your domain, you must rely on the situation and any information that will help you avoid model breakdown. In some cases, only half of the parabola will make sense in the context, so the vertex may be a good starting point or ending point for your domain.

EXAMPLE **5** DOMAIN AND RANGE OF QUADRATIC FUNCTIONS WITH NO CONTEXT

Give the domain and range for the following functions.

a. $f(x) = 2.5(x - 6)^2 + 2$

b. $f(x) = -0.25(x + 8)^2 + 10$

c. $g(x) = (x - 3)^2 - 5$

Solution

a. We graphed this function in Example 1. Looking at the graph, we can visualize the graph continuing to go upward and get wider as x increases to infinity and decreases to negative infinity.

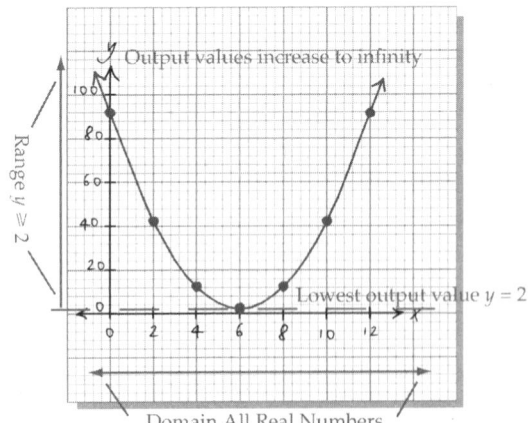

Because the input variable x can be any real number without causing the function to be undefined, the functions domain is all real numbers. The lowest point on the parabola is the vertex, so the lowest output value for the function is $y = 2$, and the outputs continue up to infinity. Thus we get

Domain: All real numbers or $(-\infty, \infty)$

Range: $y \geq 2$ or $[2, \infty)$.

b. We graphed this function in Example 2. Looking at the graph we can visualize the graph continuing to go downward and get wider as x increases to infinity and decreases to negative infinity.

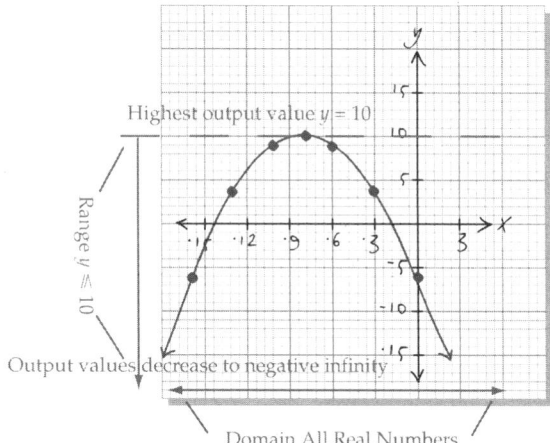

No real number will make this function undefined, so the domain of the function is all real numbers. The highest point on the parabola is the vertex, so the lowest output for the function is $y = 10$ and goes down to negative infinity from there, so we get

> Domain: All real numbers or $(-\infty, \infty)$
>
> Range: $y \le 10$ or $(-\infty, 10]$.

c. Without graphing the function, we know that no real number will make this function undefined, so the domain is all real numbers. Because the function is in vertex form, we know the vertex is $(3, -5)$ and that the parabola faces upward, Because a is positive. The vertex being the lowest point means that $y = -5$ is the lowest output value on the graph, and the outputs will continue to go upward to infinity. Thus we get

> Domain: All real numbers or $(-\infty, \infty)$
>
> Range: $y \ge -5$ or $[-5, \infty)$.

EXAMPLE ⑤ PRACTICE PROBLEM

Give the domain and range for the following:

a.

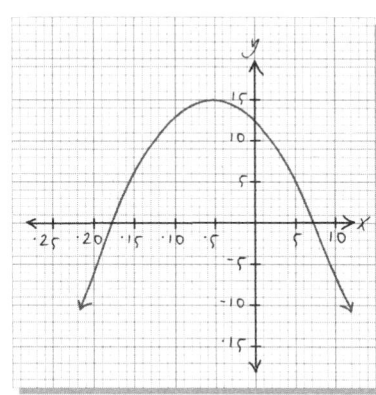

b. $f(x) = 2(x+7)^2 + 4$

c. $g(x) = -0.3(x-2.7)^2 - 8.6$

You can see from Example 5 that the domain of a quadratic function that has no context is all real numbers, because no real number will make the function undefined. The range of a quadratic function without a context will start or end with the output value of the vertex because this value is either the lowest or highest output value of the function. If the parabola faces upward, the range will go up to infinity. If the parabola faces downward, the range will go down to negative infinity.

DOMAIN AND RANGE OF QUADRATIC FUNCTIONS

Domain: all real numbers, $(-\infty, \infty)$

Range: If the vertex of the quadratic function is (h, k) then the range is

$(-\infty, k]$, or $y \leq k$ if the graph is facing downward

$[k, \infty)$, or $y \geq k$ if the graph is facing upward

4.2 Exercises

For exercises 1 through 10, refer to the values of a, h and k in the vertex form of a quadratic.

$$f(x) = a(x - h)^2 + k$$

1. Does a, h, or k shift the graph to the right? Should that value be positive or negative?

2. Does a, h, or k cause the graph to face downward? Should that value be positive or negative?

3. Does a, h, or k shift the graph down? Should that value be positive or negative?

4. Does a, h, or k affect the width of the graph? Considering a, h, and k what values will result in the graph being wider than the graph of $f(x) = x^2$?

5. Considering a, h, and k, what values will result in the graph being more narrow than the graph of $f(x) = x^2$?

6. Considering a, h, and k, what values will result in a parabola that has a vertex of $(2, 5)$?

7. Does a, h, or k cause the graph to face upward? Should that value be positive or negative?

8. Considering a, h, and k, what values will result in a downward facing parabola with a vertex $(0, 3)$?

9. Considering a, h, and k, what values will result in a parabola with a vertex at the origin?

10. Considering a, h, and k, what values will result in a parabola with a vertex on the horizontal axis?

For exercises 11 through 18, refer to the given graphs to answer the questions. Assume all functions are written in the vertex form of a quadratic.

11.

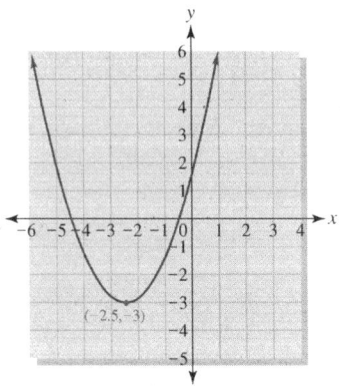

a. Where is the vertex of this graph?

b. What is the axis of symmetry for this graph?

c. What point is the symmetric pair to the point $(0.5, 4)$?

12.

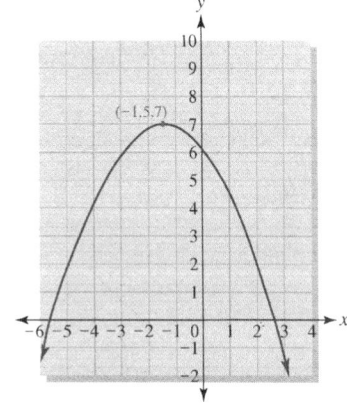

a. Where is the vertex of this graph?

b. What is the axis of symmetry for this graph?

c. What point is the symmetric pair to the point $(2, 2)$?

13.

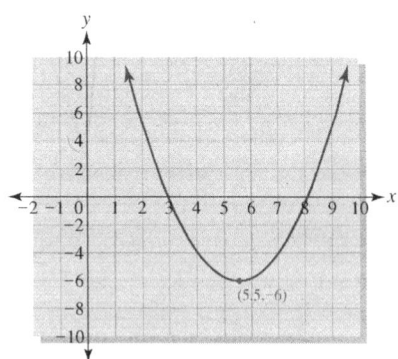

a. Where is the vertex of this graph?

b. What is the axis of symmetry for this graph?

c. What point is the symmetric pair to the point $(9, 5)$?

14.

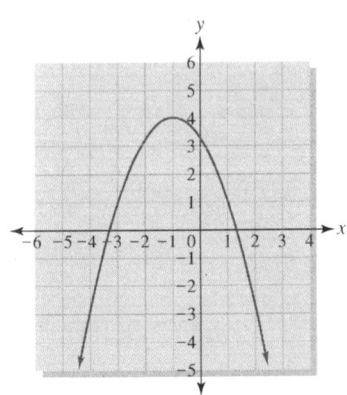

a. Where is the vertex of this graph?

b. What is the axis of symmetry for this graph?

c. What point is the symmetric pair to the point $(2, -3)$?

15.

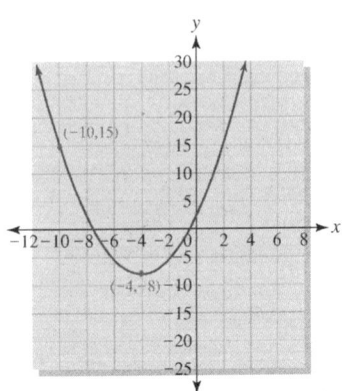

a. Where is the vertex of this graph?

b. What is the axis of symmetry for this graph?

c. Is the value of a positive or negative?

d. In vertex form, what are the values of h and k?

e. What point is the symmetric pair to the point $(-10, 15)$?

16.

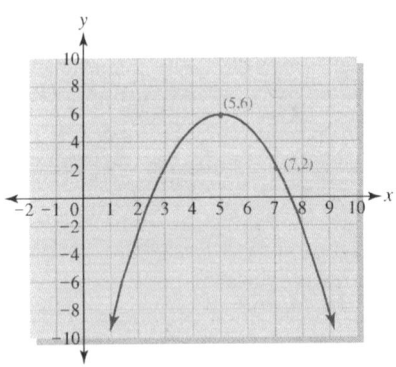

a. Where is the vertex of this graph?

b. What is the axis of symmetry for this graph?

c. Would the value of a be positive or negative?

d. In vertex form, what are the values of h and k?

e. What point is the symmetric pair to the point $(7, 2)$?

17.

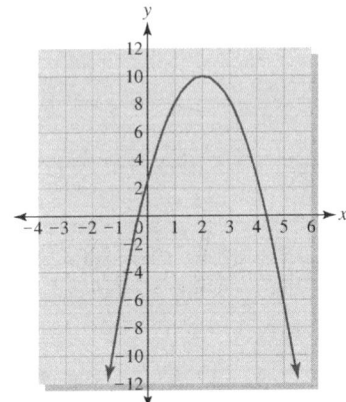

a. Where is the vertex of this graph?

b. What is the axis of symmetry for this graph?

c. Would the value of a be positive or negative?

d. In vertex form, what are the values of h and k?

e. What point is the symmetric pair to the point $(-1, -6)$?

18.

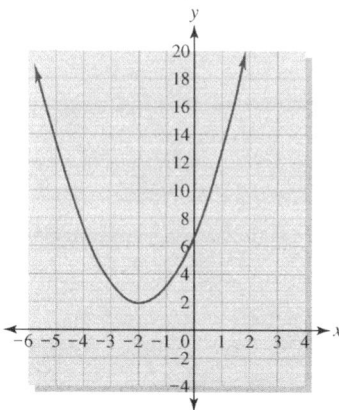

a. Where is the vertex of this graph?

b. What is the axis of symmetry for this graph?

c. Would the value of a be positive or negative?

d. In vertex form, what are the values of h and k?

e. What point is the symmetric pair to the point $(1.5, 16)$?

19. If your income level is less than the poverty threshold, you are considered to be in poverty. The poverty threshold for an individual under 65 years old can be modeled by

$$P(t) = 4.95(t - 57)^2 + 1406$$

where $P(t)$ represents the poverty threshold, in dollars, for an individual in the United States under 65 years old, t years since 1900.

Source: Model based on data from Statistical Abstract 2001.

a. What was the poverty threshold in 1990?

b. Find $P(80)$ and explain its meaning in this context.

c. Sketch a graph of this model.

d. According to this model, when did the poverty threshold reach a minimum? Does this minimum make sense?

e. Use your graph to estimate when the poverty threshold was $3000.

f. If the domain for this model is $[57, 90]$, find the range.

20. The total number of people, in thousands, who were below the poverty level can be modeled by

$$N(t) = 155(t - 9)^2 + 33417$$

where $N(t)$ represents the total number of people in the United States who were below the poverty level, in thousands, t years since 1990.

Source: Model based on data from Statistical Abstract 2007.

a. How many people were below the poverty level in 1999?

b. Find $N(15)$ and explain its meaning in this context.

c. Sketch a graph of this model.

d. Estimate the vertex of this model and explain its meaning in this context.

e. Use your graph to estimate when the number of people in the United States below the poverty level was 37 million.

f. If the domain for this model is $[3, 18]$ find the range.

21. The amount that personal households and nonprofit organizations have invested in time deposits and savings deposits can be modeled by

$$D(t) = 73.6(t - 3.5)^2 + 2176$$

where $D(t)$ represents the amount in billions of dollars the households and nonprofit organizations have invested in time and savings accounts t years since 1990.

a. Sketch a graph of this model.

b. How much did households and nonprofit organizations have invested in time and savings accounts in 1996?

c. In what year were these accounts at their lowest levels?

d. Use your graph to estimate when the amount that households and nonprofit organizations have invested in time and savings accounts reached $3000 billion.

e. If the domain for this model is $[0, 12]$, find the range.

22. The percent of the female population that experiences congestive heart failure based on age in years can be modeled by

$$H(a) = 0.004(a - 26)^2 + 0.007$$

where $H(a)$ represents the percent, as a whole number, of women a years old who experience congestive heart failure.

Source: Heart Disease and Stroke Statistics, 2004 Update, American Heart Association.

a. Sketch the graph of this model.

b. Find $H(30)$ and explain what it means in this context.

c. According to this model, at what age does the lowest percentage of women experience congestive heart failure?

d. Use your graph to estimate the age at which 5% of women would expect to experience congestive heart failure.

e. Give a reasonable domain and range for this model.

23. The height of an object dropped from the roof of a building can be found using the model

$$h(t) = -16t^2 + 256$$

where $h(t)$ is the height of the object in feet t seconds after being dropped.

a. Find $h(2)$ and explain its meaning in this context.

b. Sketch a graph of this model.

c. Use your graph to estimate when the object will hit the ground.

d. Give a reasonable domain and range for this model. (Hint: this model will not work after the ball hits the ground.)

24. A person getting ready to jump out of a plane dropped her altimeter, an instrument that gives the altitude. The height of the altimeter can be modeled by

$$h(t) = -16t^2 + 16000$$

where $h(t)$ is the height of the altimeter in feet t seconds after being dropped.

a. Find $h(10)$ and explain its meaning in this context.

b. Sketch a graph of this model.

c. Use your graph to estimate when the altimeter will hit the ground.

d. Give a reasonable domain and range for this model.

25. The monthly profit made from selling round trip airline tickets from New York City to Orlando, Florida, can be modeled by

$$P(t) = -0.000025(t - 120)^2 + 2.75$$

where $P(t)$ is the monthly profit in millions of dollars when the tickets are sold for t dollars each.

a. Find $P(100)$ and explain its meaning in this context.

b. What price will maximize the monthly profit for this route? What is the maximum profit?

c. Sketch a graph of this model.

d. Use your graph to estimate what price the company should set to have a monthly profit of 2 million dollars.

e. Give a reasonable domain and range for this model.

26. An outlet electronics store has calculated from past sales data the revenue from selling refurbished iPod nanos. The weekly revenue from selling these refurbished iPod nanos can be modeled by

$$R(n) = -1.5(n - 60)^2 + 5700$$

where $R(n)$ is the weekly revenue in dollars from selling n refurbished iPod nanos.

a. What is the weekly revenue from selling 30 refurbished iPod nanos?

b. Find the maximum weekly revenue and how many refurbished iPod nanos must be sold per week to make the maximum revenue.

c. Sketch a graph of this model.

d. Use your graph to estimate how many refurbished iPod nanos must be sold in a week to make revenue of $5400.

e. Give a reasonable domain and range of this model.

27. The Centers for Disease Control and Prevention (CDC) collect data on percentage of deaths caused by influenza (flu) and pneumonia to help alert health officials of flu epidemics that may be occurring in their areas. If the percentage of deaths grows to more than the epidemic threshold, then an epidemic is declared. The epidemic threshold for the first 6 months of 2006 can be modeled by

$$E(w) = -0.006(w - 10)^2 + 8$$

where $E(w)$ is the percentage, as a whole number, of all deaths due to pneumonia and influenza that represents the epidemic threshold during the wth week of the year.

a. Find $E(4)$ and explain its meaning in this context.

b. Sketch a graph of this model.

c. Use your graph to determine if there was a flu epidemic during the 23rd week of 2006 if 6.4% of deaths were caused by pneumonia or influenza.

d. What is the vertex of this model and what does it represent in this context?

e. What is a reasonable domain and range for this model?

28. The total fresh vegetables consumed per person per year can be modeled by

$$V(t) = 0.7(t - 1.5)^2 + 95.4$$

where $V(t)$ is the total fresh vegetables consumed per person in pounds per year t years since 2000.

Source: http://www.ers.usda.gov/data/foodconsumption/ accessed on 12/29/2006.

a. Find $V(5)$ and explain its meaning in this context.

b. Find the vertex of this model and explain its meaning in this context.

c. Sketch a graph of this model for the years 2000 through 2010.

d. Use your graph to estimate the year(s) when the total fresh vegetables consumed per person per year is 125 pounds.

e. If the domain of this model is [0, 10] find the range.

For Exercises 29 through 34, give the domain and range of the quadratic functions.

29. $f(x) = -2(x + 4)^2 + 15$

30. $g(x) = -0.25(x - 5)^2 + 17$

31. $f(x) = 5(x + 4)^2 - 58$

32. $h(x) = 0.3(x - 2.7)^2 + 5$

33.

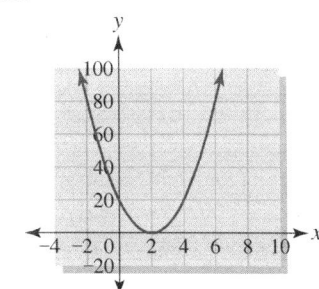

34.

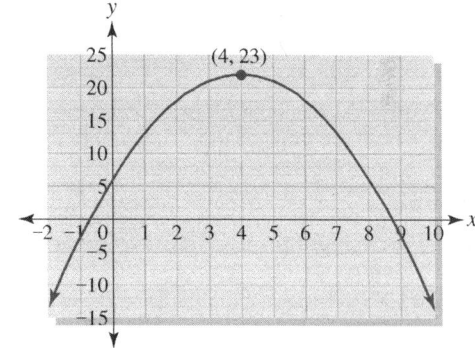

For Exercises 35 through 62, sketch the graph of the given functions and label the vertex, vertical intercept, and at least one additional symmetric pair. Give the domain and range of the function.

35. $g(x) = (x - 7)^2 + 9$

36. $f(x) = (x - 5)^2 + 3$

37. $f(x) = (x + 4)^2 + 2$

38. $w(t) = (t + 3)^2 - 8$

39. $h(t) = (t - 3)^2 - 4$

40. $f(x) = (x + 4.5)^2 - 3$

41. $k(n) = 2n^2$

42. $b(w) = -3w^2$

43. $f(x) = -1.5x^2$

44. $g(t) = 0.5t^2$

45. $h(x) = x^2 + 4$

46. $r(t) = 2t^2 - 12$

47. $f(x) = 0.4x^2 + 2$

48. $g(t) = -0.25t^2 + 6$

49. $g(x) = 4(x - 5)^2$

50. $m(c) = 0.3(c + 2)^2$

51. $f(x) = -2(x + 4)^2$

52. $r(t) = -0.15(t - 7)^2$

53. $q(a) = 3(a + 7)^2 + 2$

54. $B(n) = 4(n - 3)^2 - 4$

55. $T(z) = 0.2(z + 6)^2 - 8$

56. $g(m) = 2.4(m + 4)^2 + 20$

57. $h(x) = -5(x - 17)^2 - 15$

58. $h(d) = -0.4(d + 15)^2 - 8$

59. $s(t) = -0.25(t + 50)^2 + 25$

60. $C(u) = -0.007(u - 400)^2$

61. $p(w) = 123(w - 4)^2 - 2500$

62. $f(x) = 5(x + 13)^2 - 45$

In Exercises 63 through 72, answer the following questions:

a. Find the vertex.

b. Is the parabola wide or narrow?

c. Does the parabola face upward or downward?

d. Give a calculator window that will give a good picture of this parabola.

- X Min   X Max
- Y Min   Y Max

63. $f(x) = 5x^2 + 100$

64. $f(x) = -0.01x^2 + 200$

65. $f(x) = (x + 30)^2 - 50$

66. $f(x) = 20(x - 100)^2 + 250$

67. $f(x) = 0.002(x + 20)^2 + 50$

68. $f(x) = 0.0005(x - 1000)^2$

69. $f(x) = 0.0005(x - 1000)^2 + 1000$

70. $f(x) = -0.0005(x + 10,000)^2 + 5000$

71. $f(x) = -10(x + 25,000)^2 - 10,000$

72. $f(x) = (x - 55.8)^2 - 29.7$

Finding Quadratic Models

Now we will use all the information we have about graphing quadratics from the vertex form to find models for data that take the shape of a parabola. We will also look more at the domain and range of quadratic functions and what you should be aware of when considering model breakdown. Use the following steps to find a model for quadratic data.

> ### MODELING STEPS FOR QUADRATICS
> **1.** Define the variables and adjust the data (if needed).
> **2.** Create a scatterplot.
> **3.** Select a model type.
> **4. Quadratic Model:** Pick a vertex and substitute it for h and k in the vertex form $f(x) = a(x - h)^2 + k$.
> **5.** Pick another point and use it to find a.
> **6.** Write the equation of the model using function notation.
> **7.** Check your model by graphing it with the scatterplot.

FINDING QUADRATIC MODELS

EXAMPLE   **1**   FINDING A QUADRATIC MODEL

The average monthly temperature in Anchorage, Alaska, is given in the chart.

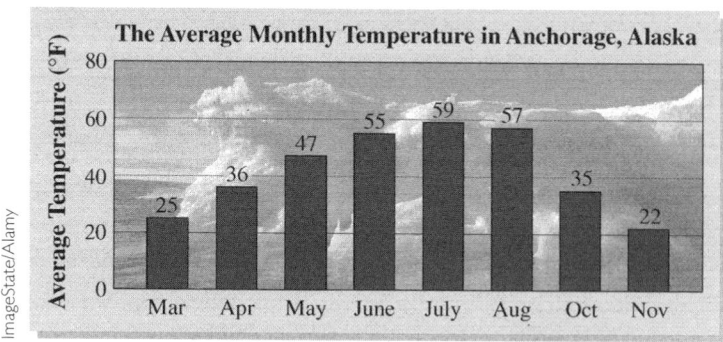

Source: Weatherbase.com.

a. Will a linear or quadratic model fit these data better? Why?

b. Find an equation for a model of these data.

c. Using your model, estimate the average temperatures during September and February. Would either of these months be a good time to travel to Alaska if you wanted to visit when temperatures were mild?

d. The actual average during September is 48°F, and that during February is 18°F. How well does your model predict these values?

e. Give a reasonable domain and range for this model.

Solution

a. A quadratic model will fit these data best because the distribution is shaped like a downward-facing parabola.

b. **Step 1** Define the variables and adjust the data (if needed).

T = Average temperature in Anchorage Alaska (in °F)

m = The month of the year; e.g., $m = 5$ represents May

The months given in the bar chart must be translated into numerical values as shown in the table.

m	T (in °F)
3	25
4	36
5	47
6	55
7	59
8	57
10	35
11	22

Step 2 Create a scatterplot.

The inputs go from 3 to 11, and the outputs go from 22 to 59, so spreading out some has the following result.

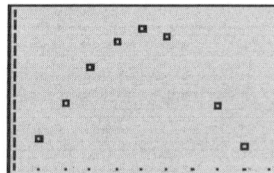 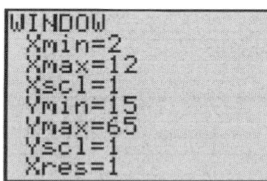

Step 3 Select a model type. We now have linear and quadratic to choose from.

Because these data have the shape of a downward-facing parabola, we should choose a quadratic model.

Step 4 **Quadratic Model:** Pick a vertex and substitute it for h and k in the vertex form.

The highest point on this scatterplot looks like a reasonable vertex, so we will use it as the vertex.

$$\text{Vertex} = (7, 59)$$
$$T(m) = a(m - 7)^2 + 59$$

Step 5 Pick another point and use it to find a.

We can choose the last point in the data set because it is farther away from the vertex and seems to follow a smooth curve.

Other point = (11, 22)

$$T(m) = a(m - 7)^2 + 59$$

$$22 = a(11 - 7)^2 + 59 \qquad \text{Substitute 11 for the month and 22 for the average temperature.}$$

$$22 = a(4)^2 + 59$$

$$22 = 16a + 59 \qquad \text{Solve for } a.$$

$$\underline{-59 \qquad\qquad -59}$$

$$-37 = 16a$$

$$\frac{-37}{16} = \frac{16a}{16}$$

$$-2.3125 = a$$

Step 6 Write the equation of the model using function notation.

You now have a, h, and k, so you can write the model.

$$T(m) = -2.3(m - 7)^2 + 59$$

Step 7 Check your model by graphing it with the scatterplot.

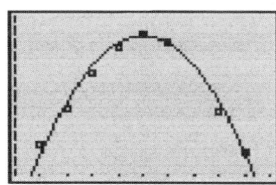

c.   September is the ninth month of the year, and February is the second month, so

$$T(m) = -2.3(m - 7)^2 + 59$$

$$T(9) = -2.3(9 - 7)^2 + 59$$

$$T(9) = 49.8$$

and

$$T(2) = -2.3(2 - 7)^2 + 59$$

$$T(2) = 1.5$$

The average temperature in Anchorage, Alaska, during September is 49.8°F, and that during February is 1.5°F. February would not be a good month to travel, and September would be questionable for most people.

d.   Our estimate for September is fairly accurate, but the estimate for February is not close to the actual value. This estimate probably represents model break-down in this situation.

e.   From what we saw in part d, the model will work well within the given data but not so well beyond the data. Therefore we can set the domain to be [3, 11]. To find the range for this model, we need to look for the lowest and highest points on this model within the chosen domain. The highest point is clearly the vertex and has a value of 59. The lowest point is on either the right or the left hand side of the domain. To determine what the lowest output is, we will substitute the two ends of the domain and find the related output values.

$$T(3) = 22.2$$
$$T(11) = 22.2$$

Because both of these values happen to be the same, the lowest output value is 22.2, so our range will be [22.2, 59].

EXAMPLE ① PRACTICE PROBLEM

The number of cable television systems in the United States during the 1990s are given below.

Year	Cable Television Systems
1992	11,075
1993	11,217
1995	11,218
1996	11,119
1997	10,950

Source: Statistical Abstract 2001.

a. Find an equation for a model of these data.

b. Explain what the vertex means in this context.

c. How many cable television systems does your model predict there will be in 1999?

d. Give a reasonable domain and range for your model.

e. Find the vertical intercept for your model and explain its meaning in this context.

f. Use a table or graph to estimate when there will only be 10,500 cable television systems.

As you can see from Example 1, the modeling process for quadratics is about the same as that for linear equations. If you can pick a reasonable vertex, the process is not difficult. If your model does not fit well, you might want to check your calculations from step 6. Students most often make mistakes when solving for the value of *a*.

DOMAIN AND RANGE

Finding a reasonable domain is exactly the same as for linear models. Try to expand the domain beyond the given data unless model breakdown occurs. You will want to use more caution when considering model breakdown, as many quadratic models are not reliable very far beyond the given range of data.

Finding the range still involves the lowest to highest output values of the function within the set domain. Unlike with linear models you must be careful not to assume that the lowest and highest outputs of the function will be the outputs from the ends of the domain. Because quadratics have a maximum or minimum point at the vertex, the range will typically include the output value of the vertex as the highest or lowest value. If the vertex is within the domain values, it will be one endpoint of the range, and one of the ends of the domain will provide the other range endpoint. Be sure to give the range stating the lowest output followed by the highest output value.

EXAMPLE 2 FINDING A QUADRATIC MODEL

The number of public branch libraries in the United States for several years is given below.

Year	Public Library Branches
1994	6223
1995	6172
1996	6205
1997	6332
1998	6435

© Blend Images/Alamy

Source: Statistical Abstract 2001.

a. Find an equation for a model of the

b. Using your model, estimate the number of branch libraries in 2000.

c. Give a reasonable domain and range for this model.

d. Use the graph or table to estimate in what year(s) there are 6500 branch libraries.

e. Use this model to determine whether it would be reasonable to assume that the number of careers in library science is going to be growing in the future.

Solution

a. **Step 1** Define the variables and adjust the data (if needed).

B = The number of public branch libraries in the United States

t = Time in years since 1990

Adjusting the years to years since 1990 give us

t	B
4	6223
5	6172
6	6205
7	6332
8	6435

Step 2 Create a scatterplot.

The inputs go from 4 to 8, and the outputs go from 6172 to 6435, so spreading out some has the following result.

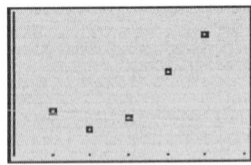 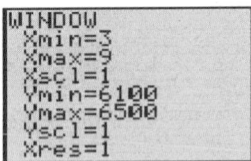

Step 3 Select a model type.

Because the data show the shape of an upward-facing parabola, we should choose a quadratic model.

Step 4 Quadratic Model: Pick a vertex and substitute it for h and k in the vertex form.

The lowest point on this scatterplot looks like a reasonable vertex, so we will use it for the model.

$$\text{Vertex} = (5, 6172)$$
$$B(t) = a(t - 5)^2 + 6172$$

Step 5 Pick another point and use it to find a.

We can choose the last data point because it is farther away from the vertex and seems to follow a smooth curve.

Other point $= (8, 6435)$

$$B(t) = a(t - 5)^2 + 6172$$
$$6435 = a(8 - 5)^2 + 6172$$
$$6435 = a(3)^2 + 6172$$
$$6435 = 9a + 6172$$
$$\underline{-6172 \qquad\quad -6172}$$
$$263 = 9a$$
$$29.22 = a$$

Substitute 8 for the year and 6435 for the number of public branch libraries.

Solve for a.

Step 6 Write the equation of the model using function notation.
You now have a, h, and k, so you can write the model.

$$B(t) = 29.22(t - 5)^2 + 6172$$

Step 7 Check your model by graphing it with the scatterplot.

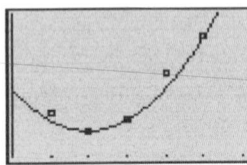

This model fits the data fairly well, but the two points that it misses are both above the model, so moving the model up slightly might make a better fit. We can adjust the model for a better fit by making k bigger. The new model becomes

$$B(t) = 29.22(t - 5)^2 + 6180$$

The graph of this adjusted model is closer overall to all the data points.

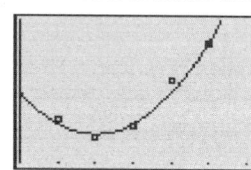

Note that the vertex of our adjusted model differs from the one we picked in step 5. There are several answers for this model that would be considered reasonable in this case. Remember that we are trying to find an "eyeball best fit" so there can be some room for differences in final models.

b. The year 2000 is represented by $t = 10$, so

$$B(10) = 29.22(10 - 5)^2 + 6180$$
$$B(10) = 6910.5$$

In 2000, there were approximately 6911 public branch libraries.

c. It is probably reasonable to assume that there will continue to be more public library branches as populations continue to grow. Therefore; we will spread out the domain to include a few years beyond the given data, giving us a domain of [3, 10]. The resulting range will have a minimum value at the vertex of 6180, and the maximum appears to be on one of the ends of the domain. $B(3) = 6296.88$ and $B(10) = 6910.5$, the maximum value is 6910.5 when $t = 10$ therefore the range will be [6180, 6910.5].

d. Using the table, we get an estimate of $t = 2$ or $t = 8$.

X	Y1	
1	6647.5	
2	6443	
3	6296.9	
4	6209.2	
6	6209.2	
7	6296.9	
8	6443	

Y1☐29.22(X-5)²+...

So there were about 6500 branch libraries in 1992 and again in 1998. Remember that you are solving a quadratic equation, and because of its symmetry, you will usually have two solutions to this type of question. In some cases, one of those solutions may not make sense in the context of the situation and, therefore, would be considered model breakdown. In this case, both answers lie within our domain and seem reasonable, we will keep them both.

e. If the number of branch libraries continues to grow, the number of library science careers should grow as well.

Some data have the same curve as a parabola but do not form both sides of the "U" shape. When data only shows half of the parabola shape, you need to use caution in determining the domain and range. Model breakdown can occur to the left or right of the vertex.

EXAMPLE   **3**   FINDING A QUADRATIC MODEL

The number of cellular telephone subscribers in the United States is given in the chart.

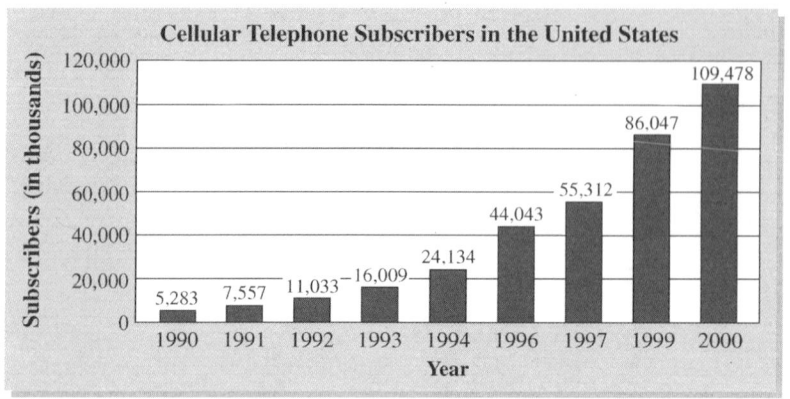

Source: Statistical Abstract 2001.

a. Find a model for these data.

b. Estimate the number of cellular phone subscribers in 1980 and 1995.

c. The actual number of cellular phone subscribers in 1995 was 33,786,000. How does your estimate from part b compare to the actual value?

d. Give a reasonable domain and range for your model.

Solution

a. **Step 1** Define the variables and adjust the data (if needed).

C = The number of cellular telephone subscribers in the United States in thousands

t = Time in years since 1990

Adjusting the years to years since 1990 gives us

Year	Cellular Telephone Subscribers *(in thousands)*
0	5,283
1	7,557
2	11,033
3	16,009
4	24,134
6	44,043
7	55,312
9	86,047
10	109,478

Step 2 Create a scatterplot.

The inputs go from 0 to 10, and the outputs go from 5,283 to 109,478, so spreading out the X and Y max and min has the following result.

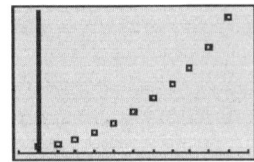 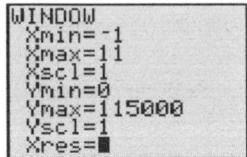

Step 3 Select a model type.

Because the distribution curves upward, the data take on the shape of the right half of an upward-facing parabola. We should choose a quadratic model.

Step 4 **Quadratic Model:** Pick a vertex and substitute it for h and k in the vertex form.

The lowest point on this scatterplot looks like a reasonable vertex, so we will use it for the model.

$$\text{Vertex} = (0, 5283)$$
$$C(t) = a(t - 0)^2 + 5283$$
$$C(t) = at^2 + 5283 \qquad \text{Simplify the function.}$$

Step 5 Pick another point and use it to find a.

We can choose the seventh point in the data because it is some distance from the vertex and seems to follow a smooth curve that will come close to all the data points.

$$\text{Other point} = (6, 44043)$$

$$C(t) = at^2 + 5283$$
$$44043 = a(6)^2 + 5283 \qquad \begin{array}{l}\text{Substitute 6 for the year and 44043 for the} \\ \text{number of cellular telephone subscribers.}\end{array}$$
$$44043 = 36a + 5283 \qquad \text{Solve for } a.$$
$$\underline{-5283 \qquad\qquad -5283}$$
$$38760 = 36a$$
$$1076.67 = a$$

Step 6 Write the equation for the model using function notation.

You now have a, h, and k, so you can write the model.

$$C(t) = 1076.67t^2 + 5283$$

Step 7 Check your model by graphing it with the scatterplot.

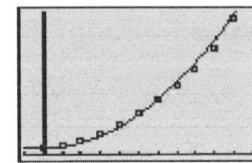

This model is not bad since it appears to be pretty balanced within the data with several points above and below the graph. Because all the points above the graph are on the left half of the graph and all the points below the graph are on the right side of the graph, we might be able to adjust it by raising the graph and then making the parabola a little wider. These adjustments will mean changing the values of a and k.

By playing with several possibilities, we might come up with the model

$$C(t) = 1000t^2 + 7000$$

and the following graph.

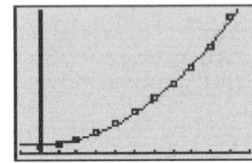

This graph comes closer to more points.

b. 1980 and 1995 are represented by $t = -10$ and $t = 5$, respectively.

$$C(-10) = 107000$$
$$C(5) = 32000$$

In 1980, there were 107,000,000 cellular telephone subscribers in the United States. This does not make sense, since that number would not decrease so drastically during the next 10 years, so we believe that this is model breakdown. The data indicated that only half the parabola would be reasonable in this situation.

In 1995 there were 32,000,000 cellular telephone subscribers.

c. Comparing the actual value in 1995 to our estimate, we are off by a little over 1 million subscribers, but that is not a very large error considering the size of these numbers. We can be satisfied with this result.

d. There were probably fewer cellular telephone subscribers before 1990. This graph trends upward to the left of the vertex that result leads us to believe that this model is not valid for years to the left of the vertex. It does appear that we could extend the curve farther to the right. So we can use the domain [0, 12].

The resulting range will start at the lowest point, the vertex, and go up until it reaches the highest point at the right-hand side of the domain.

$$\text{Vertex:} \quad (0, 7000)$$
$$C(12) = 151000$$

This gives a range of [7000, 151000].

EXAMPLE ③ PRACTICE PROBLEM

Find an equation for a model of the data given in the table below.

x	f(x)
−6	−33
−4	−45
−2	−49
0	−45
2	−33
4	−13

4.3 Exercises

For Exercises 1 through 6, graph the given data and functions on your graphing calculator and choose which of the functions best fits the given data.

1.

$f(x) = 1.5(x+5)^2 - 15 \qquad f(x) = 1.8(x+5)^2 - 15$

$f(x) = 2.0(x+5)^2 - 15 \qquad f(x) = 2.3(x+5)^2 - 15$

x	f(x)
−10	35
−7	−7
−5	−15
−3	−7
1	−57
3	−113

2.

$f(x) = -1.3(x-3)^2 + 9 \qquad f(x) = -1.4(x-3)^2 + 9$

$f(x) = -1.5(x-3)^2 + 9 \qquad f(x) = -1.6(x-3)^2 + 9$

x	f(x)
−5	−80.6
−2	−26.0
0	−3.6
2	7.6
5	3.4
7	−13.4

3.

$f(x) = 0.4(x-4)^2 + 1.2 \qquad f(x) = 0.4(x-4)^2 + 2.2$

$f(x) = 0.4(x-4)^2 + 3.2 \qquad f(x) = 0.4(x-4)^2 + 4.2$

x	f(x)
−8	60.8
−3	22.8
0	9.6
4	3.2
7	6.8
10	17.6

4.

$f(x) = -(x+5)^2 + 16.2 \qquad f(x) = -(x+5)^2 + 15.2$

$f(x) = -(x+5)^2 + 14.2 \qquad f(x) = -(x+5)^2 + 13.2$

x	f(x)
−10	−11.8
−8	4.2
−6	12.2
−4	12.2
−2	4.2
2	−35.8

5.

$f(x) = 2(x+1.0)^2 - 20 \qquad f(x) = 2(x+1.2)^2 - 20$

$f(x) = 2(x+1.5)^2 - 20 \qquad f(x) = 2(x+1.8)^2 - 20$

x	f(x)
−4	−7.5
−2	−19.5
0	−15.5
2	4.5
4	40.5
6	92.5

6.

$f(x) = -3(x-8.0)^2 + 8 \qquad f(x) = -3(x-8.4)^2 + 8$

$f(x) = -3(x-9.0)^2 + 8 \qquad f(x) = -3(x-9.4)^2 + 8$

x	f(x)
2	−156.3
4	−79.48
6	−26.68
8	2.12
10	6.92
12	−12.28

For Exercises 7 through 12, graph the data and the function on your calculator and adjust a, h and/or k to get an eye-ball best fit. Remember your answers will vary from the back of the book.

7. $f(x) = 2.7(x-5)^2 - 15$

x	f(x)
−1	88
0	59
2	15
4	−6
7	2
9	34

8. $f(x) = 0.6(x+2)^2 + 4$

x	f(x)
−5	7.6
−2	4
0	5.6
2	10.4
5	23.6
7	36.4

9. $f(x) = 2(x-7)^2 + 15$

x	f(x)
−1	−113
2	−35
4	−3
6	13
8	13
10	−3

10. $f(x) = -1.3(x-3.1)^2 + 18.6$

x	f(x)
0	−9.2
1	−0.9
2	4.8
3	8.0
4	8.5
5	6.4

11. $f(x) = 4x^2 - 10$

x	f(x)
−6	54
−4	6
−2	−10
0	6
2	54
4	134

12. $f(x) = -(x+3)^2$

x	f(x)
−9	−11
−6	16
−3	25
0	16
3	−11
6	−56

For Exercises 13 through 20, state which of a, h, and/or k in the vertex form you would change to make the parabola fit the data better.

13.

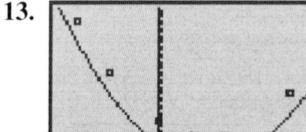

14.

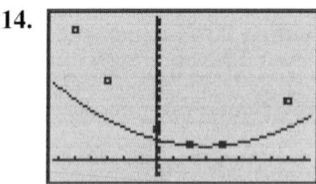

15.

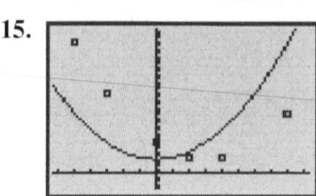

16.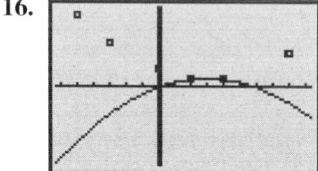

17.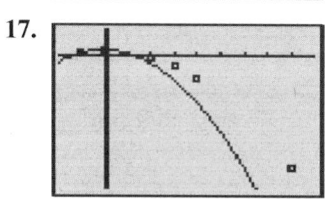

18.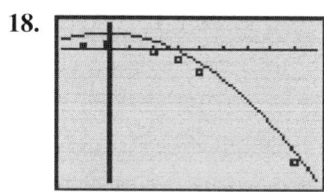

19.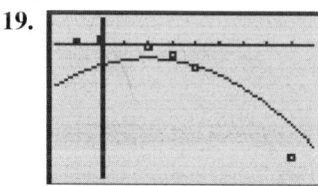

20.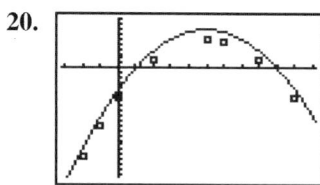

21. The average monthly low temperatures in Anchorage, Alaska for certain months are given below.

Month	Average Low Temperature (in °F)
April	28
May	39
June	47
July	51
August	49
September	41
October	28
November	15

Source: Weatherbase.com.

a. Find an equation for a model of these data.

b. Explain the meaning of the vertex for your model in this context.

c. What does your model predict the average low temperature will be in March?

d. Give a reasonable domain and range for your model.

22. The Reduced-Gravity Program operated by NASA provides astronauts with the unique "weightless" or "zero-g" environment of space flight for tests and training. The program uses specially modified KC-1 35A turbojet transport that flies on a parabolic path along the data below.

Time (seconds)	Altitude (thousands of feet)
10	26
20	31
32	32.5
45	31
60	25

Source: NASA Lyndon B. Johnson Space Center Reduced-Gravity Office

a. Find an equation for a model of these data.

b. Find the altitude of the plane after 25 seconds.

c. Give a reasonable domain and range for your model.

23. The number of Hispanic families living below the federal poverty level in the United States is given for several years in the chart.

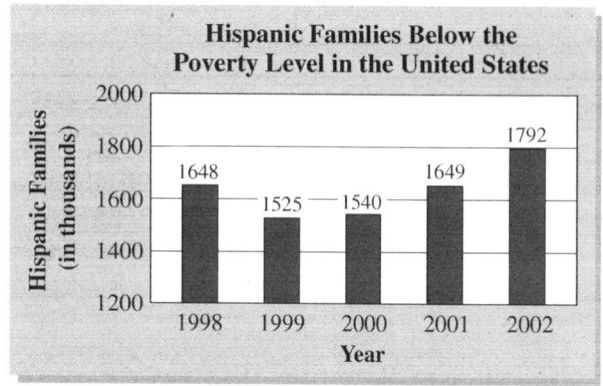

Source: Statistical Abstract 2004–2005.

a. Find an equation for a model of these data.

b. Estimate the number of Hispanic families below the poverty level in 2005.

c. Explain the meaning of the vertex in this context.

d. Give a reasonable domain and range for your model.

24. Of the total number of people who own a home computer, the percent of those who use their home computers six or seven days a week varies by age. Some of these percentages are given below.

Age	Percent
25	26.7
35	24.2
45	24.3
55	26.6
65	29.7

Source: Statistical Abstract 2001.

a. Find an equation for a model of these data.

b. Give a reasonable domain and range for this model.

c. Estimate the percent of 15 year olds who use their home computers six or seven days a week.

25. The number of hours per year an average American spent using the Internet is given in the chart.

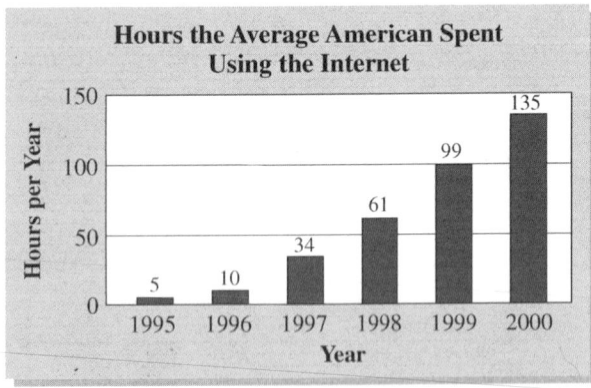

Hours the Average American Spent Using the Internet

Source: Statistical Abstract 2001.

a. Find an equation for a model of these data.

b. Determine whether model breakdown occurs near the vertex.

c. Give a reasonable domain and range for this model.

d. Estimate the number of hours per year the average person spent on the Internet in 2001.

e. Use a table or graph to estimate when the average person will spend 365 hours per year on the Internet.

26. In 2007 Mattell and other manufacturers had to recall several products because plants in China had used paint and other materials with high concentrations of lead. In the United States the use of lead in paint has been decreasing since its peak around 1920. Lead use in paint for several years is given below.

Year	Lead Use (in thousands of tons)
1940	70
1950	35
1960	10
1970	5
1980	0.01

Source: Estimated from information at AmericanScientist.org.

a. Find an equation for a model of these data.

b. Estimate the lead used in paints in 1955.

c. Estimate the lead used in paints in 2000.

d. Give a reasonable domain and range for this model.

e. Use a table or the graph to estimate when 50 thousand tons of lead were used in paints.

27. A hit at a major league baseball game flew along the following path from home plate.

Horizontal Distance (ft)	Height (ft)
10	15
75	78
110	103
190	136
230	140
310	123
400	65

a. Find an equation for the model of these data.

b. Give a reasonable domain and range for this model.

c. Find the vertex of this model and explain its meaning in this context.

d. If the center wall of the stadium is 450 feet from home plate and is 10 feet tall, will the ball make it over the wall?

28. A fly ball hit at the same major league baseball game had the following vertical heights.

Time (seconds)	Height (ft)
1	132
3	292
5	324
6	290
7	228
9	4

a. Find an equation for the model of these data.

b. Give a reasonable domain and range for this model.

c. Estimate the height of the ball after 8 seconds.

d. Use your graph to estimate how long it took for the ball to hit the ground.

e. Find the maximum height of the ball and when it reached that height.

For Exercises 29 through 42, find an equation for a model of the given data. Give the domain and range for

the model you found. (Hint: These problems do not have a context.)

29.

Input	Output
−7	−8
−5	8
−3	−8
−1	−56
1	−136
2	−188

30.

Input	Output
−20	121.2
−10	140.2
4	150.0
15	143.95
30	116.2
50	44.2

31.

Input	Output
1	12
2	5
4	−3
6	−3
8	5
9	12

32.

Input	Output
−2	−12
−1	−9.5
0	−7
3	0.5
4	3
6	8

33.

Input	Output
−100	15,300
−80	−1500
−60	−8700
−40	−6300
−30	−1500
−20	5700

37.

Input	Output
−2	−21.4
0	−1500
2	−8.6
4	−2.2
6	4.2
8	10.6

34.

Input	Output
3.5	57.6
7	42.3
10.5	33.1
17.5	33.1
21	42.3
24.5	57.6

38.

Input	Output
−3	30
−2	10
−1	−2
1	−2
2	10
3	30

35.

Input	Output
−2	−50
0	−22
2	−2
4	10
6	14
8	10

39.

Input	Output
−4	−16
−2	2
0	8
2	2
4	−16
6	−46

36.

Input	Output
−15	89.9
−12	25.1
−9	3.5
−6	25.1
−3	89.9
0	197.9

40.

Input	Output
−2	192
0	108
2	48
4	12
6	0
8	12

41.

Input	Output
−10	−10
−8	−3.6
−6	−0.4
−4	−0.4
−2	−3.6
0	−10

42.

Input	Output
−3	103.5
−2	46
−1	11.5
0	0
1	11.5
2	46

Solving Quadratic Equations by Square Root Property and Completing the Square

In this section we will learn to solve quadratic equations using both the vertex form and the standard form. We will learn several tools that can be used to solve these equations. One thing you should consider is what each tool helps you to do and when to use a tool effectively to get the job done.

Say What?

What are all these "quadratics"?

- Quadratic expression (no equals sign)
  $2x^2 + 3x - 5$

- Quadratic equation (has an equals sign)
  $2x^2 + 3x - 5 = 20$

- Quadratic function (written in function notation)
  $f(x) = 2x^2 + 3x - 5$

- Parabola: the graph of a quadratic function

SOLVING FROM VERTEX FORM

In the previous sections we were given values for the input variable and were asked to solve the function for the output variable. This has not been too difficult, since it involves simplifying one side of an equation.

When you are given a value for the output variable, solving for the input becomes more difficult. The reason for this difficulty is that the input variable is squared and therefore, harder to isolate. Isolating the variable will require removing an exponent of 2. This cannot be done by addition, subtraction, multiplication, or division. Therefore we require an operation that will undo a squared term. That operation is the square root, $\sqrt{}$.

One characteristic of numbers you need to be cautious about is that when you square a real number, squaring will effectively remove any negative sign. You will have to take extra care when you want to undo that operation.

$$x^2 = 25$$

$$(5)^2 = 25 \qquad (-5)^2 = 25$$

$$x = 5 \qquad\qquad x = -5$$

We know that 5 squared is 25. Negative 5 squared is also 25.

Both must be given as solutions to this equation.

$$x^2 = 25$$

$$x = \pm\sqrt{25}$$

$$x = 5 \qquad\qquad x = -5$$

When using a square root, we must use the plus/minus symbol to represent both solutions.

The square root that we use in mathematics always results in a positive number. This causes us to lose possible negative results, so we must use a plus/minus symbol ($\pm$) to show that there are two possible solutions. Using the square root and plus/minus is an example of the **square root property.**

DEFINITION

Square root property: If $c \geq 0$, the solutions to the equation

$$x^2 = c$$

are

$$x = \pm\sqrt{c}$$

If c is negative, then the equation has no real solutions.

With this property in mind, we are going to look at two basic ways to solve quadratic equations — the square root property and completing the square. First, we will use the square root property to solve quadratics when they are given in vertex form.

SOLVING QUADRATIC EQUATIONS IN VERTEX FORM

1. Isolate the squared part. The squared part contains the unknown.
2. Use the square root property to eliminate the exponent of 2. Use $\pm$ on the side away from the variable.
3. Write two equations and solve.
4. Check solution(s) in the original equation. Be sure that solutions are within the domain of the model and make sense in the context of the problem.

EXAMPLE   1   SOLVING QUADRATIC EQUATIONS USING THE SQUARE ROOT PROPERTY

Solve $10 = 2(x - 4)^2 - 8$

Solution

$$10 = 2(x - 4)^2 - 8$$

$$\underline{+ 8 \qquad\qquad\qquad + 8}$$ Isolate the squared part of the equation.

$$18 = 2(x - 4)^2$$

$$\frac{18}{2} = \frac{2(x - 4)^2}{2}$$

$$9 = (x - 4)^2$$

$$\pm\sqrt{9} = x - 4$$ Use the square root property. Don't forget the plus/minus symbol.

$$\pm 3 = x - 4$$

$$3 = x - 4 \quad \text{or} \quad -3 = x - 4$$
$$7 = x \quad \text{or} \quad 1 = x$$

Rewrite into two equations and solve.

$$10 = 2(7 - 4)^2 - 8$$
$$10 = 10$$

Check both answers.

$$10 = 2(1 - 4)^2 - 8$$
$$10 = 10$$

$x = 7$ and $x = 1$ are both valid solutions to this equation.

EXAMPLE 2 SOLVING APPLICATIONS USING THE SQUARE ROOT PROPERTY

In the Example 1 Practice Problem of Section 4.3 we looked at the number of cable television systems in the United States and found the model

$$C(t) = -33.3(t - 4)^2 + 11250$$

where $C(t)$ represents the number of cable systems in the United States t years since 1990.

a. Find when there were 10,000 cable systems.

b. Find the horizontal intercepts and explain their meaning in this context.

Solution

a. The quantity 10,000 cable systems is represented by $C(t) = 10000$, so we need to solve for t.

$$C(t) = -33.3(t - 4)^2 + 11250$$
$$10000 = -33.3(t - 4)^2 + 11250$$
$$\underline{-11250 \qquad\qquad\qquad -11250}$$
$$-1250 = -33.3(t-4)^2$$

Isolate the squared term.

$$\frac{-1250}{-33.3} = \frac{-33.3(t - 4)^2}{-33.3}$$
$$37.538 = (t - 4)^2$$
$$\pm\sqrt{37.538} = t - 4$$
$$\pm 6.127 = t - 4$$

Use the square root property, being sure to use a plus/minus symbol to represent both possible solutions.

$$6.127 = t - 4 \quad \text{or} \quad -6.127 = t - 4$$
$$10.127 = t \quad\quad \text{or} \quad -2.127 = t$$

Rewrite into two equations and add 4 to both sides.

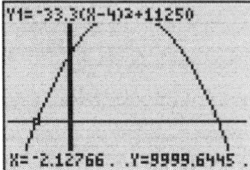 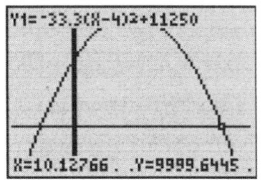

You can check your answers using the graph and trace or the table feature of your calculator.

$t = 10.127$ represents about the year 2000, and $t = -2.127$ represents about 1988. The value $t = -2.127$ is not within the domain for this model, so we must discard it, leaving us with there being about 10,000 cable systems in the year 2000.

b. To find the horizontal intercepts, we need to let the output variable equal zero and solve for t.

$$C(t) = -33.3(t - 4)^2 + 11250$$

$$0 = -33.3(t - 4)^2 + 11250$$

$$\underline{-11250 \qquad\qquad\qquad\qquad 11250} \qquad\qquad \text{Isolate the squared term.}$$

$$-11250 = -33.3(t - 4)^2$$

$$\frac{-11250}{-33.3} = \frac{-33.3(t - 4)^2}{-33.3} \qquad\qquad \text{Use the square root property, being sure to use a plus/minus symbol to represent both possible solutions.}$$

$$337.84 = (t - 4)^2$$

$$\pm\sqrt{337.84} = t - 4 \qquad\qquad \text{Rewrite into two equations and add 4 to both sides.}$$

$$\pm 18.38 = t - 4$$

$$18.38 = t - 4 \quad \text{or} \quad -18.38 = t - 4$$

$$22.38 = t \quad\quad \text{or} \quad -14.38 = t$$

X	Y1	
22.38	.44748	
-14.38	.44748	

X=

You can check your answers using the table feature of your calculator. The rounding in this solution means that the y value will not be exactly zero.

These two solutions represent the years 2012 and 1975, when according to this model there were no cable television systems in the United States. These solutions are probably model breakdown, since these inputs are not in a reasonable domain of the model.

EXAMPLE ② PRACTICE PROBLEM

In Section 4.2 we investigated the average number of hours per person per year spent watching television during the late 1990s. We used the model

$$H(t) = 9.5(t - 7)^2 + 1544$$

where $H(t)$ represents the average number of hours per person per year spent watching television t years since 1990. Determine when the average person will spend 1825 hours per year watching television.

EXAMPLE 3 SOLVING QUADRATIC EQUATIONS USING THE SQUARE ROOT PROPERTY

Solve the following.

a. $5x^2 + 10 = 255$

b. $4(2t + 5)^2 - 82 = -62.64$

c. $(x + 5)^2 - 4 = -20$

Solution

a.

$$5x^2 + 10 = 255 \qquad \text{Isolate the squared part of the equation.}$$

$$\underline{ -10 \quad -10}$$

$$5x^2 = 245$$

$$\frac{5x^2}{5} = \frac{245}{5}$$

$$x^2 = 49 \qquad \text{Use the square root property, being sure}$$
$$\text{to use a plus/minus symbol.}$$
$$x = \pm\sqrt{49}$$

$$x = \pm 7$$

$$5(7)^2 + 10 = 255 \qquad \text{Check both answers.}$$

$$255 = 255$$

$$5(-7)^2 + 10 = 255$$

$$255 = 255$$

b.

$$4(2t + 5)^2 - 82 = -62.64 \qquad \text{Isolate the squared part of the equation.}$$

$$4(2t + 5)^2 = 19.36$$

$$(2t + 5)^2 = 4.84$$

$$2t + 5 = \pm\sqrt{4.84} \qquad \text{Use the square root property, being sure to use a plus/minus symbol.}$$

$$2t + 5 = \pm 2.2$$

$$\text{Write two equations and solve.}$$

$$2t + 5 = 2.2 \qquad\qquad 2t + 5 = -2.2$$

$$2t = -2.8 \qquad\qquad 2t = -7.2$$

$$t = -1.4 \qquad\qquad t = -3.6$$

$$4(2(-1.4) + 5)^2 - 82 = -62.64 \qquad \text{Check both answers.}$$

$$-62.64 = -62.64$$

$$4(2(-3.6) + 5)^2 - 82 = -62.64$$

$$-62.64 = -62.64$$

c.

$$(x + 5)^2 - 4 = -20 \qquad \text{Isolate the squared part of the equation.}$$

$$(x + 5)^2 = -16$$

$$x + 5 = \pm\sqrt{-16} \qquad \text{Use the square root property.}$$

When we used the square root property we ended with a negative under the square root so there are no real solutions to this equation. If we look at the graph of both sides of this equation we can see that they do not cross and thus are never equal.

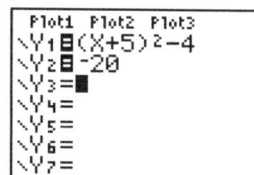
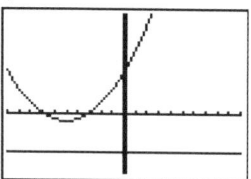

EXAMPLE PRACTICE PROBLEM

Solve the following.

a. $7x^2 + 9 = 121$

b. $-3(4x - 7)^2 + 17 = -58$

COMPLETING THE SQUARE

If a quadratic equation has both a second-degree term and a first-degree term $(ax^2 + bx + c)$, the square root property cannot be easily used to solve. If you try to get the squared part by itself you will have a variable term in the way of the square root. To handle this problem, we use a technique called completing the square. Completing the square will transform the equation so that it has a perfect square that can be solved using the square root property.

The idea behind this technique comes from the factoring process that we examined in Chapter 3. Recall from Chapter 3 that a perfect square trinomial will factor to the square of a binomial.

$$a^2 + 2ab + b^2 = (a + b)^2$$

$$a^2 - 2ab + b^2 = (a - b)^2$$

Thus, if we can get one side of a quadratic equation to have the characteristics of a perfect square trinomial we will be able to factor it into a binomial squared. First, we will consider how we can make an expression into a perfect square trinomial. The expressions

$$x^2 + 10x + 25$$

$$w^2 - 12w + 36$$

are both perfect square trinomials because they factor into a binomial squared.

$$x^2 + 10x + 25 = (x + 5)^2$$

$$w^2 - 12w + 36 = (w - 6)^2$$

If we look at these two examples, we can see a pattern that can help us complete the square of any quadratic. Consider the expressions:

$$x^2 + 10x$$

$$w^2 - 12w$$

We know from the examples above that if we add 25 to the first expression and 36 to the second expression they will become perfect square trinomials. These numbers are related to the coefficient of the first-degree term x or w. In particular, 25 is the square of half of 10, and 36 is the square of half of -12. In general, taking half of the coefficient of the first-degree term and squaring it will give us a constant that will make these expressions a perfect square trinomial.

$$x^2 + bx \qquad \text{The constant that will complete the square: } \left(\frac{b}{2}\right)^2$$

$$x^2 + bx + \left(\frac{b}{2}\right)^2$$

$$x^2 + bx + \left(\frac{b}{2}\right)^2 = \left(x + \frac{b}{2}\right)^2$$

The factored form will include half of b as one of the terms in the binomial. We will start the process of completing the square by simplifying one side of an equation. This process is easiest when the coefficient of the squared term is 1. If this coefficient is not 1 we will divide both sides of the equation by the coefficient to make it 1. Follow the steps to complete the square in the next example.

SOLVING QUADRATICS BY COMPLETING THE SQUARE

1. Isolate the variable terms on one side of the equation.

2. If the coefficient of x^2 is not 1 divide both sides of the equation by the coefficient of x^2.

3. Take half of the coefficient of x, then square it. Add this number to both sides of the equation.

4. Factor the quadratic into the square of a binomial.

5. Solve using the square root property.

6. Check your answers in the original equation.

EXAMPLE **4** SOLVING QUADRATIC EQUATIONS BY COMPLETING THE SQUARE

Solve by completing the square of $5x^2 + 30x - 35 = 0$

Solution

Step 1 Isolate the variable terms on one side of the equation.

To isolate the variable terms, we move the constant term to the other side of the equation.

$$5x^2 + 30x - 35 = 0$$
$$\underline{ \quad 35 \quad\quad 35}$$
$$5x^2 + 30x = 35$$

Step 2 If the coefficient of x^2 is not 1 divide both sides of the equation by the coefficient of x^2.

Divide both sides by 5.

$$\frac{5x^2 + 30x}{5} = \frac{35}{5}$$
$$x^2 + 6x = 7$$

Step 3 Take half of the coefficient of x, then square it. Add this number to both sides of the equation.

To complete the square on the left side of the equation we, add a number that will make the resulting trinomial a perfect square trinomial. To find this number, we take half of the coefficient of x and square it. This number is then added to both sides of the equation. In this equation, the coefficient of x is 6 so we get

$$\left(\frac{6}{2}\right)^2 = 9$$

Now, we add this constant to both sides of the equation.

$$x^2 + 6x + 9 = 7 + 9$$

$$x^2 + 6x + 9 = 16$$

Step 4   Factor the quadratic into the square of a binomial.

and the left side will factor as a perfect square.

$$x^2 + 6x + 9 = 16$$

$$(x + 3)^2 = 16$$

Step 5 Solve using the square root property.

$$(x + 3)^2 = 16$$

$$x + 3 = \pm\sqrt{16}$$

$$x + 3 = \pm 4$$

$$x + 3 = 4 \qquad x + 3 = -4$$

$$x = 1 \qquad x = -7$$

Step 6 Check your answers in the original equation.

$$5(1)^2 + 30(1) - 35 = 0$$

$$5 + 30 - 35 = 0$$

$$0 = 0$$

$$5(-7)^2 + 30(-7) - 35 = 0$$

$$5(49) - 210 - 35 = 0$$

$$245 - 210 - 35 = 0$$

$$0 = 0$$

Completing the square will allow us to solve any quadratic equation using the square root property.

EXAMPLE   5 SOLVING QUADRATIC EQUATIONS BY COMPLETING THE SQUARE

Solve the following by completing the square.

a. $x^2 + 10x + 16 = 0$

b. $3x^2 - 15x = 12$

c. $6x^2 + 8 = 10x$

Solution

a.   The coefficient of x^2 is 1, so we do not need to divide. To get the variable terms isolated on one side of the equation, we subtract 16 from both sides.

$$x^2 + 10x + 16 = 0$$

$$\underline{\qquad -16 \quad -16}$$

$$x^2 + 10x = -16$$

Now, we find the constant that will complete the perfect square trinomial. So we will take half of the coefficient of x and square it.

$$\frac{1}{2}(10) = 5$$

$$(5)^2 = 25$$

Now add this number to both sides of the equation and factor the quadratic into the square of a binomial.

$$x^2 + 10x = -16$$

$$x^2 + 10x + 25 = -16 + 25$$

$$x^2 + 10x + 25 = 9$$

$$(x + 5)^2 = 9$$

The square root property can be used to solve the equation.

$$(x + 5)^2 = 9$$

$$x + 5 = \pm\sqrt{9} \quad \text{Use the square root property.}$$

$$x + 5 = \pm 3$$

$$x + 5 = 3 \qquad x + 5 = -3$$

$$x = -2 \qquad x = -8$$

$$(-2)^2 + 10(-2) + 16 = 0 \qquad \text{Check your answers in}$$

$$4 - 20 + 16 = 0 \qquad \text{the original equation.}$$

$$0 = 0$$

$$(-8)^2 + 10(-8) + 16 = 0$$

$$64 - 80 + 16 = 0$$

$$0 = 0$$

b. The variable terms are already isolated, so we start by making the coefficient of x^2 into 1. The coefficient of x^2 is 3 so dividing both sides of the equation by 3 will make this coefficient 1.

$$3x^2 - 15x = 12$$

$$\frac{3x^2 - 15x}{3} = \frac{12}{3}$$

$$x^2 - 5x = 4$$

We want to find the constant that will complete the perfect square trinomial, so we take half of the coefficient of x and square it.

$$\frac{1}{2}(5) = \frac{5}{2}$$

$$\left(\frac{5}{2}\right)^2 = \frac{25}{4}$$

Add this number to both sides of the equation and factor the quadratic into the square of a binomial.

$$x^2 - 5x = 4$$

$$x^2 - 5x + \frac{25}{4} = 4 + \frac{25}{4}$$

$$x^2 - 5x + \frac{25}{4} = \frac{16}{4} + \frac{25}{4}$$

$$\left(x - \frac{5}{2}\right)^2 = \frac{41}{4}$$

Use the square root property to solve.

$$\left(x - \frac{5}{2}\right)^2 = \frac{41}{4}$$

$$x - \frac{5}{2} = \pm\sqrt{\frac{41}{4}}$$

$$x - \frac{5}{2} = \pm\frac{\sqrt{41}}{2}$$

$$x - \frac{5}{2} = \frac{\sqrt{41}}{2} \qquad x - \frac{5}{2} = -\frac{\sqrt{41}}{2}$$

$$x = \frac{5}{2} + \frac{\sqrt{41}}{2} \qquad x = \frac{5}{2} - \frac{\sqrt{41}}{2}$$

$$x \approx 5.702 \qquad x \approx -0.702$$

$$3(5.702)^2 - 15(5.702) = 12 \qquad \text{Check your answers in}$$
$$12.008 \approx 12 \qquad \text{the original equation.}$$

$$3(-0.702)^2 - 15(-0.702) = 12$$
$$12.008 \approx 12$$

c. First, we re-arrange the terms so the variable terms are on one side of the equation and the constant term is on the other side.

$$6x^2 + 8 = 10x$$
$$\underline{-10x \quad -10x}$$
$$6x^2 - 10x + 8 = 0$$
$$\underline{-8 \quad -8}$$
$$6x^2 - 10x = -8$$

The coefficient of x^2 is 6, so we will divide both sides of the equation by 6.

$$6x^2 - 10x = -8$$
$$\frac{6x^2 - 10x}{6} = \frac{-8}{6}$$
$$x^2 - \frac{5}{3}x = -\frac{4}{3}$$

We want to find the constant that will complete the perfect square trinomial, so we take half of the coefficient of x and square it.

$$\frac{1}{2}\left(\frac{5}{3}\right) = \frac{5}{6}$$

$$\left(\frac{5}{6}\right)^2 = \frac{25}{36}$$

Add this number to both sides of the equation and factor the quadratic into the square of a binomial

$$x^2 - \frac{5}{3}x + \frac{25}{36} = -\frac{4}{3} + \frac{25}{36}$$

$$x^2 - \frac{5}{3}x + \frac{25}{36} = -\frac{48}{36} + \frac{25}{36} \qquad \text{Get like denominators.}$$

$$\left(x - \frac{5}{6}\right)^2 = -\frac{23}{36}$$

We can use the square root property to solve the equation.

$$\left(x - \frac{5}{6}\right)^2 = -\frac{23}{36}$$

$$x - \frac{5}{6} = \pm\sqrt{-\frac{23}{36}}$$

There is a negative number under the square root, so there is no real solution to this equation. If we graph the two sides of this equation in our calculator, we can see that the two sides never touch and thus are not equal in the real number system.

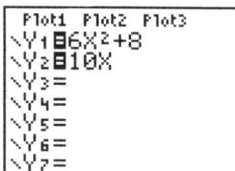 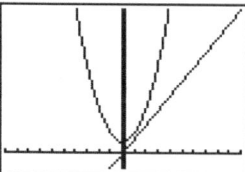

EXAMPLE ⑤ PRACTICE PROBLEM

Solve the following by completing the square.

a. $x^2 + 8x - 20 = 0$

b. $4x^2 - 20x = -8$

c. $2x^2 + 8x = -60$

CONVERTING TO VERTEX FORM

To put a quadratic function into vertex form, use the technique of completing the square. The process is basically the same, but you will change back to function notation again when you are done completing the square.

EXAMPLE CONVERTING A QUADRATIC FUNCTION INTO VER-
TEX FORM

Convert to vertex form.

a. $f(x) = 2x^2 + 20x - 6$

b. $g(x) = 5x^2 + 3x + 20$

Solution

a. To start, we will change the function notation to y so that there is less confusion as we complete the square. Then we will complete the square as before.

$$f(x) = 2x^2 + 20x - 6 \qquad \text{Get out of function notation.}$$

$$y = 2x^2 + 20x - 6$$

$$\underline{+6 \qquad\qquad\qquad +6} \qquad \begin{array}{l}\text{Move the constant to the other}\\\text{side of the equation.}\end{array}$$

$$y + 6 = 2x^2 + 20x$$

$$\frac{y+6}{2} = \frac{2x^2 + 20x}{2} \qquad \begin{array}{l}\text{Divide by the leading coefficient.}\\\text{Find the constant that will}\\\text{complete the square.}\end{array}$$

$$\frac{1}{2}y + 3 = x^2 + 10x \qquad\qquad \frac{1}{2}(10) = 5$$

$$\frac{1}{2}y + 3 + 25 = x^2 + 10x + 25 \qquad\qquad 5^2 = 25$$

$$\frac{1}{2}y + 28 = (x+5)^2 \qquad \begin{array}{l}\text{Add 25 to both sides of the}\\\text{equation and factor.}\end{array}$$

Now that the right side is factored, we want to get the y by itself again and put the equation back into function notation.

$$\frac{1}{2}y + 28 = (x+5)^2 \qquad \text{Subtract 28 from both sides.}$$

$$\underline{-28 \qquad\qquad -28}$$

$$\frac{1}{2}y = (x+5)^2 - 28 \qquad \text{Multiply both sides by 2.}$$

$$y = 2(x+5)^2 - 56$$

$$f(x) = 2(x+5)^2 - 56 \qquad \text{Re-write in function notation.}$$

b.

$$g(x) = 5x^2 + 3x + 20$$

$$y = 5x^2 + 3x + 20 \qquad \text{Get out of function notation.}$$

$$y - 20 = 5x^2 + 3x \qquad \text{Move the constant to the other side of the equation.}$$

$$\frac{y - 20}{5} = \frac{5x^2 + 3x}{5} \qquad \text{Divide by the leading coefficient.}$$

$$\text{Find the constant that will complete the square.}$$

$$\frac{1}{5}y - 4 = x^2 + \frac{3}{5}x \qquad \frac{1}{2}\left(\frac{3}{5}\right) = \frac{3}{10}$$

$$\frac{1}{5}y - 4 + \frac{9}{100} = x^2 + \frac{3}{5}x + \frac{9}{100} \qquad \left(\frac{3}{10}\right)^2 = \frac{9}{100}$$

$$\frac{1}{5}y - \frac{391}{100} = \left(x + \frac{3}{10}\right)^2 \qquad \text{Factor.}$$

$$\frac{1}{5}y = \left(x + \frac{3}{10}\right)^2 + \frac{391}{100}$$

$$y = 5\left(x + \frac{3}{10}\right)^2 + \frac{391}{20} \qquad \text{Get } y \text{ by itself.}$$

$$g(x) = 5\left(x + \frac{3}{10}\right)^2 + \frac{391}{20} \qquad \text{Re-write in function notation.}$$

EXAMPLE ⑥ PRACTICE PROBLEM

Convert into vertex form.

a. $f(x) = 0.25x^2 + 10x - 3$

b. $h(x) = -3x^2 + 4x + 10$

GRAPHING FROM VERTEX FORM WITH X INTERCEPTS

Now that we know how to solve a quadratic in vertex form, we can find the horizontal intercepts of a quadratic in vertex form and these can be part of our graphs.

STEPS TO GRAPHING A QUADRATIC FROM THE VERTEX FORM

1. Determine whether the graph opens up or down.
2. Find the vertex and axis of symmetry.
3. Find the vertical intercept.
4. Find the horizontal intercepts.
5. Plot the points, plot their symmetric pairs, and sketch the graph. (Find an additional symmetric pair if needed.)

EXAMPLE ▮7▮ GRAPHING A QUADRATIC FUNCTION IN VERTEX FORM

Sketch the graph of $f(x) = -1.5(x - 2.5)^2 + 45.375$

Solution

Step 1 Determine whether the graph opens up or down.

The value of a is -1.5. Because a is negative, the graph will open downward.

Step 2 Find the vertex and axis of symmetry.

Because the quadratic is given in vertex form, the vertex is $(2.5, 45.375)$. The axis of symmetry is the vertical line through the vertex $x = 2.5$.

Step 3 Find the vertical intercept.

To find the vertical intercept, we make the input variable zero.

$$f(x) = -1.5(x - 2.5)^2 + 45.375$$
$$f(0) = -1.5(0 - 2.5)^2 + 45.375$$
$$f(0) = -1.5(-2.5)^2 + 45.375$$
$$f(0) = -1.5(6.25) + 45.375$$
$$f(0) = 36$$

Therefore, the vertical intercept is $(0, 36)$.

Step 4 Find the horizontal intercepts.

Horizontal intercepts occur when the output variable is equal to zero, so substitute zero for the output variable and solve.

$$f(x) = -1.5(x - 2.5)^2 + 45.375$$
$$0 = -1.5(x - 2.5)^2 + 45.375$$

$$-45.375 = -1.5(x - 2.5)^2 \qquad \text{Isolate the squared part.}$$

$$\frac{-45.375}{-1.5} = \frac{-1.5(x - 2.5)^2}{-1.5}$$

$$30.25 = (x - 2.5)^2 \qquad \text{Use the square root property.}$$

$$\pm\sqrt{30.25} = x - 2.5$$

$$\pm 5.5 = x - 2.5$$

$$5.5 = x - 2.5 \qquad -5.5 = x - 2.5 \qquad \text{Write two equations and solve.}$$

$$8 = x \qquad -3 = x$$

Therefore, we have the horizontal intercepts $(-3, 0)$ and $(8, 0)$.

Step 5 Plot the points, plot their symmetric pairs, and sketch the graph. (Find an additional symmetric pair if needed.)

The vertical intercept $(0, 36)$ is 2.5 units to the left of the axis of symmetry $x = 2.5$ so its symmetric point will be 2.5 units to the right of the axis of symmetry at $(5, 36)$.

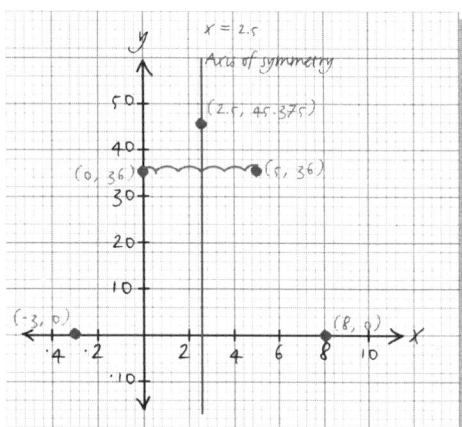

Now connect the points with a smooth curve.

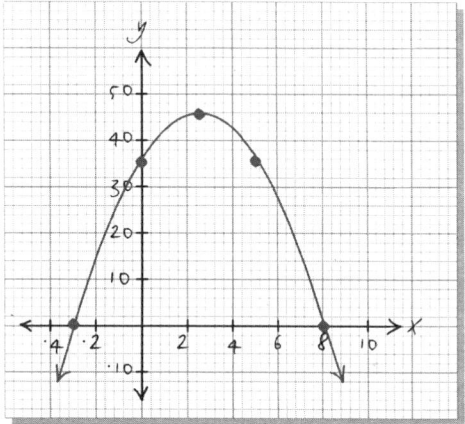

EXAMPLE (7) PRACTICE PROBLEM

Sketch the graph of $f(x) = 2(x + 1)^2 - 32$

4.4 Exercises

For exercises 1 through 16 solve the equation using the square root property.

1. $x^2 = 100$

2. $x^2 = 70$

3. $2x^2 = 162$

4. $5x^2 = 180$

5. $x^2 + 12 = 181$

6. $4x^2 - 43 = 1764$

7. $(x-5)^2 = 49$

8. $(x+4.5)^2 = 156.25$

9. $4(x+7)^2 = 400$

10. $-9(x-5)^2 = -369$

11. $3(x-7)^2 + 8 = 647.48$

12. $-2(x+3)^2 + 11 = -151$

13. $-5(x-6)^2 - 15 = -30$

14. $2.5(x-6.5)^2 - 20 = 42.5$

15. $-0.5(x+4)^2 - 9 = -20$

16. $-0.25(x-8)^2 + 18 = 6.25$

17. The number of cassette singles shipped by major recording media manufacturers can be modeled by

$$C(t) = -2.336(t - 13)^2 + 85.6$$

where $C(t)$ is the number of cassette singles in millions shipped t years since 1980.

Source: Model based on data from Statistical Abstract 2001.

a. Find how many cassette singles were shipped in 1998.

b. Find when 50 million cassette singles were shipped.

c. What is the vertex of this model and what does it represent in this context?

d. When do you think model breakdown occurs for this model?

18. The poverty threshold for an individual under 65 years old can be modeled by

$$P(t) = 4.95(t - 57)^2 + 1406$$

where $P(t)$ represents the poverty threshold, in dollars, for an individual in the United States under 65 years old t years since 1900.

Source: Model based on data from Statistical Abstract 2001.

a. What was the poverty threshold in 1995?

b. What year was the poverty threshold $11,500?

c. What year was the poverty threshold $3,000?

d. What is the vertex of this model and what does it represent in this context?

e. When do you think model breakdown occurs for this model?

19. The amount that personal households and nonprofit organizations have invested in time and savings deposits can be modeled by

$$D(t) = 73.6(t - 3.5)^2 + 2176$$

where $D(t)$ represents the amount in billions of dollars the households and nonprofit organizations have deposited in time and savings accounts t years since 1990.

a. How much did personal households and nonprofit organizations have invested in time and savings deposits in 2000?

b. Find what year personal households and nonprofit organizations had $7500 billion invested in time and savings deposits.

c. What is the vertex of this model, and what does it represent in this context?

20. The total number of people, in thousands, age 65 years old and over who were below the poverty level can be modeled by

$$N(t) = -59(t - 12)^2 + 3890$$

where $N(t)$ represents the total number of people 65 years old and over in the United States who were below the poverty level, in thousands, t years since 1980.

Source: Model based on data from Statistical Abstract 2001.

a. How many people 65 years old and over were below the poverty level in 2000?

b. When were there 1 million people 65 years old and over below the poverty level? (Hint: 1 million = 1000 thousand)

c. What is the vertex of this model, and what does it represent in this context?

21. The poverty threshold for a family of four is given below.

Year	Poverty Threshold (in $)
1965	3022
1970	3223
1975	3968
1980	5500
1985	8414

Source: Statistical Abstract 2001.

a. Find an equation for a model of these data.

b. Estimate the poverty threshold for a family of four in 1990.

c. Give a reasonable domain and range for your model.

d. Find what year the poverty threshold was $5000.

22. A basketball player shoots a basket from 12 ft away. The ball flies along a parabolic path given by the data.

Distance from Basket (feet)	Height of Ball (feet)
12	6.5
10	10.3
9	11.7
5	14.2
4	14
0	10

a. Find an equation for a model of these data.

b. What is the height of the ball when it was 8 ft from the basket?

c. Where did the ball reach its maximum height?

d. When was the ball 11 ft high?

Evangelista Torricelli
(1608 - 1647)

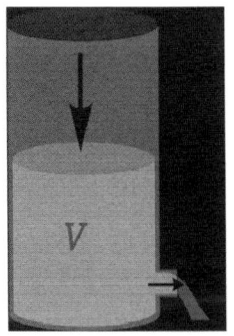

23. If a tank holds 2000 gallons of water, which drains from the bottom of the tank in 20 minutes, then Torricelli's law gives the volume in gallons of water remaining in the tank after t minutes as

$$V(t) = 2000\left(1 - \frac{t}{20}\right)^2$$

a. Find the volume of water remaining in the tank after 5 minutes.

b. When will there be only 500 gallons of water remaining in the tank?

c. When will there be only 20 gallons of water remaining in the tank?

24. If a tank holds 5000 gallons of water, which drains from the bottom of the tank in 80 minutes, then Torricelli's law gives the volume in gallons of water remaining in the tank after t minutes as

$$V(t) = 5000\left(1 - \frac{t}{80}\right)^2$$

a. Find the volume of water remaining in the tank after 1 hour.

b. When will there be only 2450 gallons of water remaining in the tank?

c. When will there be only 100 gallons of water remaining in the tank?

For exercises 25 through 28 use the information and formula below.

Money in a savings account that earns interest at r percent compounded annually will grow according to the formula

$$A = P(1 + r)^t$$

where

A = The account balance in dollars

t = time in years

P = The principal in dollars (amount deposited)

r = The interest rate compounded annually.

25. Find the interest rate that will turn a deposit of $800 into $900 in 2 years.

26. Find the interest rate that will turn a deposit of $4000 into $4400 in 2 years.

27. Find the interest rate that will turn a deposit of $2000 into $2150 in 2 years.

28. Find the interest rate that will turn a deposit of $15,000 into $17,500 in 2 years.

29. The KVLY-TV mast is a television transmitter in North Dakota used by Fargo station KVLY channel 11. At 2,063 ft, it is currently the world's tallest supported structure on land. The mast is held in place by several guy wires that are attached on the sides of the mast. One of the guy wires is anchored in the ground 1,000 ft from the base of the mast and it attaches to the mast 2,000 ft up from the base. How long is the guy wire? (Hint: use the Pythagorean Theorem).

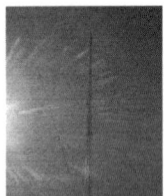

KVLY-TV mast

Guy wire securing TV antenna.
Source: Wikipedia.org

30. Scott has a 30 foot ladder and puts the base of the ladder 12 feet from the house. How far will the ladder reach?

31. Mike is a painter and has a 25 foot ladder that he wants to lean against the house 20 feet up. How far from the house should he put the base of the ladder?

32. Joan installs awnings above windows on homes and needs her 30 foot ladder to reach 22 feet up the side of a home. How far from the house should she put the base of the ladder?

For exercises 33 through 50 solve by .completing the square.

33. $x^2 + 6x = 7$

34. $t^2 + 10x = 24$

35. $k^2 - 16x = 3$

36. $m^2 - 12x = -27$

37. $x^2 + 4x + 7 = 0$

38. $h^2 + 2x - 5 = 0$

39. $t^2 + 11x = 4$

40. $p^2 + 7p - 12 = 0$

41. $x^2 - 9x + 14 = 0$

42. $x^2 - 20x + 51 = 0$

43. $m^2 + 8m + 20 = 0$

44. $k^2 - 10k + 40 = 5$

45. $3x^3 + 12x - 15 = 0$

46. $4t^2 + 80t + 20 = -56$

47. $5x^2 + 7x = 0$

48. $3h^2 + 9h - 14 = 0$

49. $-7x^2 + 4x + 20 = 0$

50. $3x^2 - 10x - 7 = 0$

For exercises 51 through 68 convert the function into vertex form.

51. $f(x) = x^2 + 6x + 8$

52. $g(x) = x^2 - 10x - 7$

53. $g(t) = t^2 - 8t - 20$

54. $s(p) = p^2 + 14p + 35$

55. $f(x) = x^2 - 7x + 10$

56. $k(m) = m^2 + 5m - 12$

57. $h(x) = 3x^2 + 12x + 24$

58. $m(d) = 4d^2 + 24d - 30$

59. $f(t) = 2t^2 - 16t - 12$

60. $k(p) = 5p^2 - 50p + 20$

61. $f(x) = 4x^2 + 5x - 20$

62. $m(p) = 6p^2 + 10p + 21$

63. $g(x) = 0.5x^2 + 7x - 30$

64. $d(m) = 0.25m^2 - 9m + 13$

65. $f(x) = 0.2x^2 - 7x - 10$

66. $h(t) = \frac{1}{3}t^2 + 6t + 8$

67. $c(p) = \frac{2}{7}p^2 - 5p - \frac{3}{7}$

68. $g(x) = \frac{3}{8}x^2 + \frac{1}{8}x - \frac{7}{8}$

For Exercises 69 through 78, sketch the graph of the given functions and label the vertex, vertical intercept, and horizontal intercepts (if any). Give the domain and range of the function.

69. $f(x) = (x - 5)^2 - 16$

70. $g(x) = (x - 3)^2 - 9$

71. $h(x) = (x + 2)^2 - 5$

72. $f(x) = (x + 6)^2 - 11$

73. $g(x) = -2(x - 4)^2 + 18$

74. $h(x) = -0.5(x + 3)^2 + 32$

75. $f(x) = \frac{1}{3}(x - 5)^2 - 12$

76. $g(x) = \frac{1}{5}(x + 7)^2 + 10$

77. $h(x) = -0.3(x - 10)^2 - 5$

78. $f(x) = 5(x + 1.5)^2 - 6$

Solving Quadratic Equations by Factoring

SOLVING BY FACTORING

In many situations, you may be given an equation that is easy to factor. Factoring can be used to solve some quadratic equations. We saw in chapter 3 that the basis of factoring is to take a polynomial and break it into simpler pieces that are multiplied together.

Standard Form   Factored Form

$$f(x) = x^2 + 8x + 15 \qquad f(x) = (x + 3)(x + 5)$$

If we multiply the factored form out and simplify, it will become the standard form.

$$f(x) = (x + 3)(x + 5)$$
$$f(x) = x^2 + 5x + 3x + 15$$
$$f(x) = x^2 + 8x + 15$$

The main reason factoring an expression is useful is the **product property of zero,** which says that when you multiply any number by zero, the answer is zero.

DEFINITIONS
Product property of zero: $a \cdot 0 = 0$.
Zero factor property: If $ab = 0$, then $a = 0$, $b = 0$, or both.
Where a and b are real numbers.

The **zero factor property** follows from the product property of zero and states that if two numbers are multiplied together and the answer is zero, then one or both of those factors must be zero. To understand this simple property, do the following arithmetic in your head.

- First pick any number.
- Now multiply that number by zero.
- You should now have zero as the result.

Okay. Start over. This time:

- Pick any number.
- Now multiply by any number you want, but the result must be zero.
- You should have multiplied by zero.

If you do these two operations with any non-zero number, you will get the same results every time. This statement seems obvious, but it is very helpful in algebra. Let's see why.

CONCEPT INVESTIGATION HOW DOES FACTORING HELP US SOLVE? ···

I. Solve the following equations for x.

a. $3x = 0$ **b.** $-9x = 0$

c. $5(x + 2) = 0$ **d.** $-13(x - 6) = 0$

Note that all of these problems include multiplication.

II. For each of the following equations, list the two factors being multiplied together, and tell which one must equal zero for the product to equal zero.

a. $3x = 0$ 3 and x are the factors.
                                               3 cannot equal zero, so x must equal zero.

b. $-9x = 0$

c. $5(x + 2) = 0$ 5 and $(x + 2)$ are the factors
                                      $5 \neq 0$, so $(x + 2)$ must equal zero.

d. $-13(x - 6) = 0$

Note that in all of the equations given, there has been only one factor that could have equaled zero.

III. In the following equations, list the factors being multiplied together and state which can equal zero. Then solve.

a. $(x + 5)(x - 9) = 0$ $(x + 5)$ and $(x - 9)$; both can equal zero.
$$x + 5 = 0 \quad \text{or} \quad x - 9 = 0$$
$$x = -5 \quad \text{or} \quad x = 9$$

b. $(x - 3)(x - 12) = 0$

c. $(x + 17)(x + 2) = 0$

d. $(2x - 5)(x - 4) = 0$

e. $3(x - 7)(x + 15) = 0$

f. $(x + 3)(x - 7)(x + 15) = 0$

···

    As you can see from this investigation, equations that involve a product that is set equal to zero can be solved by setting each factor equal to zero one at a time rather than the entire equation at once. If we factor a quadratic into simpler parts, we will be able to solve the simpler equations.

It is very important to note that we can use the factored form to solve an equation only if it is set equal to zero. If the quadratic equation is not equal to zero, you must first move everything to one side so that it equals zero and then factor.

EXAMPLE ❚ APPLICATIONS OF SOLVING BY FACTORING

The profit a local photographer makes from selling n copies of a photograph can be modeled by

$$P(n) = n(40 - 2n)$$

where $P(n)$ represents the amount of profit in dollars from selling n copies of a photograph. Find the number of copies the photographer must sell of a photo to make a profit of $150.

© images-of-france/Alamy

Solution

The $150 is a desired profit, so it must take the place of $P(n)$.

$$150 = n(40 - 2n)$$ Substitute 150 for P.
$$150 = 40n - 2n^2$$ Distribute the n.

$$\underline{+ 2n^2 \qquad\qquad + 2n^2}$$

$$2n^2 + 150 = 40n$$ Get everything to one side and make it equal to zero.

$$\underline{-40n \qquad\qquad -40n}$$

$$2n^2 - 40n + 150 = 0$$ Factor using AC method. Step 1

$$2(n^2 - 20n + 75) = 0$$

$$\frac{2(n^2 - 20n + 75)}{2} = \frac{0}{2}$$ Because this is an equation, we can divide both sides by 2.

Skill Connection

AC METHOD OF FACTORING QUADRATICS

1. Factor out any common factors.

2. Multiply a and c together.

3. Find factors of ac that add up to b.

4. Rewrite the middle (bx) term using the factors from step 3.

5. Group and factor out what is in common.

$$n^2 - 20n + 75 = 0$$
$$n^2 - 15n - 5n + 75 = 0 \quad \text{Step 4}$$
$$(n^2 - 15n) + (-5n + 75) = 0 \quad \text{Step 5}$$
$$n(n - 15) - 5(n - 15) = 0$$
$$(n - 15)(n - 5) = 0$$

$$n - 15 = 0 \qquad n - 5 = 0$$
$$n = 15 \qquad n = 5$$

Set the two factors equal to zero
using the zero factor property.

Step 2 1(75)

	1	75
	3	25
	5	15
	-1	-75
	-3	-25

Step 3| -5 | -15 |

To make a profit of \$150, the photographer must sell 5 or 15 photographs. If he sells between 5 and 15, he will make more than \$150 in profit.

EXAMPLE ① PRACTICE PROBLEM

The average cost for building a small plane at Private Planes 101 depends on the number of planes built in a year. This average cost can be modeled by

$$A(p) = p(31 - 5p) + 194$$

where $A(p)$ represents the average cost in thousands of dollars per plane to build p planes in a year. How many planes does Private Planes 101 have to build in a year to have an average cost per plane of \$200,000?

SOLVING QUADRATICS USING FACTORING

1. Put the quadratic into standard form, $ax^2 + bx + c = 0$
2. Factor the quadratic completely.
3. Set each of the factors equal to zero and solve.
4. Check your answers in the original equation.

EXAMPLE 2 SOLVING QUADRATICS BY FACTORING

Solve the following by factoring.

a. $x^2 + 7x - 40 = 20$

b. $2x^2 + 27 = 21x$

c. $x^3 + 8x^2 + 15x = 0$

Solution

a.

$$x^2 + 7x - 40 = 20 \quad \text{Set the quadratic equal to zero and factor.}$$

$$x^2 + 7x - 60 = 0$$

$$x^2 - 5x + 12x - 60 = 0$$

$$(x^2 - 5x) + (12x - 60) = 0$$

$$x(x - 5) + 12(x - 5) = 0$$

$$(x - 5)(x + 12) = 0$$

$$x - 5 = 0 \qquad x + 12 = 0$$

$$x = 5 \qquad x = -12$$

$$(5)^2 + 7(5) - 60 = 0 \quad \text{Check both answers.}$$

$$0 = 0$$

$$(-12)^2 + 7(-12) - 60 = 0$$

$$0 = 0$$

	$1(-60) = -60$		
1	-60	-1	60
2	-30	-2	30
3	-20	-3	20
4	-15	-4	15
5	-12	-5	12
6	-10	-6	10

Set the factors equal to zero and finish solving.

b.

$$2x^2 + 27 = 21x \quad \text{Set the quadratic equal to zero and factor.}$$

$$2x^2 - 21x + 27 = 0$$

$$2x^2 - 3x - 18x + 27 = 0$$

$$(2x^2 - 3x) + (-18x + 27) = 0$$

$$x(2x - 3) - 9(2x - 3) = 0$$

$$(2x - 3)(x - 9) = 0$$

$$2x - 3 = 0 \qquad x - 9 = 0$$

$$x = \frac{3}{2} \qquad x = 9$$

$$2\left(\frac{3}{2}\right)^2 + 27 = 21\left(\frac{3}{2}\right)$$

$$31.5 = 31.5$$

$$2(9)^2 + 27 = 21(9)$$

$$189 = 189$$

	$2(27) = 54$		
1	54	-1	-54
2	27	-2	-27
3	18	-3	-18
6	9	-6	-9

Set the factors equal to zero and finish solving.
Check both answers.

c.

$$x^3 + 8x^2 + 15x = 0 \quad \text{Take out the } x \text{ in common, and factor the remaining quadratic.}$$

$$x(x^2 + 8x + 15) = 0$$

$$x(x^2 + 3x + 5x + 15) = 0 \quad \text{The factored-out } x \text{ will remain at the front of each step until the factoring is complete.}$$

$$x[(x^2 + 3x) + (5x + 15)] = 0$$

$$x[x(x + 3) + 5(x + 3)] = 0$$

$$x(x + 3)(x + 5) = 0$$

$$x = 0 \qquad x + 3 = 0 \qquad x + 5 = 0 \quad \text{Now rewrite into three equations and solve each separately.}$$

$$x = 0 \qquad x = -3 \qquad x = -5$$

$$(0)^3 + 8(0)^2 + 15(0) = 0$$
$$0 = 0$$
$$(-3)^3 + 8(-3)^2 + 15(-3) = 0$$
$$0 = 0$$
$$(-5)^3 + 8(-5)^2 + 15(-5) = 0$$
$$0 = 0$$

Check all three answers in the original equation.

EXAMPLE 2 PRACTICE PROBLEM

Solve the following by factoring.

a. $x^2 + 13x = -36$

b. $2x^3 + 11x^2 + 21x - 5 = 6x - 5$

In some contexts, the factored form can be the easiest way to model a situation. In business, when a company changes the price of an item, it will affect both the price and the quantity of that item that is sold. For example if a company can sell 2000 bikes at a price of $1500.00, the company may only be able to sell 1900 bikes if they raise the price to $1700.00.

Through experience and research, companies can get an idea of how price changes will increase or decrease the quantity they sell and therefore also increase or decrease their revenue. Remember that revenue can be calculated by multiplying the price by the quantity sold.

EXAMPLE 3 APPLICATIONS OF THE FACTORED FORM OF A QUADRATIC

A movie theatre sells an average of 35,000 tickets per week when the price per ticket averages $7. From researching other theatres in the area, the managers believe that for each $0.50 increase in the ticket price they will lose about 1400 ticket sales per week.

a. What will the weekly revenue for this theatre be if management increases the ticket price to $8.

b. Find a model for the weekly revenue at this theatre if management increases the ticket price $0.50 x times.

c. Use your model to determine the weekly revenue if the ticket price is $9.

d. Use your model to determine the weekly revenue if the ticket price is $12.

Solution

a. Let R be the weekly revenue in dollars for this theatre.

If the theatre increases the ticket price to $8, that change in price, will represent two $0.50 increases, so the quantity sold will decrease by $2(1400) = 2800$.

Therefore they will expect to sell $35000 - 2800 = 32200$ tickets at $8 each for a revenue of:

$$\text{Revenue} = (\text{Price})(\text{Quantity})$$
$$R = (8)(32200)$$
$$R = 257600$$

Therefore at $8 per ticket the theatre expects to have a weekly revenue of $257,600.

b. Let x be the number of $0.50 increases in price.

The new price will be represented by $7 + 0.50x$. We also know that the quantity sold will decrease by 1400 per $0.50 increase. Therefore the quantity sold can be represented by $35000 - 1400x$. Thus the weekly revenue in dollars can be represented by

$$R(x) = (7 + 0.5x)(35000 - 1400x)$$

c. If the ticket price is $9, there were four $0.50 increases, so $x = 4$, we get:

$$R(4) = (7 + 0.5(4))(35000 - 1400(4))$$
$$R(4) = 264600$$

If the theater increases the price to $9 per ticket, weekly revenue will be $264,600.00.

d. If the ticket price is $12, the increase was $5 or ten $0.50 increases. Therefore $x = 10$, we get:

$$R(10) = (7 + 0.5(10))(35000 - 1400(10))$$
$$R(10) = 252000$$

If the theatre increases the ticket price to $12 per ticket, weekly revenue will be $252,000. At this price, the theatre is making less revenue because fewer people are buying tickets.

EXAMPLE ③ PRACTICE PROBLEM

A small restaurant is planning to increase the fee it charges for valet parking. Currently, the restaurant charges $3 per car, and the valet parks about 700 cars per week. For every dollar the restaurant raises the price, the restaurant expects about 100 fewer cars to use the valet parking service.

a. What will the weekly revenue from valet parking be if the restaurant increases the fee to $4?

b. Find a model for the weekly revenue from valet parking at this restaurant if the owners increase the fee x dollars.

c. Use your model to determine the weekly revenue if the fee is $5.

d. Use your model to determine the weekly revenue if the fee is $6.

e. What fee brings the restaurant the highest weekly revenue from valet parking?

FINDING AN EQUATION FROM THE GRAPH

The zero factor property can also be used to help find equations by looking at their graphs. As you solved different equations by factoring, you always set the equation equal to zero and then factor. If we set a function equal to zero and solve, we are looking for the horizontal intercepts of the function. When we solve using factoring, then, we are actually finding the horizontal intercepts of the graph related to that equation. We can use the connection between the horizontal intercepts and the factors of an expression to find an equation using its graph.

The horizontal intercepts of a graph also represent the zeros of the function. In this graph, we can see that the horizontal intercepts are $(-2, 0)$ and $(-8, 0)$, so the zeros of the equation are $x = -2$ and $x = -8$. Taking the zeros and working backwards through the process of solving by factoring, we will get an equation for the graph.

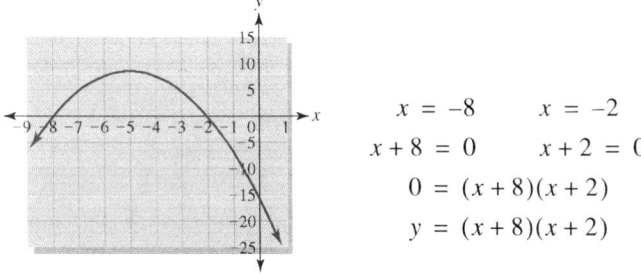

$$x = -8 \qquad x = -2$$
$$x + 8 = 0 \qquad x + 2 = 0$$
$$0 = (x + 8)(x + 2)$$
$$y = (x + 8)(x + 2)$$

This equation in factored form could be missing a constant that was factored-out. To find this constant, we can use another point from the graph and solve for the missing constant. If we let a represent the missing constant and use the point $(-3, 5)$, we can solve for a.

$$y = a(x + 8)(x + 2)$$
$$5 = a(-3 + 8)(-3 + 2)$$
$$5 = a(5)(-1)$$
$$5 = -5a$$
$$\frac{5}{-5} = \frac{-5a}{-5}$$
$$-1 = a$$

Substituting in the -1 for a and simplifying, we get;

$$y = -1(x + 8)(x + 2)$$
$$y = -x^2 - 10x - 16$$

We can see in the graph that a negative a value makes sense because the graph is facing downward. After simplifying, we can see that the equation also matches the y-intercept of $(0, -16)$ found in the graph. The horizontal intercepts cannot tell us everything about the equation, but once we have the factors, we can use another point on the graph to find any missing constant factors.

FINDING AN EQUATION FROM THE GRAPH

1. Find the horizontal intercepts of the graph and write them as zeros of the equation.

2. Undo the solving process to find the factors of the equation.

3. Use another point from the graph to find any constant factors.

4. Multiply out the factors and simplify.

EXAMPLE   **4**   FINDING A QUADRATIC EQUATION FROM THE GRAPH

Use the graph to find an equation for the quadratic.

a.

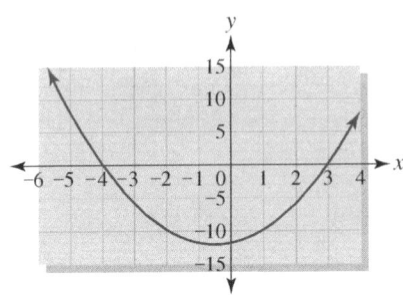

b.

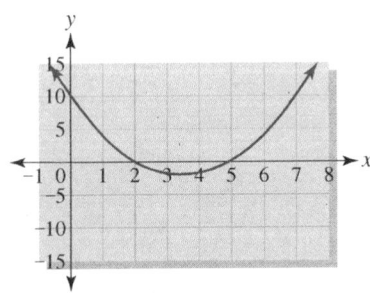

c.

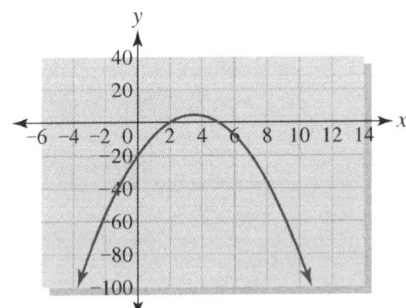

Solution

a.

> **Step 1** Find the horizontal intercepts of the graph and write them as zeros of the equation.
>
> The horizontal intercepts for this graph are located at $(-4, 0)$ and $(3, 0)$, so we know that $x = -4$ and $x = 3$ are the zeros of the quadratic equation.
>
> **Step 2** Undo the solving process to find the factors of the equation.

Therefore, we have:

$$x = -4 \qquad\qquad x = 3$$
$$x + 4 = 0 \qquad\qquad x - 3 = 0$$

This gives us the factors $(x + 4)$ and $(x - 3)$ and the factorization

$$y = (x + 4)(x - 3)$$

Step 3 Use another point from the graph to find any constant factors.
Using the point $(1, -10)$, we get:

$$y = a(x + 4)(x - 3)$$
$$-10 = a(1 + 4)(1 - 3)$$
$$-10 = a(5)(-2)$$
$$-10 = -10a$$
$$1 = a$$

So the constant factor is 1.

Step 4 Multiply out the factors and simplify.
Therefore we have the equation

$$y = 1(x + 4)(x - 3)$$
$$y = x^2 + x - 12$$

The y-intercept for this quadratic would be $(0, -12)$, which agrees with the graph, so we have the equation correct.

b.

Step 1 Find the horizontal intercepts of the graph and write them as zeros of the equation.
The horizontal intercepts for this graph are $(2, 0)$ and $(5, 0)$, so we know that $x = 2$ and $x = 5$ are the zeros of the quadratic equation.

Step 2 Undo the solving process to find the factors of the equation.
Therefore, we have:

$$x = 2 \qquad\qquad x = 5$$
$$x - 2 = 0 \qquad\qquad x - 5 = 0$$

This gives us the factors $(x - 2)$ and $(x - 5)$ and the factorization

$$y = (x - 2)(x - 5)$$

Step 3 Use another point from the graph to find any constant factors.
Using the point $(7, 10)$, we get:

$$y = a(x - 2)(x - 5)$$
$$10 = a(7 - 2)(7 - 5)$$
$$10 = a(5)(2)$$
$$10 = 10a$$
$$1 = a$$

So the constant factor is 1.

Step 4 Multiply out the factors and simplify.
Therefore, we have the equation

$$y = 1(x - 2)(x - 5)$$
$$y = x^2 - 7x + 10$$

The y-intercept for this quadratic would be $(0, 10)$, which agrees with the graph, so we have the equation correct.

c.

Step 1 Find the horizontal intercepts of the graph and write them as zeros of the equation.
The horizontal intercepts for this graph are $(2, 0)$ and $(5, 0)$, so we know that $x = 2$ and $x = 5$ are the zeros of the quadratic equation.

Step 2 Undo the solving process to find the factors of the equation.
Therefore we have:

$$x = 2 \qquad x = 5$$
$$x - 2 = 0 \qquad x - 5 = 0$$

This gives us the factors $(x - 2)$ and $(x - 5)$ and the factorization

$$y = (x - 2)(x - 5)$$

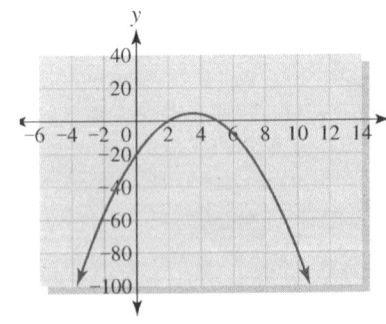

Step 3 Use another point from the graph to find any constant factors.
Using the point $(10, -80)$, we get:

$$y = a(x - 2)(x - 5)$$
$$-80 = a(10 - 2)(10 - 5)$$
$$-80 = 40a$$
$$-2 = a$$

So the constant factor is -2.

Step 4 Multiply out the factors and simplify.
Therefore, we have the equation

$$y = -2(x - 2)(x - 5)$$
$$y = -2x^2 + 14x - 20$$

The y-intercept for this quadratic would be $(0, -20)$, which agrees with the graph, so we have the equation correct.

EXAMPLE ④ PRACTICE PROBLEM

Use the graph to find an equation for the quadratic.

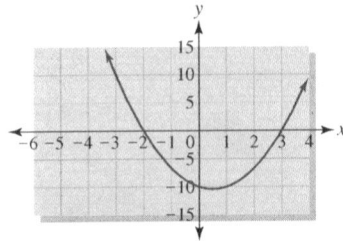

4.5 Exercises

In Exercises 1-10 solve and check.

1. $(x + 3)(x - 2) = 0$

2. $(x - 8)(x - 4) = 0$

3. $2(w + 7)(3w + 10) = 0$

4. $8(k - 46)(k + 23) = 0$

5. $x(x - 4)(x + 5) = 0$

6. $x(x + 1)(5x + 6) = 0$

7. $4x(x - 9)(x + 7) = 0$

8. $-3x(2x + 5)(4x - 9) = 0$

9. $4.5w(3w - 8)(7w + 5)(w - 6) = 0$

10. $-2m(m + 3.8)(m - 7.5)(2m - 9)(m + 8) = 0$

11. A math professor vacationing in Kauai jumped off a 20 foot waterfall into a pool of water below. The height of the math professor above the pool of water can be modeled by

    $$H(t) = -16t^2 + 4t + 20$$

    where $H(t)$ represents the height of the professor in feet t seconds after jumping off the waterfall.

    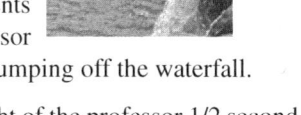

    a. What was the height of the professor 1/2 second after jumping off the waterfall?

    b. Use factoring to determine how many seconds it took for the professor to hit the pool of water below.

    c. Use factoring to determine how many seconds after jumping the professor was 8 feet above the pool of water.

12. The Grand Canyon National Park recently opened a new attraction called the Skywalk. The Skywalk is a glass-bottomed platform that hangs over the edge of the Grand Canyon and floats 4000 feet above the floor of the canyon. If a tourist shot an arrow straight up in the air from the observation platform, the arrow's height could be modeled by

    $$H(t) = -16t^2 + 240t + 4000$$

    where $H(t)$ is the height of the arrow above the canyon floor in feet t seconds after being shot.

    a. Find $H(5)$ and explain its meaning in this context.

    b. Use factoring to determine how many seconds before the arrow was at a height of 2400 feet.

    c. Use factoring to determine after how many seconds the arrow would hit the canyon floor.

13. The average profit made by a local company that manufactures custom car parts for racing teams can be modeled by

    $$A(p) = p(80 - 4p)$$

    where $A(p)$ represents the average profit in dollars per part when p parts are produced. Use factoring to find the following. Check your answers using the table.

    a. Find the number of parts that have to be produced for the company to earn $144 profit per part.

    b. Find the number of parts that have to be produced for the company to earn $300 profit per part.

    c. Find the number of parts that have to be produced for the company to earn $400 profit per part.

14. The average profit for mixing specialized metallic paints for a small manufacturing company can be modeled by

    $$P(q) = q(105 - 7q)$$

    where $P(q)$ represents the average profit in dollars per quart mixed when q quarts are mixed for one order. Use factoring to find the number of quarts that must be ordered and mixed for the average profit to be $350 per quart. Check your answer with the table.

15. The average cost to manufacture custom tooling machines can be modeled by

$$C(m) = m(108 - 6m)$$

where $C(m)$ represents the average cost in dollars per machine when m machines are made.

a. When will the average cost per machine be $390? Check your answer with the table.

b. When will the average cost per machine be $480? Check your answer with a graph.

16. An open box can be made from a rectangular piece of cardboard that is 36 by 30 inches by cutting a square from each corner and folding up the sides. The volume of the open box can be modeled by

$$V(x) = 4x^3 - 132x^2 + 1080x$$

where $V(x)$ is the volume of the open box in cubic inches when squares of length x inches are cut from each corner.

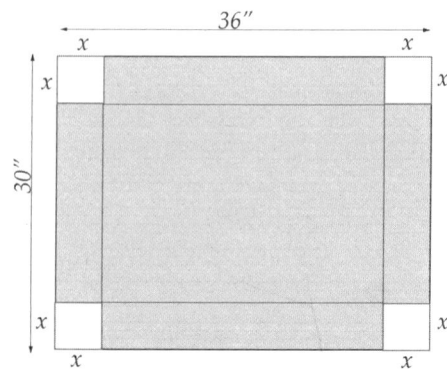

a. Completely factor the volume formula.

b. Find $V(4)$ and explain its meaning in this context.

c. Use the graph to estimate the size of square that should be cut out of each corner to get a box with the greatest volume.

In Exercises 17 through 54, solve by factoring.

17. $x^2 - 4x - 21 = 0$

18. $x^2 + 6x + 5 = 0$

19. $h^2 + 12h + 27 = 0$

20. $p^2 + 3p - 28 = 0$

21. $x^2 - 36 = 0$

22. $t^2 - 64 = 0$

23. $5x^2 - 80 = 0$

24. $-3m^2 + 75 = 0$

25. $x^2 + 50 = 150$

26. $4x^2 - 20 = 5$

27. $5x^2 + 20x = 0$

28. $2x^2 - 9x = 0$

29. $6x^2 = 10x$

30. $8x^2 = 6x$

31. $x^2 + 9x + 20 = 0$

32. $w^2 - 7w - 18 = 0$

33. $t^2 - 11t + 21 = -7$

34. $2x^2 + 13x + 10 = -5$

35. $7m^2 - 25m + 18 = 6$

36. $3t^2 - 7t - 5 = 5$

37. $4x^2 - 31x - 45 = 0$

38. $6w^2 + 29w + 28 = 0$

39. $-41t + 10t^2 = 18$

40. $11x + x^2 = -28$

41. $x^2 + 7x - 9 = 7x + 16$

42. $x^2 - 3x + 8 = 24 - 3x$

43. $28p^2 + 3p + 60 = 100$

44. $12t^2 - 43t + 40 = -16$

45. $55 + 55h = 10h^2 - 50$

46. $6w^2 - 15w = 4w - 10$

47. $x^3 - 4x^2 - 21x = 0$

48. $x^3 - 4x^2 - 21x = 0$

49. $w^3 - 7w^2 + 15w = 5w$

50. $w^3 - 3w^2 - 3w + 9 = 7w + 9$

51. $10x^3 + 30x^2 + 40x + 15 = -5x^2 + 10x + 15$

52. $14x^3 + 35x^2 - 20x + 10 = 6x^2 - 5x + 10$

53. $x^4 - 16 = 0$

54. $x^5 = 81x$

55. A school sells candy as a fund-raiser for their sports programs. They sell about 5500 candy bars at a profit of $0.75 each. For each $0.50 increase in the price, and therefore the profit of each candy bar, they will sell about 1000 fewer candy bars.

    a. What is the profit for this school if the profit is $0.75 per candy bar?

    b. Find a model for the profit this school makes if they increase the price of each candy bar by $0.50 x times.

    c. Use your model to determine the profit if the price per candy bar is increased by $1.

    d. Use your model to determine the profit if the price per candy bar is increased by $2.

    e. What price will bring the school the highest profit?

56. A church is bringing a group of its members to a "clean comedians" show. If they bring 15 people, it will cost $10 each. For each additional person the church brings, the price per ticket will go down $0.20. Groups are limited to 35 people because of the seating capacity of the venue.

    a. What will it cost the church to bring 20 members?

    b. Find a model for the total cost of bringing a group to the show. (Define x as the number of members beyond 15.)

    c. Use your model to find the total cost of bringing a group of 25 people.

    d. If the church can only afford to pay $200 for this trip, how many people can the church bring?

57. A sporting goods store sells about 100 boxes of golf balls a week when the price is $15 per box. If the store decreases the price by $1, they will sell an additional 20 boxes.

    a. What is the store's weekly revenue from selling boxes of golf balls for $15?

    b. What is the store's weekly revenue from selling boxes of golf balls for $17?

    c. Find a model for the store's weekly revenue from selling boxes of golf balls if the store lowers the price by x.

    d. Use your model to find the weekly revenue from selling boxes of golf balls if the price is $8.

    e. What price will maximize the store's weekly revenue from selling boxes of golf balls?

58. A local college has 12,000 students who each pay $800 per semester for tuition. For every $25 increase in tuition, the college will lose 300 students.

    a. How much revenue will the college make per semester if the tuition is raised to $850?

    b. Find a model for the college's revenue per semester when the tuition is raised $25 x times.

    c. Use your model to find the college's revenue if their tuition is raised to $1000.

    d. What should the tuition be increased to so that the college makes the most revenue?

59. A nature conservancy group wants to protect a rectangular nesting ground next to a river to save an endangered bird species. The group used 140 feet of fencing and only fenced three sides of the rectangular area, since the river made the fourth side. What are the dimensions of the enclosure if they were able to protect a total of area of 2400 square feet?

60. A walkway is being installed around a rectangular swimming pool. The pool is 30 feet by 12 feet, and the total area of the pool and the walkway is 1288 square feet. What is the width of the walkway?

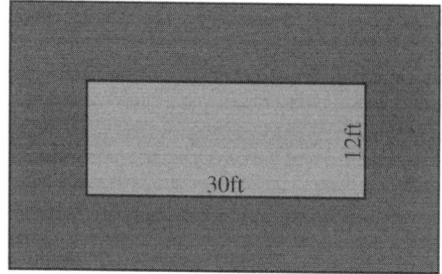

30ft

12ft

61. A framed painting has overall dimensions of 32 inches by 29 inches. The area of the painting itself is 550 square inches. Find the width of the frame.

32 inches

29 inches

62. A framed painting has overall dimensions of 36 inches by 30 inches. The area of the painting itself is 832 square inches. Find the width of the frame.

In Exercises 63 through 70, use the graph to find the equation for the quadratic.

63.

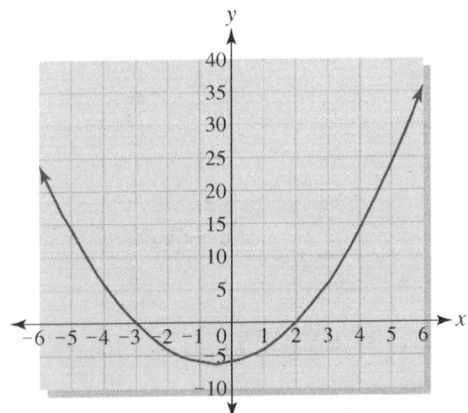

64.

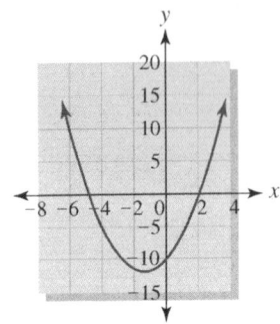

65.

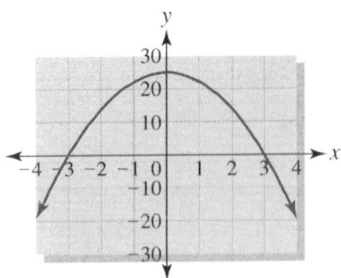

66.

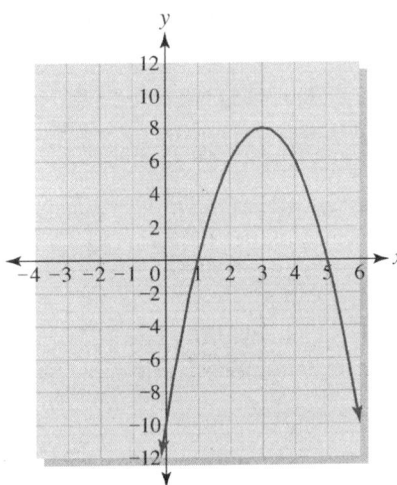

67.

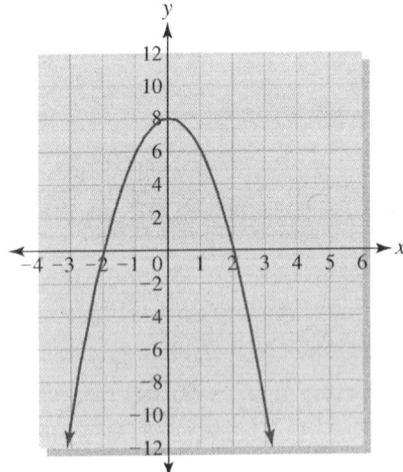

68.

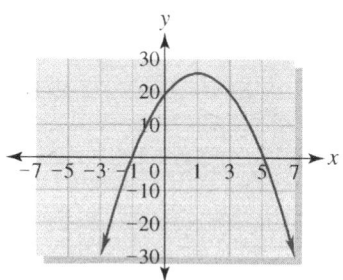

69.

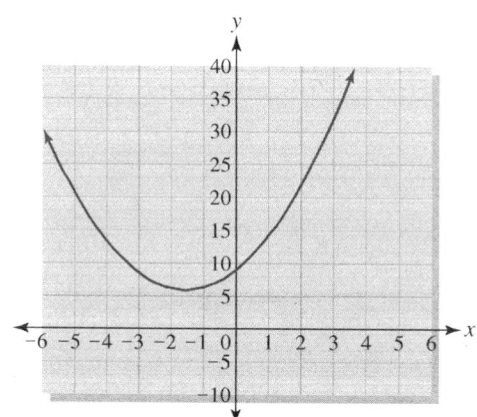

70.

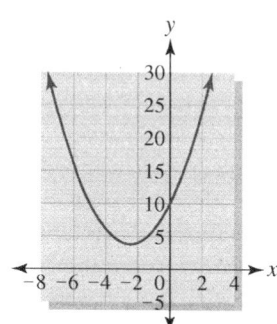

For exercises 71 through 82 use the given information to find the equation of the quadratic function.

71. The zeros of the function are $x = 3$ and $x = -7$, and there is no constant multiplier.

72. The zeros of the function are $x = -5$ and $x = -9$, and there is no constant multiplier.

73. The zeros of the function are $x = 2$ and $x = 4$, and there is no constant multiplier.

74. The zeros of the function are $x = \frac{1}{2}$ and $x = -8$, and there is no constant multiplier.

75. The zeros of the function are $x = \frac{2}{3}$ and $x = 4$, and there is no constant multiplier.

76. The zeros of the function are $x = 6$ and $x = -1$. Use the fact that $f(2) = -36$ to find the constant multiplier.

77. The zeros of the function are $x = 4$ and $x = 2$. Use the fact that $f(0) = 40$ to find the constant multiplier.

78. The zeros of the function are $x = 15$ and $x = -12$. Use the fact that $f(3) = -45$ to find the constant multiplier.

79. The zeros of the function are $x = \frac{1}{4}$ and $x = -2$. Use the fact that $f(2) = -21$ to find the constant multiplier.

80. The zeros of the function are $x = \frac{2}{5}$ and $x = 9$. Use the fact that $f(4) = 13.5$ to find the constant multiplier.

81. The zeros of the function are $x = -\frac{1}{3}$ and $x = \frac{9}{2}$. Use the fact that $f(3) = 20$ to find the constant multiplier.

82. The zeros of the function are $x = -\frac{2}{5}$ and $x = \frac{3}{4}$. Use the fact that $f(2) = 18$ to find the constant multiplier.

Solving Quadratic Equations Using the Quadratic Formula

SOLVING BY THE QUADRATIC FORMULA

In this section, we will discuss a third method for solving quadratics. Many models that we may encounter have numbers that are not going to be easy to factor. We are now going to generate a formula based on completing the square and the square root property. If we start with a standard quadratic form we can solve for x while leaving the coefficients and constants unknown. The only requirement is that $a \neq 0$ or the division would be undefined and the equation would not be a quadratic.

Say What?

Be careful with the difference between the following:

• Quadratic formula: a formula that is used to solve quadratic equations.

$$x = \frac{-b \pm \sqrt{b^2 - 4ac}}{2a}$$

• Quadratic equation (what is being solved)

$$2x^2 + 3x - 5 = 20$$

$ax^2 + bx + c = 0$	We start with the standard quadratic equal to zero.
$ax^2 + bx = -c$	Move the constant to the other side of the equal sign. Divide by a to make the leading coefficient 1. Note: a cannot equal zero so that division by a is defined.
$x^2 + \dfrac{b}{a}x = \dfrac{-c}{a}$	
$x^2 + \dfrac{b}{a}x + \dfrac{b^2}{4a^2} = \dfrac{-c}{a} + \dfrac{b^2}{4a^2}$	Add a constant that will complete the square. (half of the coefficient of x squared).
$\left(x + \dfrac{b}{2a}\right)^2 = \dfrac{b^2 - 4ac}{4a^2}$	Factor and simplify.
$x + \dfrac{b}{2a} = \pm\sqrt{\dfrac{b^2 - 4ac}{4a^2}}$	Use the square root property.
$x + \dfrac{b}{2a} = \pm\dfrac{\sqrt{b^2 - 4ac}}{2a}$	Simplify the denominator of the radical.
$x = \dfrac{-b}{2a} \pm \dfrac{\sqrt{b^2 - 4ac}}{2a}$	Get x by itself.
$x = \dfrac{-b \pm \sqrt{b^2 - 4ac}}{2a}$	

Because we solved the quadratic equation for any values for a, b and c, the mathematics will hold for any real values of $a \neq 0$, b and c. Therefore in general we have solved any quadratic from standard form. The final result is called the **quadratic formula**.

$$x = \frac{-b \pm \sqrt{b^2 - 4ac}}{2a} \qquad a \neq 0$$

This formula can be used to solve any quadratic equation in standard form as long as the equation equals zero.

SOLVING QUADRATICS USING THE QUADRATIC FORMULA

1. Set the quadratic equal to zero.

2. Put the quadratic into standard form.

3. Substitute the values of a, b, and c into the quadratic formula.

$$x = \frac{-b \pm \sqrt{b^2 - 4ac}}{2a}$$

4. Simplify the quadratic formula.

5. Check solutions in the original equation.

EXAMPLE ▮ USING THE QUADRATIC FORMULA

Solve the following quadratic equations. Round your answers to three decimal places.

a. $3x^2 - 6x - 24 = 0$

b. $4t^2 - 8t + 5 = 50$

c. $3.4x^2 + 4.2x - 7.8 = 0$

Solution

a.

$$3x^2 - 6x - 24 = 0$$

$$a = 3 \qquad b = -6 \qquad c = -24$$

The equation is in standard form and equal to zero, so use the quadratic formula.

$$x = \frac{-(-6) \pm \sqrt{(-6)^2 - 4(3)(-24)}}{2(3)}$$

$$x = \frac{6 \pm \sqrt{324}}{6}$$

```
(-6)²-4(3)(-24)
                324
√(Ans)
                 18
```

$$x = \frac{6 \pm 18}{6}$$

Separate into two equations.
Simplify.

$$x = \frac{6 + 18}{6} \qquad x = \frac{6 - 18}{6}$$

$$x = \frac{24}{6} \qquad x = \frac{-12}{6}$$

$$x = 4 \qquad x = -2$$

$$3(4)^2 - 6(4) - 24 = 0$$

$$48 - 24 - 24 = 0$$

Check both solutions.

$$0 = 0$$

$$3(-2)^2 - 6(-2) - 24 = 0$$

$$12 + 12 - 24 = 0$$

$$0 = 0$$

b.

$$4t^2 - 8t + 5 = 50$$ 　　　Set the equation equal to zero.

$$4t^2 - 8t - 45 = 0$$

$$a = 4 \qquad b = -8 \qquad c = -45$$ 　　Use the quadratic formula.

$$t = \frac{-(-8) \pm \sqrt{(-8)^2 - 4(4)(-45)}}{2(4)}$$

$$t = \frac{8 \pm \sqrt{784}}{8}$$

$$t = \frac{8 \pm 28}{8}$$

$$t = \frac{8 + 28}{8} \qquad t = \frac{8 - 28}{8}$$ 　　Separate into two equations.

$$t = \frac{36}{8} \qquad t = \frac{-20}{8}$$ 　　Simplify

$$t = \frac{9}{2} \qquad t = \frac{-5}{2}$$

Check both solutions using the table.

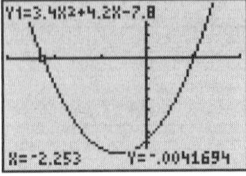

c.

$$3.4x^2 + 4.2x - 7.8 = 0$$

$$a = 3.4 \qquad b = 4.2 \qquad c = -7.8$$

$$x = \frac{-(4.2) \pm \sqrt{(4.2)^2 - 4(3.4)(-7.8)}}{2(3.4)}$$

$$x = \frac{-4.2 \pm \sqrt{123.72}}{6.8}$$

$$x \approx \frac{-4.2 \pm 11.1229}{6.8}$$

Separate into two equations and simplify.

$$x \approx \frac{-4.2 + 11.1229}{6.8} \qquad x \approx \frac{-4.2 - 11.1229}{6.8}$$

$$x \approx 1.018 \qquad x \approx -2.253$$

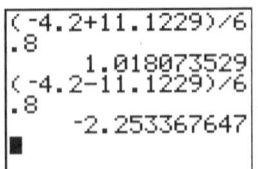

Check both solutions using the graph.

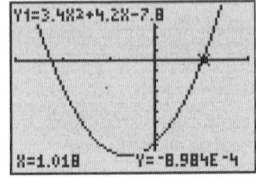

EXAMPLE PRACTICE PROBLEM

Solve the following quadratic equations. Round your answers to three decimal places.

a. $2x^2 + 13x + 15 = 0$

b. $1.5x^2 + 2 = -6.5x$

The quadratic formula will find all possible solutions of every quadratic, making it a very powerful tool. When using the quadratic formula, be very careful with your calculations. Doing each calculation one step at a time on the calculator will lessen the number of arithmetic errors you make. In the formula itself you should note that it starts with the opposite of b and has a plus/minus sign in front of the square root. A common mistake is to put the $2a$ in the denominator only under the square root and not under the b as well.

EXAMPLE 2 USING THE QUADRATIC FORMULA

In Section 4.1 we considered data for the number of public branch libraries in the United States and were given the model

$$B(t) = 28.7t^2 - 286.2t + 6899.3$$

where $B(t)$ represents the number of public branch libraries in the United States t years since 1990. In what year were there 6500 public branch libraries in the United States?

Solution

6500 represents a number of branch libraries, so it will be substituted into $B(t)$. Because the numbers in this model are large, we will use the quadratic formula to solve.

$$6500 = 28.7t^2 - 286.2t + 6899.3$$

| -6500 | | -6500 | Set the equation equal to zero. |

$$0 = 28.7t^2 - 286.2t + 399.3$$

$$a = 28.7 \qquad b = -286.2 \qquad c = 399.3$$ Use the quadratic formula.

$$t = \frac{-b \pm \sqrt{b^2 - 4ac}}{2a}$$

Remember -b means the "opposite of b."

$$t = \frac{-(-286.2) \pm \sqrt{(-286.2)^2 - 4(28.7)(399.3)}}{2(28.7)}$$

$$t = \frac{286.2 \pm \sqrt{36070.8}}{57.4}$$

```
(-286.2)²-4(28.7
)(399.3)
           36070.8
√(Ans)
         189.9231423
■
```

$$t \approx \frac{286.2 \pm 189.92}{57.4}$$

$$t \approx \frac{286.2 - 189.92}{57.4} \qquad t \approx \frac{286.2 + 189.92}{57.4}$$ Separate into two equations and simplify.

$$t \approx \frac{96.28}{57.4} \qquad t \approx \frac{476.12}{57.4}$$

$$t \approx 1.68 \qquad t \approx 8.29$$

Check both solutions using the table.

In 1992 and 1998 there were approximately 6500 public branch libraries in the United States. Both of these years are reasonable due to the fall in branch libraries during the early 90's and the increase in the mid to late 90's.

EXAMPLE ② PRACTICE PROBLEM

The number of people who were unemployed in Hawaii during the 1990s can be modeled by

$$U(t) = -726.25t^2 + 9824.85t + 3763.21$$

where $U(t)$ represents the number of people unemployed in Hawaii t years since 1990. According to this model, when were there 30,000 people unemployed in Hawaii?

© BananaStock/Alamy

DETERMINING WHICH ALGEBRAIC METHOD TO USE

When you are faced with an equation to solve, you should first decide which type of equation you are trying to solve (linear, quadratic, or higher degree polynomial) and then choose a method that is best suited to solve that equation. We have looked at four algebraic methods to solve a quadratic

- Square Root Property
- Completing the Square
- Factoring
- Quadratic Formula

Determining what method is best depends on the characteristics of the equation you are trying to solve.

Square Root Property
The square root property works best when the quadratic is in vertex form or when there is a squared variable term but no other variable terms.

$$(x + 5)^2 - 9 = 16$$
$$x^2 + 13 = 49$$
$$6x^2 - 4 = 18$$

In all of these equations, the square root property would be a good way to solve. Remember when solving using the square root property to isolate the squared part of the equation on one side before you use the square root property. Also do not forget to use the plus/minus symbol to indicate all possible answers.

Completing the Square

Completing the square works well for equations that have both a squared term and a first degree term. This method is usually easiest if the numbers are not too large and the leading coefficient is 1.

$$x^2 + 4x + 9 = 0$$

$$x^2 - 5x = 20$$

After completing the square, you will again use the square root property to solve.

Factoring

Factoring works best when the numbers are not too large or you have terms higher than second degree. Always remember first factor out the greatest common factor. Although factoring does not work with all quadratics, if the equation factors easily, it can be one of the fastest solution methods. Remember that for factoring to be used when solving, the equation must be equal to zero, so that the zero product property can be used.

$$x^2 + 5x + 6 = 0$$

$$x^3 + x^2 - 6x = 0$$

Equations that cannot be factored may still have solutions, so use one of the other methods such as completing the square or the quadratic formula. Occasionally, an equation can be factored partially and then the separate pieces can be solved using another method.

Quadratic Formula

The quadratic formula will work with any quadratic but is easiest if the quadratic starts out in standard form. Because the quadratic formula basically requires you to substitute values for a, b and c and then simplify an arithmetic expression, the formula will work just as well for large or small numbers. Whenever a quadratic equation has decimals or fractions, the quadratic formula is probably the best method for you to choose.

$$5x^2 + 16x - 85 = 0$$

$$0.25x^2 - 3.4x + 9 = 0$$

Remember that some quadratic equations will have no real solutions. When solving with the square root property or the quadratic formula, you will end up having a negative under the square root indicating no real solutions.

EXAMPLE **3** CHOOSING A METHOD TO SOLVE EQUATIONS

The five parts of this example go through the different processes to choose the right method to solve the equation. Solve each part, keeping the four methods in mind. Check your answers using the calculator table or graph.

a. $5x^3 + 37x^2 - 5x = -19x$

b. $x^2 + 12x + 7 = 0$

c. $5x^2 - 30 = 40$

d. $2x(x + 5) = 4x$

e. $x^3 + 8x^2 + 5x = 0$

Solution

a. This equation has a third-degree term, so we will try to factor it and break it up into smaller pieces that will be easier to solve. The first step will be to set one side of the equation equal to zero.

$$5x^3 + 37x^2 - 5x = -19x$$

$$5x^3 + 37x^2 + 14x = 0$$ Set the equation equal to zero.

$$x(5x^2 + 37x + 14) = 0$$ Factor.

$$x(5x^2 + 2x + 35x + 14) = 0$$

$$x[(5x^2 + 2x) + (35x + 14)] = 0$$

$$x[x(5x + 2) + 7(5x + 2)] = 0$$

$$x(x + 7)(5x + 2) = 0$$

$$x = 0 \qquad x + 7 = 0 \qquad 5x + 2 = 0$$ Separate into three equations and solve.

$$x = 0 \qquad x = -7 \qquad x = -\frac{2}{5}$$

Check all three solutions using the table.

Plot1 Plot2 Plot3		X	Y1	Y2
\Y1**█**5X^3+37X²-5X		0	0	0
		-7	133	133
\Y2**█**-19X		-.4	7.6	7.6
\Y3=				
\Y4=				
\Y5=				
\Y6=		X=		

b. This equation does not factor (try it!), so we must use another method. Since the equation has a squared term and a first-degree term, we will not be able to use the square root property directly. Therefore, we could complete the square and then use the square root property.

$$x^2 + 12x + 7 = 0$$ Complete the square.

$$x^2 + 12x = -7$$

$$x^2 + 12x + 36 = -7 + 36$$

$$(x + 6)^2 = 29$$ Use the square root property.

$$x + 6 = \pm\sqrt{29}$$

$$x + 6 = \sqrt{29} \qquad x + 6 = -\sqrt{29}$$ Separate into two equations and solve.

$$x = -6 + \sqrt{29} \qquad x = -6 - \sqrt{29}$$

$$x \approx -0.615 \qquad x \approx -11.385$$

Check both solutions using the graph.

Y1=X²+12X+7	Y1=X²+12X+7
X=-11.38298 Y=-.0235401	X=-.615 Y=-.001775

c. This equation has no first-degree term, so the square root property will be a good method to choose. First isolate the squared part of the equation.

$$5x^2 - 30 = 40 \qquad \text{Isolate the squared part.}$$

$$5x^2 = 70$$

$$x^2 = 14$$

$$x = \pm\sqrt{14} \qquad \text{Use the square root property.}$$

$$x = \sqrt{14} \qquad x = -\sqrt{14} \qquad \text{Separate into two}$$
equations and solve.

$$x \approx 3.742 \qquad x \approx -3.742$$

Plot1 Plot2 Plot3	X	Y1	Y2
\Y1■5X²-30	3.742	40.013	40
\Y2■40	-3.742	40.013	40
\Y3=			
\Y4=			
\Y5=			
\Y6=			
\Y7=	X=		

Check both solutions using the table.

d. This equation looks like it is already factored, but since it does not equal zero, we will have to multiply it out and set it equal to zero before we can factor and solve.

$$2x(x + 5) = 4x \qquad \text{Multiply out using the}$$
distributive property.

$$2x^2 + 10x = 4x$$

$$2x^2 + 6x = 0 \qquad \text{Set the equation equal to}$$
zero.

$$2x(x + 3) = 0 \qquad \text{Factor.}$$

$$2x = 0 \qquad x + 3 = 0 \qquad \text{Separate into two}$$
equations and solve.

$$x = 0 \qquad x = -3$$

Plot1 Plot2 Plot3	X	Y1	Y2
\Y1■2X(X+5)	0	0	0
\Y2■4X	-3	-12	-12
\Y3=			
\Y4=			
\Y5=			
\Y6=			
\Y7=■	X=		

Check both solutions using the table.

e. This equation has a third-degree term, so it should be factored.

$$x^3 + 8x^2 + 5x = 0 \qquad \text{Factor out the common}$$
term.

$$x(x^2 + 8x + 5) = 0$$

The equation cannot be factored any further. We will set each factor equal to zero and then solve using the quadratic formula.

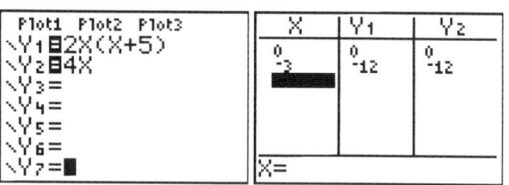

$$x = 0 \quad \text{or} \quad x^2 + 8x + 5 = 0 \qquad \text{Separate into two}$$
equations and solve.

$$a = 1 \qquad b = 8 \qquad c = 5$$

Use the quadratic formula.

$$x = \frac{-(8) \pm \sqrt{(8)^2 - 4(1)(5)}}{2(1)}$$

$$x = \frac{-8 \pm \sqrt{44}}{2}$$

$$x = \frac{-8 \pm 2\sqrt{11}}{2}$$ Simplify the radical.

$$x = -4 \pm \sqrt{11}$$

$$x = -4 + \sqrt{11} \qquad x = -4 - \sqrt{11}$$

$$x = 0 \quad \text{or} \quad x \approx -0.683 \quad \text{or} \quad x \approx -7.317$$

This equation appears to have three solutions. We can check them in the table.

```
Plot1 Plot2 Plot3
\Y1■X^3+8X²+5X
\Y2■0
\Y3=
\Y4=
\Y5=
\Y6=
\Y7=
```

X	Y1	Y2
0	0	0
-.683	-.0017	0
-7.317	-.0182	0

X=

EXAMPLE ③ PRACTICE PROBLEM

Solve the following equations.

a. $6x^3 - 7x^2 = 20x$

b. $4x^2 + 10x - 40 = 2(5x + 12)$

SOLVING SYSTEMS OF EQUATIONS WITH QUADRATICS

Now that we can solve quadratics in several ways, we will consider systems of equations that contain quadratics. In Chapter 2 we learned that we could solve systems using three different methods: graphing, substitution and elimination.

Using graphs and tables to estimate solutions to systems of equations that involve functions other than lines is the same process as with linear systems. When using the graph, you look for the place(s) where the two graphs intersect. When using a table, you need to find input value(s) that make the outputs for both equations equal. When looking for these input value(s), notice when one equation changes from being smaller than the other equation to being larger than the other. The intersection of the two equations must be between these input values.

EXAMPLE 4 USING GRAPHS AND TABLES TO SOLVE SYSTEMS

Use the given graph or table to estimate the solutions to the systems of equations.

a.

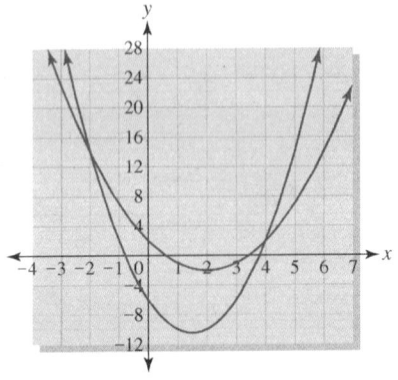

b.

X	Y1	Y2
0	4	-6
1	4	0
2	10	10
3	22	24
4	40	42
5	64	64
6	94	90

X=6

c.

X	Y1	Y2
-4.4	.36	.24
-4.375	.26563	.35938
-4.35	.1725	.4275
-.13	2.3669	2.5031
-.12	2.4144	2.4656
-.11	2.4621	2.4279
-.1	2.51	2.39

X=-.1

Solution

a. From the graph, we can see that the two graphs intersect at about the points $(4, 2)$ and $(-2, 14)$.

b. The table shows that the two equations are equal at the points $(2, 10)$ and $(5, 64)$.

c. This table does not show any exact places where the equations are equal. The equation in Y1 is greater than Y2 at $x = -4.4$, but Y1 is less than Y2 at $x = -4.375$. So there must be an intersection between these values. One estimate for this intersection could be $(-4.38, 0.3)$. The two equations must also intersect between $x = -0.12$ and $x = -0.11$, so we might estimate this intersection at about $(-0.115, 2.48)$. Remember that these are only estimates. Many other answers could be reasonable.

Although graphs and tables can be used to find solutions to these systems we can also use algebraic methods to solve. When solving systems that contain quadratics algebraically, substitution is often the best choice.

EXAMPLE **5** SOLVING SYSTEMS ALGEBRAICALLY

Solve the following systems of equations.

a. $y = 4x + 9$
$y = x^2 + 5x + 3$

b. $y = 5x^2 + 2x + 7$
$y = 2x^2 - 3x + 10$

Solution

a.
$$y = 4x + 9$$
$$y = x^2 + 5x + 3$$

$$4x + 9 = x^2 + 5x + 3 \qquad \text{Substitute for } y.$$

$$0 = x^2 + x - 6$$ Set the equation equal to zero.

$$0 = (x + 3)(x - 2)$$ Factor.

$$x + 3 = 0 \qquad x - 2 = 0$$ Separate into two equations and solve.

$$x = -3 \qquad\qquad x = 2$$

$$y = 4(-3) + 9$$ Use the values of x to find the

$$y = -3$$ corresponding values of y.

$$y = 4(2) + 9$$

$$y = 17$$ These two points are the solutions

$$(-3, -3) \qquad (2, 17)$$ to the system of equations.

These solutions can be checked by using either the table feature or the graph feature on your calculator. First, put both equations into your Y= screen, and then graph them using a window that will include both solutions or by going to the table and inputting both x values to see whether the y values are the same. Remember that some rounding error can occur if the solutions were not exact.

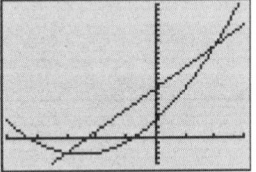

b.
$$y = 5x^2 + 2x + 7$$
$$y = 2x^2 - 3x + 10$$

$$5x^2 + 2x + 7 = 2x^2 - 3x + 10$$ Substitute for y and simplify.

$$3x^2 + 5x - 3 = 0$$ You are left with a quadratic

$$a = 3 \qquad b = 5 \qquad c = -3$$ that can be solved by using the quadratic formula.

$$x = \frac{-(5) \pm \sqrt{(5)^2 - 4(3)(-3)}}{2(3)}$$

$$x = \frac{-5 \pm \sqrt{61}}{6}$$

$$x \approx \frac{-5 \pm 7.81}{6}$$

$$x \approx 0.468 \qquad x \approx -2.135$$

Use the x values to find the corresponding y values;

$$y = 5(0.468)^2 + 2(0.468) + 7$$

$$y = 9.031$$

$$y = 5(-2.135)^2 + 2(-2.135) + 7$$

$$y = 25.521$$ There are two solutions to this

$$(0.468, \ 9.031) \qquad (-2.135, \ 25.521)$$ system.

Check the solutions using the table or graph.

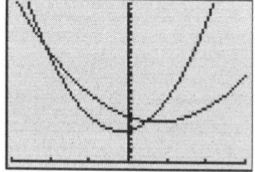

EXAMPLE ⑤ PRACTICE PROBLEM

Solve the following system of equations.

$$y = 3x^2 + 5x - 74$$
$$y = 6x^2 + 60x + 126$$

You may be asked to find a solution of a quadratic equation that does not exist in the real numbers. A quadratic equation can have no real solutions when you are asked to find a value that the function will never reach. No real solutions will occur in an upward-facing parabola when the output value you are looking for is below the vertex or in a downward-facing parabola when the output value is above the vertex.

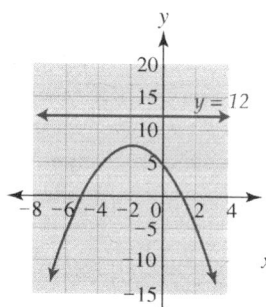

$$y = -x^2 - 4x + 5$$
$$12 = -x^2 - 4x + 5$$

Neither of these will have real number solutions.

$$y = (x + 2)^2 + 6$$
$$3 = (x + 2)^2 + 6$$

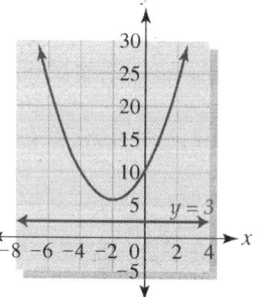

In using the quadratic formula, the value of the **discriminant,** $b^2 - 4ac$, determines whether the equation will have real solutions or not. When the discriminant is negative, the quadratic will have no real solutions, because in the real number system, we cannot take the square root of negative numbers. Using the quadratic formula to solve the equation above, we get;

$$12 = -x^2 - 4x + 5$$
$$0 = -x^2 - 4x - 7$$
$$a = -1 \qquad b = -4 \qquad c = -7$$
$$x = \frac{-(-4) \pm \sqrt{(-4)^2 - 4(-1)(-7)}}{2(-1)}$$
$$x = \frac{4 \pm \sqrt{-12}}{-2}$$

For our work in this chapter, when we get a negative under the square root, we will simply state that there are no real solutions.

4.6 Exercises

For exercises 1 through 10 solve each equation using the quadratic formula.

1. $x^2 + 8x + 15 = 0$

2. $x^2 + 5x + 6 = 0$

3. $x^2 - 7x + 10 = 0$

4. $x^2 - 12x - 45 = 0$

5. $3x^2 + 9x - 12 = 0$

6. $5x^2 + 11x - 42 = 0$

7. $2x^2 - 9x = 5$

8. $7x^2 + 15x - 9 = 11$

9. $-4x^2 + 7x - 8 = -20$

10. $-3x^2 - 8x + 1000 = 360$

11. In the United States, the use of lead in paint has been decreasing since its peak around 1920. The amount of lead used in paint can be modeled by

$$L(t) = 0.06t^2 - 8.56t + 320.32$$

where $L(t)$ represents lead use in thousands of tons t years since 1900.

Source: Model based on data from Statistical Abstract 2001.

   a. How much lead was used in 1955?

   b. When were 5500 tons of lead used in paints? (Hint: 5500 tons = 5.5 thousand tons.)

   c. When were 51 thousand tons of lead in paints used?

12. The number of Americans who participate in aerobic exercise in thousands can be modeled by

$$E(a) = -7.5a^2 + 523.1a - 3312.3$$

where $E(a)$ represents the number of people in thousands who participated in aerobic exercise at age a years old.

   a. Find the number of 20-year-old Americans who participate in aerobic exercise.

   b. At what age(s) do 3 million Americans participate in aerobic exercise? (Hint: 3 million = 3000 thousand.)

13. The amount of electricity flowing through a surface is called the current I and is measured in amperes (A). The current through a particular wire can be modeled by

$$I(t) = 3t^2 - 4t + 5$$

where $I(t)$ is the current in amperes after t seconds.

   a. Find the current in the wire after 0.5 seconds.

   b. After how many seconds would the current reach 20 amperes?

   c. After how many seconds would the current reach 60 amperes?

14. The current through a particular wire can be modeled by

$$I(t) = 2t^2 - 6t + 7$$

where $I(t)$ is the current in amperes after t seconds.

   a. Find the current in the wire after 0.25 seconds.

   b. After how many seconds would the current reach 10 amperes?

   c. After how many seconds would the current reach 50 amperes?

The **marginal cost** of a product is the cost to produce one more item. For example, the marginal cost of 100 items is the cost to produce the 101st item.

15. The marginal cost of producing a particular kind of dress shoe can be modeled by

$$M(n) = 0.00045n^2 + 0.02n + 5$$

where $M(n)$ is the marginal cost in dollars per pair of dress shoes when n dress shoes are produced.

   a. Find the marginal cost of producing the 201st pair of shoes. (Hint: Find $M(200)$)

   b. Find when the marginal cost will be $20 per pair of shoes.

16. The marginal cost of producing a cell phone can be modeled by

$$M(n) = 0.0000015n^2 + 0.0001n + 8$$

where $M(n)$ is the marginal cost in dollars per phone when n phones are produced.

a. Find the marginal cost of producing the 5001st phone. (Find $M(5000)$)

b. Find when the marginal cost will be $30 per phone.

17. The marginal cost in dollars of producing a bike can be modeled by

$$M(n) = 0.0005n^2 + 0.07n + 50$$

where $M(n)$ is the marginal cost in dollars per bike when n bikes are produced.

a. Find the marginal cost of producing the 401st bike. (Find $M(400)$)

b. Find when the marginal cost will be $700 per bike.

18. The marginal cost in dollars of producing a laptop computer can be modeled by

$$M(n) = 0.0007n^2 + 0.04n + 85$$

where $M(n)$ is the marginal cost in dollars per laptop when n laptops are produced.

a. Find the marginal cost of producing the 801st laptop. (Find $M(800)$)

b. Find when the marginal cost will be $825 per laptop.

19. A local rocketry club held a competition for the highest launch of a model rocket. From the power produced by the motor, the height t seconds after launch is modeled by

$$h(t) = -16t^2 + 200t + 2$$

where $h(t)$ represents the height in feet t seconds after the rocket is launched.

a. Find the height of the rocket 1 second after launch.

b. When will the rocket first reach a height of 450 feet?

c. When will the rocket first reach a height of 600 feet?

d. When will the rocket first reach a height of 700 feet?

20. The height of a ball thrown straight up into the air can be modeled by

$$h(t) = -16t^2 + 30t + 3.5$$

where $h(t)$ is the height of the ball in feet t seconds after it is thrown.

a. What is the height of the ball after 0.5 second?

b. When will the ball be at a height of 15 feet?

c. Will the ball ever reach a height of 20 feet?

21. The number of hours per year an average person spent using the Internet can be modeled by

$$I(t) = 3.8t^2 - 29.5t + 56.0$$

where $I(t)$ represents the number of hours per year an average person spent using the Internet t years since 1990.

Source: Model based on data from Statistical Abstract 2001.

a. How many hours did the average person spend using the Internet in 2000?

b. When will/did the average person spend 100 hours per year on the Internet?

c. When will the average person spend an average of 1 hour per day on the Internet?

22. The number of cable television systems in the United States during the 1990s can be modeled by

$$C(t) = -34.6t^2 + 285.5t + 10649.6$$

where $C(t)$ represents the number of cable television systems in the United States t years since 1990.

a. Find the number of cable television systems in the United States in 1995.

b. Find when there were 11,000 cable television systems in the United States.

In Exercises 23 through 68, solve the given equations using any of the methods you have learned thus far. Round all answers to two decimal places when necessary. Check your solutions graphically or with a table.

23. $a^2 + 2a = 15$

24. $x^2 + 8x = 30$

25. $5t^2 - 14 = 0$

26. $-6m^2 + 56 = 0$

27. $\frac{1}{7}x^2 - \frac{5}{7}x = 0$

28. $\frac{2}{5}x^2 + \frac{3}{5}x = 0$

29. $4.7x^2 - 2.6x = 0$

30. $3.5w^2 + 8.2w = 0$

31. $5x - 12 = 80$

32. $4x + 45 = 5$

33. $5x^2 + 3x = 0$

34. $4t^2 - 4t = 0$

35. $(x - 9)^2 + 8 = 24$

36. $(p + 5)^2 - 4 = 0$

37. $3(4x - 5)^2 + 20 = 47$

38. $7(2k - 7)^2 - 11 = 17$

39. $33 = -7(4 - w)^2 + 59$

40. $47 = -4(6 - x)^2 + 63$

41. $r^2 + 1.4r - 14.9 = 0$

42. $k^2 + 2.3k - 11.4 = 0$

43. $d^2 + 2d - 35 = 0$

44. $x^2 + 11x + 18 = 0$

45. $b^2 - 3b = 28$

46. $v^2 + 14v = -48$

47. $23 = -2(s + 9)^2 + 5$

48. $105 = 5(17 + d)^2 - 20$

49. $\frac{2}{3}c^2 - \frac{5}{6} = \frac{1}{6}c - 2$

50. $\frac{3}{4}x^2 + \frac{1}{2}x - 7 = \frac{1}{4}x - 3$

51. $120 = -28f + 7f^2 - 939$

52. $9.9x^2 + 57.4x - 134.8 = 0$

53. $(3p - 4)(p + 3) = 15$

54. $(z + 9)(5z - 4) = 18$

55. $3x^2 + 4x + 20 = 0$

56. $-2x^2 + 7x - 18 = 0$

57. $\frac{3}{2}(x - 8)^2 + 10 = 1$

58. $\frac{2}{3}(x - 5)^2 - 7 = 10$

59. $3x^3 - 15x^2 = 252x$

60. $1.75x^3 + 2.5x^2 = 6.25x$

61. $7(x + 3) - 8 = 20$

62. $-3(r - 8) + 7 = -15$

63. $-1.5(d + 5) - 7 = 2d + 8$

64. $-2.3(p + 7) + 2 = 4.3p + 20$

65. $x^2 + 6x + 25 = 0$

66. $w^2 - 12x - 16 = 0$

67. $\frac{1}{4}x^2 - \frac{3}{4}x + 7 = 13$

68. $\frac{11}{21}x^2 - \frac{9}{21}x - \frac{4}{21} = 0$

For Exercises 69 through 76, estimate the solutions to the systems using the graphs.

69.

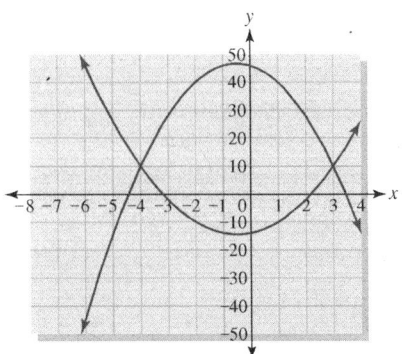

70.

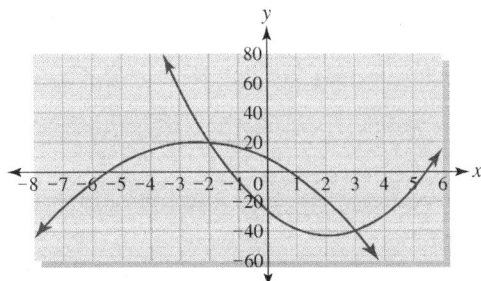

71.

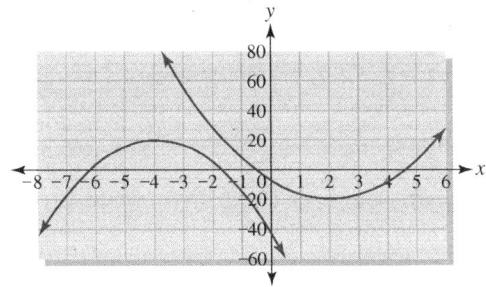

72.

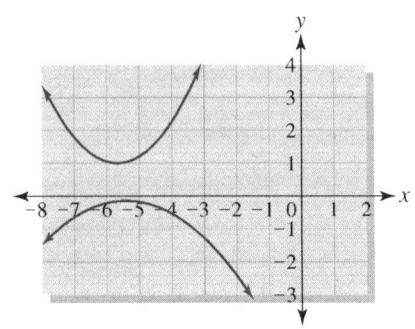

73.

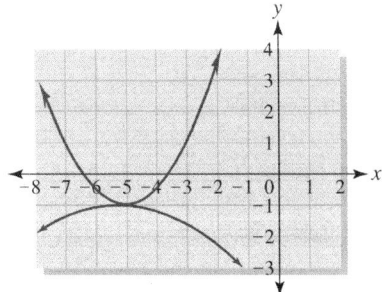

74.

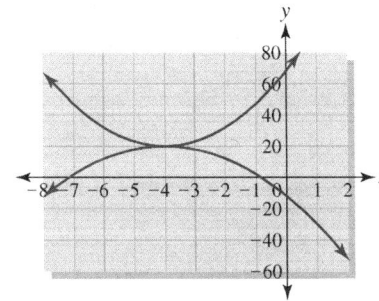

75.

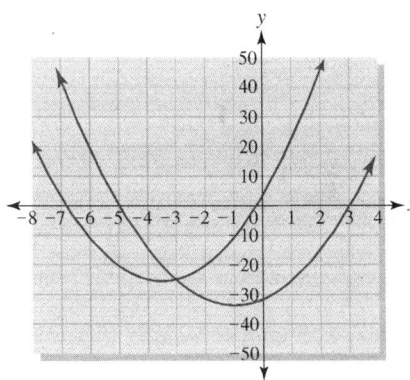

76.

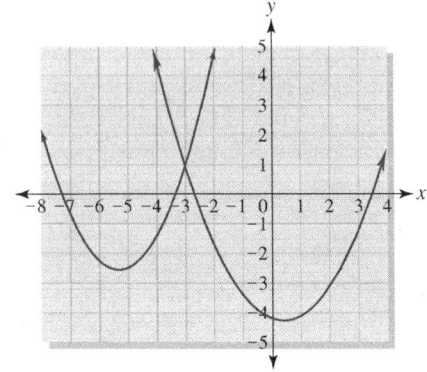

For Exercises 77 through 80, estimate the solution(s) to the system using the given tables.

77.

X	Y1	Y2
-6	48	-42
-4	12	12
-2	-8	42
0	-12	48
2	0	30
3	12	12
4	28	-12

Y1◻2X²+2X-12

78.

X	Y1	Y2
2.176	17.292	17.797
2.188	17.525	17.499
2.203	17.819	17.125
0	-6	48
-2.735	18.451	18.804
-2.743	18.61	18.609
-2.765	19.051	18.069

Y1◻4X²+2X-6

79.

X	Y1	Y2
-3.9	1.41	.1
-3.8	.84	.2
-3.7	.29	.3
-3.6	-.24	.4
2.6	5.96	6.6
2.7	6.69	6.7
2.8	7.44	6.8

Y1◻X²+2X-6

80.

X	Y1	Y2
2	27	9
3	17	9
4	11	9
5	9	9
6	11	9
7	17	9
8	27	9

Y1◻2(X-5)²+9

In Exercises 81 through 98, solve the system of equations algebraically. Check your solution(s) graphically or with a table.

81. $y = x^2 + 3x - 9$

$y = 5x - 8$

82. $y = 0.25x^2 + 5x - 3.4$

$y = -4.5x + 7.5$

83. $y = x^2 + 5x - 3$

$y = 2x - 10$

84. $y = 2x^2 - 3x + 7$

$y = x - 12$

85. $y = 3x^2 + 5x - 9$

$y = -0.5x^2 - 3x + 15$

86. $y = 3x^2 + 4x - 20$

$y = 2x^2 + 6x + 15$

87. $y = x^2 - 4x + 11$

$y = -x^2 + 7x - 4$

88. $y = 4x^2 + 2x - 7$

$y = -2x^2 + 5x + 12$

89. $y = -1.8x^2 - 2.3x + 4.7$

$y = 2.5x^2 + 3.4x - 8.5$

90. $y = -0.3x^2 + 5x - 2.6$

$y = 0.5x^2 - 3x - 7.5$

91. $y = x^2 + 6x - 20$

$y = -x^2 - 6x - 38$

92. $y = 5x^2 + 20x - 40$

$y = -3x^2 - 12x - 72$

93. $y = x^2 + 2x - 8$

$y = -2x^2 + 2x - 18$

94. $y = -x^2 + 4x + 1$

$y = x^2 + 3x + 14$

95. $y = x^2 - 10x + 30$

$y = -x^2 + 6x - 15$

96. $y = -0.5x^2 + 4x - 8$

$y = 0.3x^2 + 5x + 7$

97. $y = 6x^2 + 2x - 9$

$y = 9x^2 + 2x - 15$

98. $y = 4x^2 + 9x - 12$

$y = -2x^2 + 20x + 23$

Graphing from Standard Form

GRAPHING FROM STANDARD FORM

Because quadratic functions are often written in standard form rather than in vertex form, it is important that you learn to graph from the standard form. Since all quadratics have a vertex and a vertical intercept, these two will be the starting points for graphing.

A quadratic function in standard form $f(x) = ax^2 + bx + c$ gives us several key pieces of information about the graph. The value of a in the standard form affects the graph of the quadratic the same way that it affects the vertex form. If a is positive, the parabola will open upward, and if a is negative, the parabola will open downward. Also a will affect whether the graph is wide or narrow. The value c will also give us information about the graph. When you substitute zero into the input variable for the function, the output will be c,

$$\text{Let } f(x) = ax^2 + bx + c \quad \text{and } x = 0$$
$$f(0) = a(0)^2 + b(0) + c$$
$$f(0) = c$$

which means that $(0, c)$ will be the vertical intercept for the graph.

$$f(x) = 4x^2 + 3x - 12$$
$$f(0) = 4(0)^2 + 3(0) - 12$$
$$f(0) = -12$$
$$(0, -12) \quad \text{The vertical intercept.}$$

Another important relationship for parabolas is between the vertex and the standard form of a quadratic function. There are several ways to see this relationship, one being to consider the quadratic formula and symmetry.

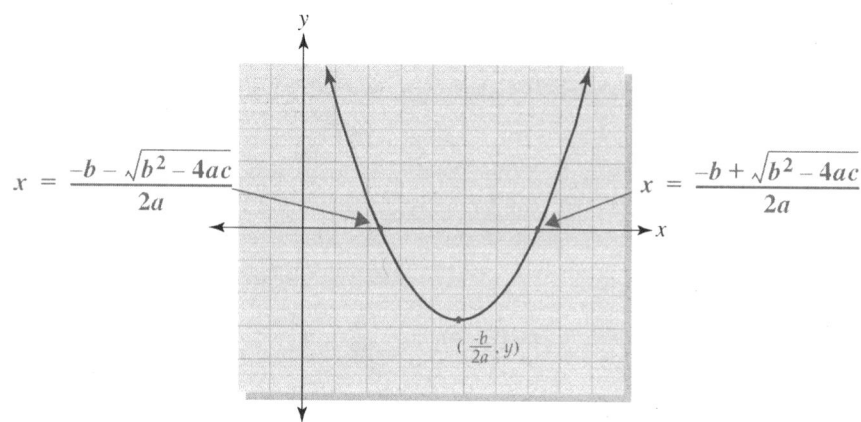

Concept Connection

The a-value in the vertex form is the same as the a-value in standard form. If we multiply out the vertex form and simplify into the standard form we can see that the a-value does not change.

We can best see this with an example.

$$f(x) = 5(x-2)^2 + 9$$
$$f(x) = 5(x-2)(x-2) + 9$$
$$f(x) = 5(x^2 - 4x + 4) + 9$$
$$f(x) = 5x^2 - 20x + 29$$

The fact that the a value is the same in both forms of a quadratic means that it will affect the graph of the parabola in the same ways.

Positive a: upward facing

Negative a: downward facing

$|a| < 1$: wide parabola

$|a| > 1$: narrow parabola

From the graph, we can see the two x-intercepts are located at the x values given by the quadratic formula. The symmetry of the graph indicates that the x value of the vertex must be midway between these two points. If you take the two x values and add them together, the square root portions will cancel out because one is positive and the other is negative. Then you divide by 2 to get the average.

$$x = \frac{-b - \sqrt{b^2 - 4ac}}{2a} \qquad x = \frac{-b + \sqrt{b^2 - 4ac}}{2a} \qquad \text{Horizontal intercepts } x \text{ values.}$$

$$x = \frac{-b}{2a} - \frac{\sqrt{b^2 - 4ac}}{2a} \qquad x = \frac{-b}{2a} + \frac{\sqrt{b^2 - 4ac}}{2a}$$

$$\frac{\dfrac{-b}{2a} - \dfrac{\sqrt{b^2 - 4ac}}{2a} + \dfrac{-b}{2a} + \dfrac{\sqrt{b^2 - 4ac}}{2a}}{2} \qquad \text{Add them and divide by 2}$$

$$\frac{2\left(\dfrac{-b}{2a}\right)}{2} \qquad \text{The radical parts cancel.}$$

$$\frac{-b}{2a} \qquad \text{Reduce.}$$

This result indicates that the x-value for the vertex will be $x = \dfrac{-b}{2a}$.

Another way to discover this relationship is to consider that the only point on a parabola without a symmetric point is the vertex. With this information in mind, the only way to come to a single solution from the quadratic formula is to have the square root after the plus/minus symbol be zero. This way, you are not adding or subtracting anything to or from the remaining parts of the formula. If we consider the square root to be zero, we are left with:

$$x = \frac{-b \pm \sqrt{0}}{2a}$$

$$x = \frac{-b}{2a}$$

This x value will then represent the one input value that does not have a symmetric pair. Hence, the vertex has the input value $x = \dfrac{-b}{2a}$. Remember that the vertex is a point on the graph, so you must find the y-value of the point. To find the y-value of the vertex you simply evaluate the equation at $x = \dfrac{-b}{2a}$.

$$f(x) = 4x^2 + 24x - 20$$

$$x = \frac{-b}{2a} = \frac{-(24)}{2(4)} = \frac{-24}{8} = -3 \qquad \text{Find the } x\text{-part.}$$

$$f(-3) = 4(-3)^2 + 24(-3) - 20 \qquad \text{Substitute the } x\text{-part into the equation to}$$
$$f(-3) = -56 \qquad \text{find the } y\text{-part.}$$
$$(-3, -56) \qquad \text{The vertex.}$$

EXAMPLE ▮ FINDING THE VERTICAL INTERCEPT AND VERTEX OF QUADRATICS

Find the vertical intercept and vertex of the following quadratics. State if the vertex is a minimum or maximum point on the graph.

a. $f(x) = 3x^2 + 24x + 8$

b. $g(t) = t^2 - 5t - 11$

c. $h(m) = 3(m + 4)^2 - 8$

Solution

a. This function is a quadratic in standard form. When $x = 0$, the function equals c, which is 8.

$$f(x) = 3x^2 + 24x + 8$$
$$f(0) = 3(0)^2 + 24(0) + 8$$
$$f(0) = 8$$
$$(0, 8)$$

Therefore, the vertical intercept is $(0, 8)$.

The input value of the vertex can be found using the formula $x = \dfrac{-b}{2a}$.

$$a = 3 \qquad b = 24$$
$$x = \frac{-(24)}{2(3)}$$
$$x = -4$$

Because the input value of the vertex is $x = -4$, we can find the output value by substituting -4 for x.

$$f(-4) = 3(-4)^2 + 24(-4) + 8$$
$$f(-4) = -40$$
$$(-4, -40)$$

Therefore, the vertex is the point $(-4, -40)$. The a value of this quadratic function is positive, so this parabola faces upward. Therefore this vertex is a minimum point.

b. This function is a quadratic in standard form. When $t = 0$, the function equals c, which is -11. Therefore the vertical intercept is $(0, -11)$.

The t-value of the vertex can be found using the formula $t = \dfrac{-b}{2a}$.

$$a = 1 \qquad b = -5$$
$$t = \frac{-(-5)}{2(1)}$$
$$t = 2.5$$

Because the input value of the vertex is $t = 2.5$, we can find the output value by substituting 2.5 for t.

$$f(2.5) = (2.5)^2 - 5(2.5) - 11$$
$$f(2.5) = -17.25$$
$$(2.5, -17.25)$$

Therefore, the vertex is the point $(2.5, -17.25)$. The a value of this quadratic function is positive, so this parabola faces upward. Therefore, this vertex is a minimum point.

c.   This function is a quadratic in vertex form. The vertical intercept must be calculated carefully by substituting in zero for m.

$$h(m) = -3(m + 4)^2 - 8$$
$$h(0) = -3(0 + 4)^2 - 8$$
$$h(0) = -3(16) - 8$$
$$h(0) = -56$$
$$(0, -56)$$

Therefore, the vertical intercept is $(0, -56)$. Note the value we calculated is not the constant in the function. Instead, you must substitute zero into the equation and calculate to get the correct vertical intercept for this quadratic function.

Since the function is in vertex form, we do not have to calculate the vertex. The vertex will be (h, k), so the vertex is $(-4, -8)$. The a value is negative, so this parabola faces downward. The vertex is the maximum point on the graph.

EXAMPLE   ⓘ PRACTICE PROBLEM

Find the vertical intercept and vertex of the following quadratics. State if the vertex is a minimum or maximum point on the graph.

a. $f(x) = 4x^2 + 40x + 7$

b. $k(x) = -2(x - 3)^2 + 12$

c. $h(t) = -7t^2 - 15t - 6$

The vertical intercept, vertex and symmetry will all help us to graph quadratic functions. The process will be basically the same as graphing from vertex form. We will find the vertex and vertical intercept, and use the axis of symmetry to find symmetric pairs. Now that we know how to solve quadratic equations, we will also find the horizontal intercepts, if there are any, and use them as an additional symmetric pair. In general at least 5 points should be plotted for a graph of a parabola. If you want more details you can always find more symmetric pairs by plugging in values for x and finding y.

GRAPHING QUADRATICS FROM STANDARD FORM

1. Determine whether the graph opens up or down. Find the vertical intercept.

2. Find the vertex.

3. Find the horizontal intercepts (if any).

4. Find the axis of symmetry.

5. Plot the points, plot their symmetric pairs, and sketch the graph. (Find an additional symmetric pair if needed.)

EXAMPLE 2 GRAPHING QUADRATIC FUNCTIONS

Sketch a graph of the following:

a. $f(x) = 2x^2 - 12x + 5$.

b. $f(x) = -0.5x^2 + 4x - 10$.

Solution

a.

Step 1 Determine whether the graph opens up or down. Find the vertical intercept.

Because a is positive in this quadratic, we know that the parabola will open upward. Because this quadratic is in standard form, we know that the vertical intercept has an output value equal to the constant c. In this case the vertical intercept is $(0, 5)$.

Step 2 Find the vertex.

This function is in standard form for a quadratic, so we can use the formula $x = \dfrac{-b}{2a}$ to find the x-value of the vertex.

$$x = \frac{-(-12)}{2(2)}$$

$$x = 3$$

Now that we have the input value for the vertex, we can substitute it into the function and find the output value for the vertex.

$$f(x) = 2x^2 - 12x + 5$$

$$f(3) = 2(3)^2 - 12(3) + 5$$

$$f(3) = 18 - 36 + 5$$

$$f(3) = -13$$

So the vertex is $(3, -13)$.

Step 3 Find the horizontal intercepts (if any).

Horizontal intercepts will happen when the output variable is equal to zero, so substitute zero for the output variable and solve.

$$f(x) = 2x^2 - 12x + 5$$

$$0 = 2x^2 - 12x + 5$$

$$a = 2 \qquad b = -12 \qquad c = 5 \qquad \text{Use the quadratic formula.}$$

$$x = \frac{-b \pm \sqrt{b^2 - 4ac}}{2a}$$

$$x = \frac{-(-12) \pm \sqrt{(-12)^2 - 4(2)(5)}}{2(2)}$$

$$x = \frac{12 \pm \sqrt{144 - 40}}{4}$$

$$x = \frac{12 \pm \sqrt{104}}{4}$$

$$x = \frac{12 \pm 10.2}{4}$$

$$x = \frac{12 + 10.2}{4} \qquad x = \frac{12 - 10.2}{4}$$

$$x = 5.55 \qquad\qquad x = 0.45$$

Therefore, the horizontal intercepts are (5.55, 0) and (0.45, 0).

Step 4 Find axis of symmetry.

The axis of symmetry is the vertical line through the vertex. Therefore, the axis of symmetry is $x = 3$.

Step 5 Plot the points, plot their symmetric pairs, and sketch the graph. (Find an additional symmetric pair if needed.

We now have the vertex $(3, -13)$, the vertical intercept $(0, 5)$, and the horizontal intercepts $(5.55, 0)$ and $(0.45, 0)$. Plotting these points and using the axis of symmetry to plot their symmetric pairs, we get the following:

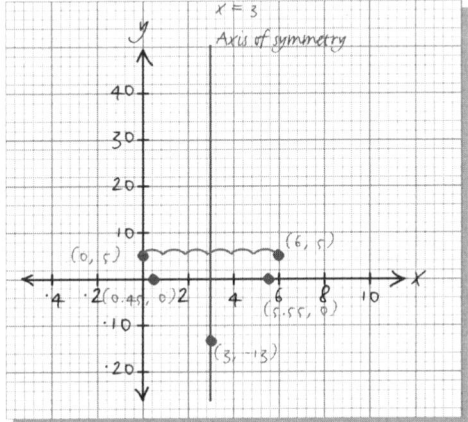

Sketching a smooth curve through the points gives us the following graph.

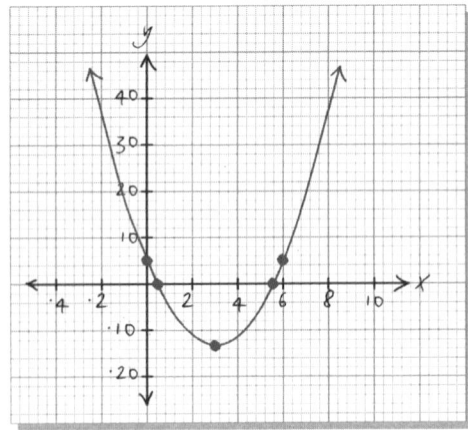

b.

Step 1 Determine whether the graph opens up or down. Find the vertical intercept.

Because a is negative in this quadratic, we know that the parabola will open downward. The vertical intercept is $(0, -10)$.

Step 2 Find the vertex.

This function is in standard form for a quadratic, so we can use the formula $x = \dfrac{-b}{2a}$ to find the x-value of the vertex.

$$x = \frac{-(4)}{2(-0.5)}$$

$$x = 4$$

Now that we have the input value for the vertex, we can substitute it into the function and find the output value for the vertex.

$$f(x) = -0.5x^2 + 4x - 10$$
$$f(4) = -0.5(4)^2 + 4(4) - 10$$
$$f(4) = -8 + 16 - 10$$
$$f(4) = -2$$

So the vertex is $(4, -2)$.

Step 3 Find the horizontal intercepts (if any).

Horizontal intercepts will happen when the output variable is equal to zero, so substitute zero for the output variable and solve.

$$f(x) = -0.5x^2 + 4x - 10$$
$$0 = -0.5x^2 + 4x - 10 \qquad \text{Set the function equal to zero.}$$
$$a = -0.5 \qquad b = 4 \qquad c = -10$$

Use the quadratic formula.

$$x = \frac{-(4) \pm \sqrt{(4)^2 - 4(-0.5)(-10)}}{2(-0.5)}$$

$$x = \frac{4 \pm \sqrt{-4}}{-1} \qquad \text{A negative discriminant indicates no real solutions.}$$

There are no real solutions when this function is set equal to zero, so there are no horizontal intercepts.

Step 4 Find axis of symmetry.

The axis of symmetry is the vertical line through the vertex. Therefore, the axis of symmetry is $x = 4$.

Step 5 Plot the points, plot their symmetric pairs, and sketch the graph. (Find an additional symmetric pair if needed.

We now have the vertex $(4, -2)$ and the vertical intercept $(0, -10)$. There are no horizontal intercepts, so we will want an additional symmetric pair to give us more points to sketch our curve through. To find another point, we can pick an additional input value on one side of the axis of symmetry and find the output value.

$$x = 2$$
$$f(2) = -0.5(2)^2 + 4(2) - 10 \qquad \text{Pick any input value we don't already have and substitute to find the output value.}$$
$$f(2) = -4$$
$$(2, -4)$$

Plotting all the points that we have found and using the axis of symmetry to plot their symmetric pairs, we get the following:

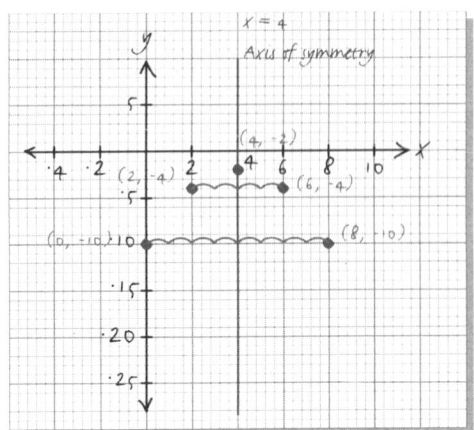

We now have the vertex $(4, -2)$ and two sets of symmetric pairs: $(2, -4)$, $(6, -4)$ and $(0, -10)$, $(8, -10)$. Sketching a smooth curve through these points gives us the final graph.

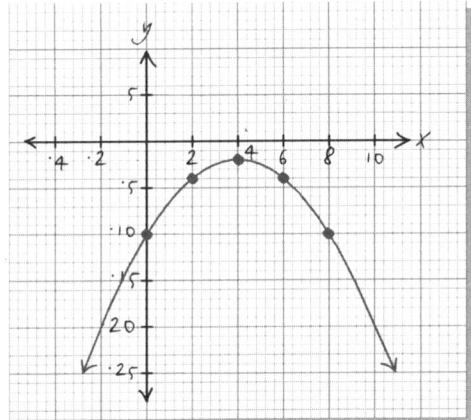

EXAMPLE ② PRACTICE PROBLEM

Sketch a graph of the following:

a. $f(x) = 0.25x^2 + 6x - 15$.

b. $f(x) = 3x^2 + 6x + 7$.

In Example 2 part b the parabola had no horizontal intercepts. If we thought ahead about the fact that the vertex is below the x-axis and the parabola is facing downward we could have known that there would be no horizontal intercepts. After finding the vertex and deciding if the graph faces upward or downward you can determine if parabola has any horizontal intercepts and therefore if you need to solve for them.

EXAMPLE 3 APPLICATIONS OF THE VERTEX

The monthly profit for an amusement park can be modeled by

$$P(t) = -0.181t^2 + 13.767t - 235.63$$

where $P(t)$ represents the monthly profit in millions of dollars when tickets are sold for t dollars each.

a. What is the amusement park's monthly profit if it sells tickets for $30 each?
b. How much should the amusement park charge for tickets if it wants a monthly profit of $25 million?
c. Find the vertex for this model and explain its meaning in this context.

Solution

a. $30 is a ticket price, so it can be substituted into t.

$$P(t) = -0.181t^2 + 13.767t - 235.63$$
$$P(30) = -0.181(30)^2 + 13.767(30) - 235.63$$
$$P(30) = 14.48$$

If the amusement park sells tickets for $30 each, it will have a monthly profit of about $14.48 million.

b. The $25 million is a monthly profit, which can be substituted into P.

$$P(t) = -0.181t^2 + 13.767t - 235.63$$

$$25 = -0.181t^2 + 13.767t - 235.63 \qquad \text{Substitute 25 for } P.$$
$$\underline{-25 \qquad\qquad\qquad\qquad\qquad -25}$$

Set the equation equal to zero and use the quadratic formula.

$$0 = -0.181t^2 + 13.767t - 260.63$$
$$t \approx 35.51 \qquad t \approx 40.55$$

If the amusement park charges $35.51 or $40.55 per ticket, it will have a monthly profit of about $25 million. We hope that the park would choose the $35.51 to make more people happy.

c. The vertex will have an input of

$$t = \frac{-b}{2a} = \frac{-13.767}{2(-0.181)}$$
$$t = 38.03$$

Using a ticket price of $38.03, we get

$$P(t) = -0.181t^2 + 13.767t - 235.63$$
$$P(38.03) = -0.181(38.03)^2 + 13.767(38.03) - 235.63$$
$$P(38.03) = 26.15$$
$$(38.03, \ 26.15)$$

If the amusement park charges $38.03 per ticket, it will make its maximum profit of $26.15 million per month.

EXAMPLE PRACTICE PROBLEM

The net sales for computer manufacturer Gateway, Inc. can be modeled by
$$S(t) = 75.85t^2 - 1204.22t + 4725.95$$
where $S(t)$ represents the annual net sales in millions of dollars for Gateway, Inc. t years since 1980.

Source: Model derived from data found in Gateway, Inc. annual reports at Gateway.com.

a. Using this model, estimate the annual net sales for Gateway, Inc. in 1995.

b. When were Gateway's net sales $9000 million?

c. Find the vertex for this model and explain its meaning in this context.

GRAPHING QUADRATIC INEQUALITIES IN TWO VARIABLES

Graphing quadratic inequalities can be done using the same techniques as graphing linear inequalities. First graph the inequality as if it were an equality using a solid curve if there is an "equal to" part and a dashed curve if it is only "less than" or "greater than" but not "equal to." This separates the graph into two sections, points that are inside the parabola and points that are outside the parabola. Then decide which section satisfies the inequality and, therefore, should be shaded. So pick a point that is not on the graph and check to see whether it satisfies the inequality. Finally, shade the side of the curve that does satisfy the inequality.

EXAMPLE 4 GRAPHING QUADRATIC INEQUALITIES

Graph the following inequalities.

a. $y < x^2 + 3x - 10$

b. $y \geq -2x^2 + 11x - 12$

Solution

a. First, we will sketch the graph of the quadratic using a dashed curve because there is no "equal to" part.

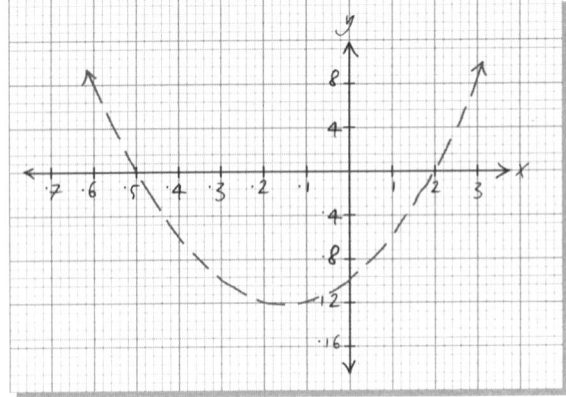

Doing so separates the grid into two sections, the points above the parabola and the points below the parabola.

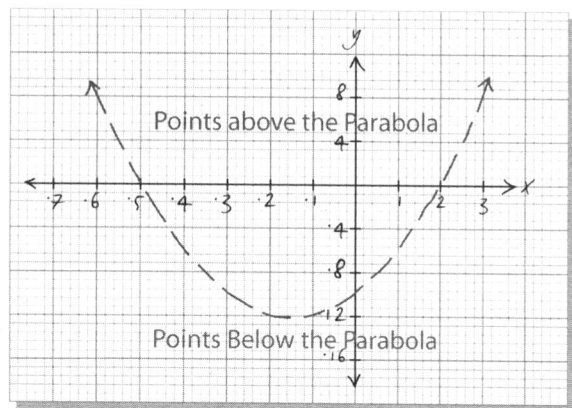

Now if we pick (0, 0) as a test point, we find that it does not satisfy the inequality, so we will shade the points below the parabola.

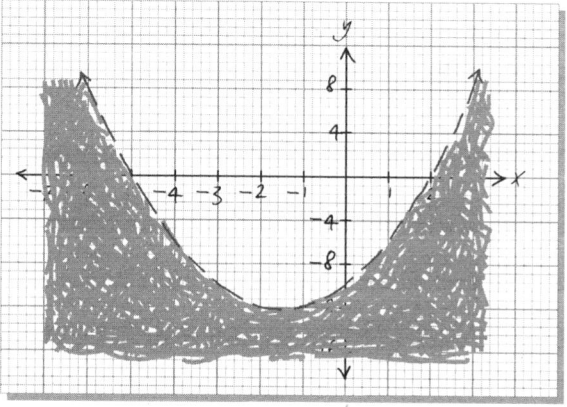

b. First, we will sketch the graph using a solid line because it is a "greater than or equal to" inequality.

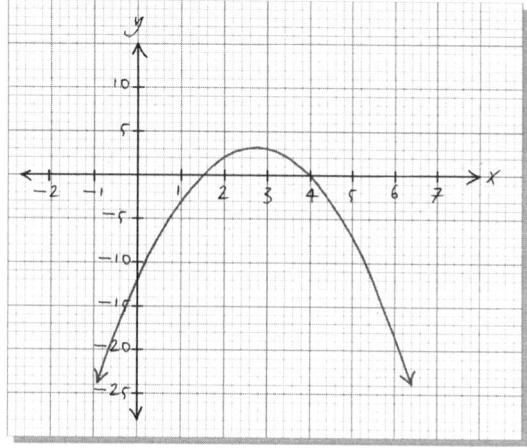

If we pick (0, 0) for our test point, we find that it does satisfy the inequality, so we should shade above the curve.

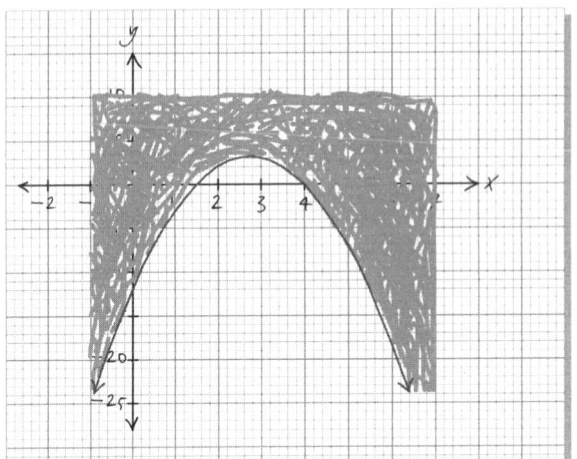

EXAMPLE ④ PRACTICE PROBLEM

Graph the following inequalities.

a. $y > 2x^2 - 12x - 8$

b. $y \leq -0.25x^2 - 3x + 5$

4.7 Exercises

For exercises 1 through 4, without using your calculator, match the given equations with the appropriate graph. State which equation does not match a given graph.

1. $y = -2x^2 - 6x - 7$

$y = -0.5x^2 - x + 7$

$y = x^2 + 4x + 4$

$y = x^2 - 5x - 6$

Graph A   Graph B   Graph C

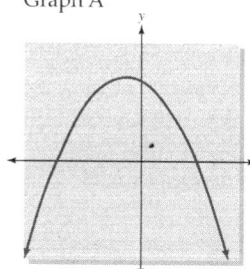 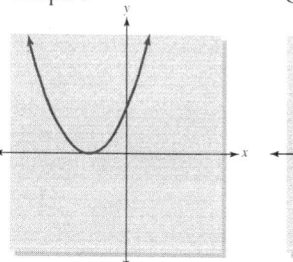 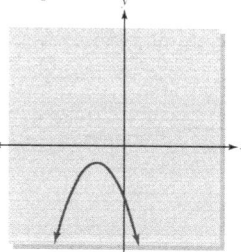

2. $y = x^2 - 6x + 10$

$y = x^2 - 3x - 4$

$y = -0.3x^2 + 2x + 3$

$y = -3x^2 + 12x - 8$

Graph A   Graph B   Graph C

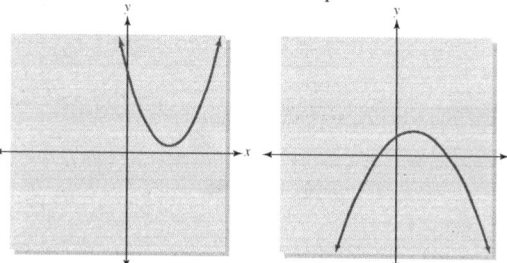 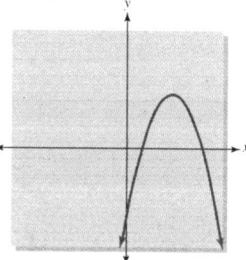

3. $y = -2(x - 3)^2 - 9$

$y = 2(x - 3)^2 - 9$

$y = 2(x + 4)^2 + 1$

$y = 2(x + 4)^2 - 9$

Graph A   Graph B   Graph C

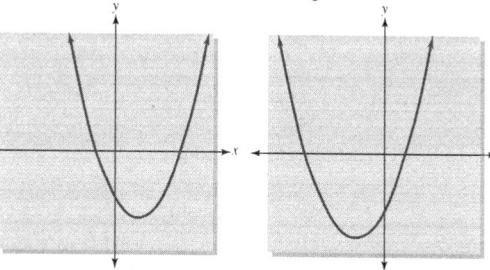 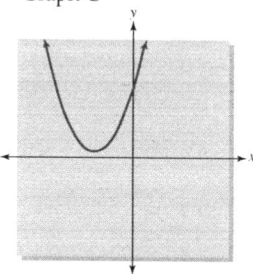

4. $y = x^2$

$y = x^2 - 7$

$y = -x^2 + 20$

$y = -x^2$

Graph A   Graph B   Graph C

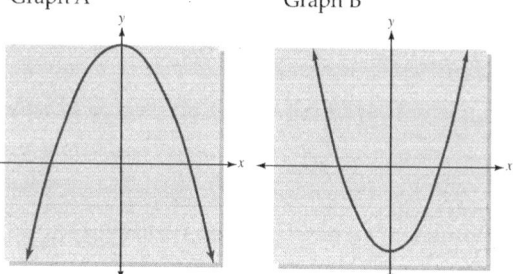 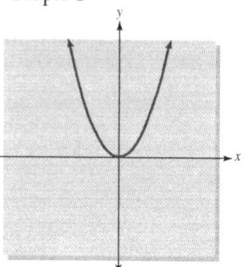

For exercises 5 through 14 find the vertex, vertical intercept and horizontal intercept(s) of each quadratic function.

5. $f(x) = x^2 + 6x + 8$

6. $g(x) = x^2 + 12x - 28$

7. $h(x) = 3x^2 - 18x + 15$

8. $f(x) = -4x^2 + 10x + 20$

9. $h(x) = 4(x - 8)^2 - 20$

10. $f(x) = -2(x + 10)^2 + 15$

11. $g(x) = 5x^2 + 12x + 10$

12. $h(x) = -3x^2 + 11x - 20$

13. $f(x) = 1.5x^2 - 6x + 4$

14. $g(x) = 0.75x^2 + 10x - 15$

15. The cost to produce football uniforms for a school can be modeled by

$$C(u) = \frac{1}{4}u^2 - 25u + 3500$$

where $C(u)$ represents the cost in dollars to produce u football uniforms.

   a. Find the cost to produce 30 uniforms.

   b. Find the vertex and describe its meaning in this context.

   c. How many uniforms can a school get with a budget of $1600?

16. The cost to produce backpacks with a school's logo can be modeled by

$$C(b) = \frac{2}{5}b^2 - 12b + 500$$

where $C(b)$ represents the cost in dollars to produce b backpacks.

   a. Find the cost to produce 20 backpacks.

   b. Find the vertex and describe its meaning in this context.

   c. How many backpacks can a school get with a budget of $3000?

17. The revenue from selling digital cameras can be modeled by

$$R(c) = -3c^2 + 90c$$

where $R(c)$ represents the revenue in thousands of dollars from selling c thousand digital cameras.

   a. Find the revenue from selling 5 thousand digital cameras.

   b. How many cameras must the company sell to have a revenue of $600,000?

   c. How many digital cameras must it sell to maximize its revenue?

18. The revenue from selling sunglasses can be modeled by

$$R(s) = -1.5s^2 + 30s$$

where $R(s)$ represents the revenue in hundreds of dollars from selling s hundred pairs of sunglasses.

   a. Find the revenue from selling 9 hundred pairs of sunglasses.

   b. How many pairs of sunglasses must the company sell to have a revenue of $10,000? (Hint: $10,000 = $100 hundred)

   c. How many pairs of sunglasses must the company sell to maximize its revenue?

19. The net sales for Home Depot can be modeled by

$$N(t) = 376.5t^2 + 548.1t + 2318.4$$

where $N(t)$ represents the net sales in millions of dollars for Home Depot t years since 1990.

Source: Model derived from data found in Home Depot annual reports.

   a. Using this model estimate the annual net sales for Home Depot in 1995.

   b. When were Home Depot's net sales 30,000 million dollars?

   c. Find the vertex for this model and explain its meaning in this context.

20. The net income for Dell Corporation can be modeled by

$$I(t) = -145.86t^2 + 3169.342t - 15145.2$$

where $I(t)$ represents the net income in millions of dollars t years since 1990.

Source: Model derived from data found in Dell's 2002 annual report.

  a. Using this model, estimate the net income for Dell in 2000.

  b. When did Dell's net income reach 1500 million dollars?

  c. Find the vertex for this model and explain its meaning in this context.

21. The net income for clothing and gear retailer Quiksilver can be modeled by

$$I(t) = -1.5t^2 + 32.3t - 138.8$$

where $I(t)$ represents the net income in millions of dollars t years since 1990.

Source: Model derived from data found in Quiksilver annual report 2001.

  a. Using this model, estimate the net income for Quiksilver in 2001.

  b. When did Quiksilver's net income reach 18 million dollars?

  c. Find the vertex for this model and explain its meaning in this context.

22. The average fuel consumption of vehicles driving the roads can be modeled by

$$F(t) = 0.425t^2 - 16.431t + 840.321$$

where $F(t)$ represents the gallons per vehicle per year t years since 1970.

  a. What was the average fuel consumption in 1975?

  b. What was the average fuel consumption in 1985?

  c. Find the vertex for this model and explain its meaning in this context.

23. A baseball is hit so that its height in feet t seconds after it is hit can be represented by

$$h(t) = -16t^2 + 40t + 4$$

  a. What is the height of the ball when it is hit?

  b. When does the ball reach a height of 20 feet?

  c. When did the ball reach its maximum height?

  d. What is the ball's maximum height?

  e. If the ball does not get caught, when does it hit the ground?

24. A baseball is hit so that its height in feet t seconds after it is hit can be represented by

$$h(t) = -16t^2 + 56t + 4$$

  a. What is the height of the ball when it is hit?

  b. When does the ball reach a height of 25 feet?

  c. When does the ball reach its maximum height?

  d. What is the ball's maximum height?

25. A baseball is hit so that its height in feet t seconds after it is hit can be represented by

$$h(t) = -16t^2 + 60t + 4.2$$

  a. What is the height of the ball when it is hit?

  b. When does the ball reach a height of 40 feet?

  c. What is the ball's maximum height?

d. If it is not caught, when does the ball hit the ground?

26. A baseball is hit so that its height in feet t seconds after it is hit can be represented by

$$h(t) = -16t^2 + 65t + 3.5$$

a. What is the height of the ball when it is hit?

b. When does the ball reach a height of 50 feet?

c. What is the ball's maximum height?

d. If it is not caught, when does the ball hit the ground?

27. The average high temperature in Melbourne, Australia, can be modeled by

$$H(m) = 0.9m^2 - 13m + 104$$

where $H(m)$ represents the average high temperature (°F) in Melbourne during month m of the year.

a. Find $H(6)$ and explain its meaning in this context.

b. Find the vertex for this model and explain its meaning in this context.

c. During what month(s) is the average temperature in Melbourne 60°F?

28. The average high temperature in Paris, France, can be modeled by

$$H(m) = -1.5m^2 + 21.6m - 3.2$$

where $H(m)$ represents the average high temperature (°F) in Paris during month m of the year.

a. Find $H(4)$ and explain its meaning in this context.

b. Find the vertex for this model and explain its meaning in this context.

c. During what month(s) is the average temperature in Paris 70°F?

In Exercises 29 through 50:

    a. *Find the vertex.*

    b. *Find the vertical and horizontal intercept(s).*

    c. *Sketch a graph of the function.*

    d. *Give the domain and range of the function.*

29. $f(x) = x^2 + 2x - 15$

30. $g(x) = x^2 + 6x + 3$

31. $m(b) = -b^2 + 11b - 24$

32. $D(z) = -z^2 - 12z + 43$

33. $g(s) = 2s^2 - 62s + 216$

34. $h(t) = 4t^2 + 30t - 17$

35. $f(x) = 2x^2 + 5$

36. $b(w) = -3w^2 + 8$

37. $d(p) = -1.5p^2 - 3$

38. $g(n) = \frac{2}{3}n^2 + 1$

39. $h(x) = -\frac{1}{4}x^2 - 5$

40. $s(m) = -\frac{1}{2}x^2 + 12$

41. $p(k) = -5k^2 - 17.5k - 12.5$

42. $f(x) = -3x^2 - 14.6x - 11.2$

43. $h(w) = 0.4w^2 - 3.6w - 44.8$

44. $f(x) = 0.25x^2 + 4x + 6.5$

45. $p(x) = \frac{2}{5}x^2 - 2x - \frac{3}{5}$

46. $h(t) = \frac{1}{3}t^2 - 2t + 4$

47. $Q(p) = -0.3p^2 - 2.4p + 82$

48. $M(a) = -0.25a^2 - 3.5a - 27.5$

49. $W(g) = -0.3(g + 2)^2 + 17$

50. $f(x) = 3(x - 4)^2 - 18$

For Exercises 51 through 64, graph the inequalities.

51. $y < x^2 + 8x + 15$

52. $y > x^2 - 4x - 21$

53. $y \leq -2(x + 3)^2 + 10$

54. $y \leq -3(x - 4)^2 + 7$

55. $y \geq 1.5x^2 + 6x + 10$

56. $y < 0.5x^2 - 8x - 11$

57. $y \geq -0.25x^2 + 3x - 2$

58. $y \leq 4x^2 - 20x - 16$

59. $y > 0.3(x - 4)^2 - 6$

60. $y < (x + 8)^2 - 9$

61. $y \leq 2.5x^2$

62. $y \geq -1.75x^2 + 5$

63. $y > 2(x + 7)(x - 4)$

64. $y < 0.5(x + 2)(x - 6)$

Chapter 4 Summary

Section 4.1 Introduction to Quadratics

- The **vertex** of a parabola is the maximum or minimum point on the graph and is the point where the graph changes directions.

- The graph of a quadratic function is called a **parabola** and is either an upward- or downward-facing "U" shape.

- A **quadratic function** can be written in two forms:
$$f(x) = ax^2 + bx + c$$

or
$$f(x) = a(x - h)^2 + k$$
where $a, b, c, h,$ and k are real numbers and $a \neq 0$.

EXAMPLE 1

Create a scatterplot for the given data and estimate the vertex.

x	0	1	2	3	4	5	6	7
y	14	10.5	8	6.5	6	6.5	8	10.5

Solution

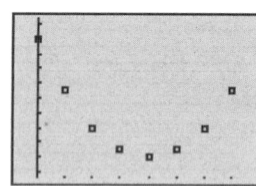

This distribution takes the shape of a parabola, and the vertex is a minimum point at about $(4, 6)$.

EXAMPLE 2

Use the graph to estimate the following.

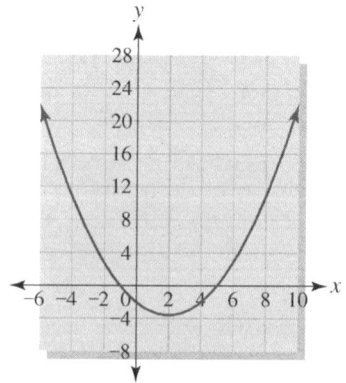

a. Estimate the vertex.

b. Estimate the vertical intercept.

c. Estimate the horizontal intercept(s).

Solution

a. Vertex: $(2, -4)$

b. Vertical intercept: $(0, -2)$

c. Horizontal intercepts: $(-1, 0)$ and $(5, 0)$

Section 4.2 Graphing Quadratics in Vertex Form

- The **vertex form of a quadratic** is
$$f(x) = a(x - h)^2 + k,$$
where a, h, and k are real numbers and $a \neq 0$.

- The **value of a** will determine if the parabola faces upward or downward as well as how wide or narrow the graph is.

- The **value of h** will determine how far the vertex is moved to the left or right from the origin.

- The **value of k** will determine how far the vertex is moved up or down from the origin.

- The **vertex** of the parabola will be (h, k).

- The **axis of symmetry** is the vertical line through the vertex and has the equation $x = h$.

- The **symmetry** of the parabola helps us sketch the graph more accurately and more quickly.

- To **graph a quadratic from the vertex form** follow these 5 steps:

  1. Determine whether the graph opens up or down.
  2. Find the vertex and axis of symmetry.
  3. Find the vertical intercept.
  4. Find an extra point by picking an input value on one side of the axis of symmetry and calculating the output value.
  5. Plot the points, plot their symmetric pairs and sketch the graph. (Find an additional symmetric pair if needed.)

- The **domain of a quadratic model** will only be restricted by the context. Avoid model breakdown.

- The **range of a quadratic model** is the output values of the model that come from the domain.

- The **domain** for a quadratic function with no context will be all real numbers.

- The **range** for a quadratic function will be either $(-\infty, k]$ if $a < 0$ or $[k, \infty)$ if $a > 0$.

EXAMPLE 3

Sketch the graph of $f(x) = -2(x + 5)^2 + 10$. Label the vertex, vertical intercept, and at least one other symmetric pair. Give the domain and range of the function.

Solution

The vertex is $(-5, 10)$.
$\quad f(0) = -40$, so the vertical intercept is $(0, -40)$.
The graph faces downward because a is negative.
Another symmetric pair is $(-2, -8)$ and $(-8, -8)$.

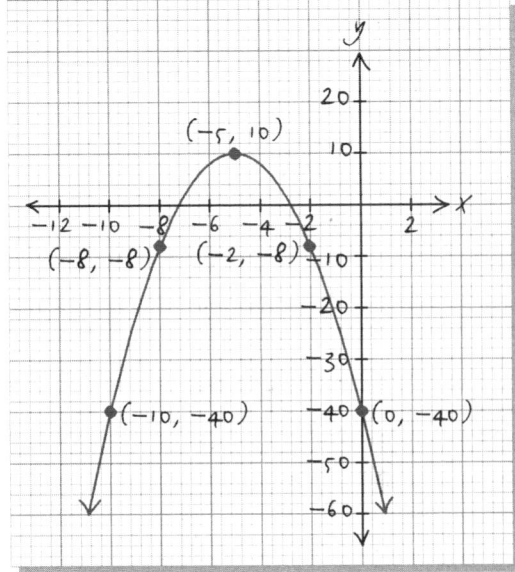

Domain: All real numbers
Range: $y \leq 10$

EXAMPLE 4

Give the domain and range of $f(x) = 4(x + 7)^2 - 45$.

Solution

Domain: All real numbers
Range: $[-45, \infty)$.

Section 4.3 Finding Quadratic Models

- To find a **quadratic model,** you need to follow these seven steps.

  1. Define the variables and adjust the data (if needed).
  2. Create a scatterplot.
  3. Select a model type.
  4. **Quadratic Model:** Pick a vertex and substitute it for h and k in the vertex form $f(x) = a(x - h)^2 + k$.
  5. Pick another point and use it to find a.
  6. Write the equation of the model using function notation.
  7. Check your model by graphing it with the scatterplot.

- In the **context of an application** the **domain** of a quadratic model expands beyond the data. Be sure to avoid inputs that will cause model breakdown to occur.

- In the **context of an application** the **range** of a quadratic model is the lowest to highest points on the graph within the domain.

EXAMPLE 5

The number of solar flares each year appears to be cyclical. The data from the latest cycle are given in the table.

Year	Number of Solar Flares
1997	790
1998	2423
1999	3963
2000	4474
2002	3223
2003	1552

a. Find a model for the data.

b. What is the vertex, and what does it represent in this context?

c. According to your model, how many solar flares were there in 2001?

d. Give a reasonable domain and range for your model.

Solution

a. S = the number of solar flares during the year

t = the years since 1995.

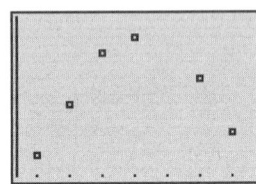

The vertex could be (5, 4474), so we have
$$S(t) = a(t - 5)^2 + 4474.$$

Using (2, 790), we get $a = -409$ and
$$S(t) = -409(t - 5)^2 + 4474.$$

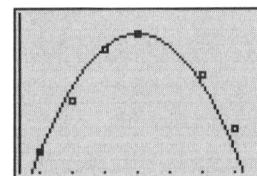

This model could be moved over to the right to make a better fit, so we can adjust the model to $S(t) = -409(t - 5.2)^2 + 4474$.
This give us the graph below.

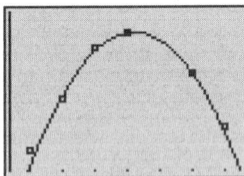

b. The vertex is (5.2, 4474), so the maximum number of solar flares occurred in about 2000 with 4474 solar flares.

c. $S(6) = 4212.2$, so in 2003 there were about 4212 solar flares.

d. Domain: [2, 8], range: [286, 1267]. Note that if we go beyond this domain, model breakdown occurs because the model predicts a negative number of solar flares.

Section 4.4 Solving Quadratic Equations by Square Root Property and Completing the Square

- When **solving** quadratic equations in vertex form, isolate the squared part of the equation and use the **square root property** to remove the exponent. Don't forget the plus/minus symbol.

- When **solving** quadratic equations using **completing the square** you can use these 6 steps.

  1. Get all the variable terms along on one side of the equation.
  2. If the coefficient of x^2 is not 1, divide both sides by the coefficient of x^2.
  3. Take half of the coefficient of x and then square it. Add this number to both sides of the equation.
  4. Factor the quadratic into the square of a binomial.
  5. Solve using the square root property.
  6. Check your answers in the original equation.

- When **converting to vertex form,** use the same steps as completing the square but end with function notation.

- When solving a quadratic, if you get a negative under the square root, the quadratic will have **no real solutions**.

- When **graphing a quadratic in vertex form**, you should include the horizontal intercepts. You can use these steps.

  1. Determine whether the graph opens up or down.
  2. Find the vertex and axis of symmetry.
  3. Find the vertical intercept.
  4. Find the horizontal intercepts.
  5. Plot the points, plot their symmetric pairs and sketch the graph. (Find an additional symmetric pair if needed.)

EXAMPLE 6

Solve the following equations.

a. $15 = 2(x + 5)^2 - 17$

b. $x^2 - 6x + 20 = 47$

Solution

a. Because this equation is in vertex form, we would use the square root property to solve.

$$15 = 2(x + 5)^2 - 17$$
$$32 = 2(x + 5)^2$$
$$16 = (x + 5)^2$$
$$\pm 4 = x + 5$$
$$4 = x + 5 \qquad -4 = x + 5$$
$$x = -1 \qquad x = -9$$

Check your answers.

$$15 = 2((-1) + 5)^2 - 17 \quad 15 = 2((-9) + 5)^2 - 17$$
$$15 = 2(4)^2 - 17 \qquad\qquad 15 = 2(-4)^2 - 17$$
$$15 = 32 - 17 \qquad\qquad 15 = 32 - 17$$
$$15 = 15 \qquad\qquad 15 = 15$$

b. This equation is not in vertex form, so we will use completing the square.

$$x^2 - 6x + 20 = 47$$
$$x^2 - 6x = 27$$
$$x^2 - 6x + 9 = 36$$
$$(x - 3)^2 = 36$$
$$x - 3 = \pm 6$$
$$x - 3 = 6 \qquad x - 3 = -6$$
$$x = 9 \qquad\quad x = -3$$

Check your answers.

$$(9)^2 - 6(9) + 20 = 47 \qquad (-3)^2 - 6(-3) + 20 = 47$$
$$47 = 47 \qquad\qquad\qquad 47 = 47$$

EXAMPLE 7

Convert $f(x) = 3x^2 + 12x - 21$ into vertex form.

Solution

$$y = 3x^2 + 12x - 21$$
$$y + 21 = 3x^2 + 12x$$
$$\frac{1}{3}y + 7 = x^2 + 4x$$

$$\frac{1}{3}y + 11 = x^2 + 4x + 4$$

$$\frac{1}{3}y + 11 = (x+2)^2$$

$$f(x) = 3(x+2)^2 - 33$$

Section 4.5 Solving Quadratic Equations by Factoring

- To solve a quadratic using **factoring**, the equation must be set equal to zero and put into standard form. Then factor, set each of the factors equal to zero, and solve.

- When finding an equation from the graph, set x equal to the zeros and undo the solving process to find the factors. Use another point to find any constant multiplier.

EXAMPLE 8

Solve the following equations.

a. $2x^2 - x - 4 = 11$

b. $2x^3 - 26x^2 + 80x = 0$

Solution

a.
$$2x^2 - x - 4 = 11$$
$$2x^2 - x - 15 = 0$$
$$2x^2 - 6x + 5x - 15 = 0$$
$$(2x^2 - 6x) + (5x - 15) = 0$$
$$2x(x-3) + 5(x-3) = 0$$
$$(2x+5)(x-3) = 0$$
$$2x+5 = 0 \qquad x-3 = 0$$
$$x = -\frac{5}{2} \qquad x = 3$$

Check your answers.

$$2x^2 - x - 4 = 11 \qquad 2x^2 - x - 4 = 11$$
$$2\left(-\frac{5}{2}\right)^2 - \left(-\frac{5}{2}\right) - 4 = 11 \qquad 2(3)^2 - (3) - 4 = 11$$
$$2\left(\frac{25}{4}\right) + \frac{5}{2} - 4 = 11 \qquad 2(9) - 3 - 4 = 11$$

$$15 - 4 = 11 \qquad 18 - 3 - 4 = 11$$
$$11 = 11 \qquad 11 = 11$$

b.
$$2x^3 - 26x^2 + 80x = 0$$
$$2x(x^2 - 13x + 40) = 0$$
$$2x[x^2 - 5x - 8x + 40] = 0$$
$$2x[(x^2 - 5x) + (-8x + 40)] = 0$$
$$2x[x(x-5) - 8(x-5)] = 0$$
$$2x(x-5)(x-8) = 0$$
$$2x = 0 \qquad x-5 = 0 \qquad x-8 = 0$$
$$x = 0 \qquad x = 5 \qquad x = 8$$

Check your answers.
$$2(0)^3 - 26(0)^2 + 80(0) = 0$$
$$0 = 0$$

$$2(5)^3 - 26(5)^2 + 80(5) = 0$$
$$1024 - 1664 + 640 = 0$$
$$0 = 0$$

$$2(8)^3 - 26(8)^2 + 80(8) = 0$$
$$1024 - 1664 - 640 = 0$$
$$0 = 0$$

EXAMPLE 9

Use the graph to find an equation for the quadratic.

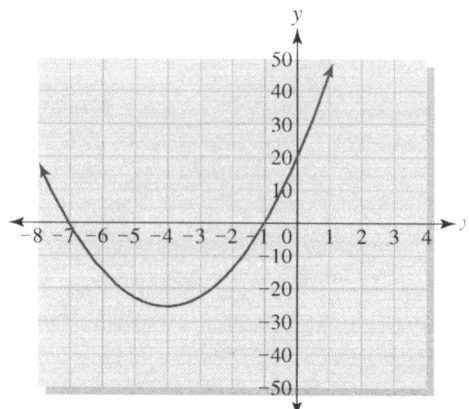

$$x = -1 \qquad x = -7$$
$$x + 1 = 0 \qquad x + 7 = 0$$
$$y = a(x + 1)(x + 7)$$
$$20 = a(0 + 1)(0 + 7)$$
$$20 = 7a$$
$$\frac{20}{7} = a$$
$$y = \frac{20}{7}x^2 + \frac{160}{7}x + 20$$

Section 4.6 Solving Quadratic Equations Using the Quadratic Formula

- The **quadratic formula** can be used to solve any quadratic equation that is in standard form:
$$ax^2 + bx + c = 0$$
$$x = \frac{-b \pm \sqrt{b^2 - 4ac}}{2a}$$

- Be sure that your equation is set equal to zero before using the quadratic formula.

- A **system of equations with quadratics** can be solved **graphically** by looking for the intersections of the graphs.

- A **system of equations with quadratics** can be solved **numerically** by looking for the input values that give the same output values in both equations.

- A **system of equations with quadratics** can be solved **algebraically** by using the substitution method.

- Whenever the quadratic formula results in a negative under the square root, there will be **no real solutions** to the equation.

EXAMPLE

Solve the following quadratics.

a. $5x^2 + 3x - 20 = 10$

b. $4x^2 - 2x + 15 = -30$

Solution

a. $5x^2 + 3x - 20 = 10$
$$5x^2 + 3x - 30 = 0$$
$$a = 5 \qquad b = 3 \qquad c = -30$$
$$x = \frac{-(3) \pm \sqrt{(3)^2 - 4(5)(-30)}}{2(5)}$$
$$x = \frac{-3 \pm \sqrt{609}}{10}$$
$$x = \frac{-3 \pm 24.678}{10}$$
$$x = \frac{-3 + 24.678}{10} \qquad x = \frac{-3 - 24.678}{10}$$
$$x = 2.168 \qquad x = -2.768$$

Verify with the table.

X	Y₁	Y₂
2.168	10.005	10
-2.768	10.005	10

X=

b. $4x^2 - 2x + 15 = -30$
$$4x^2 - 2x + 45 = 0$$
$$a = 4 \qquad b = -2 \qquad c = 45$$

$$x = \frac{-(-2) \pm \sqrt{(-2)^2 - 4(4)(45)}}{2(4)}$$

$$x = \frac{2 \pm \sqrt{-716}}{8}$$

Verify with the graph.

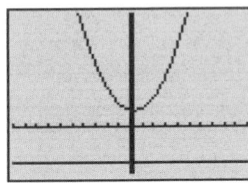

$y = -30$

Because the number under the square root is negative, there are no real solutions to this equation. This result is verified by the graph, which shows that the parabola never gets down to -30.

EXAMPLE

Estimate the solutions to the system of equations given in the graph.

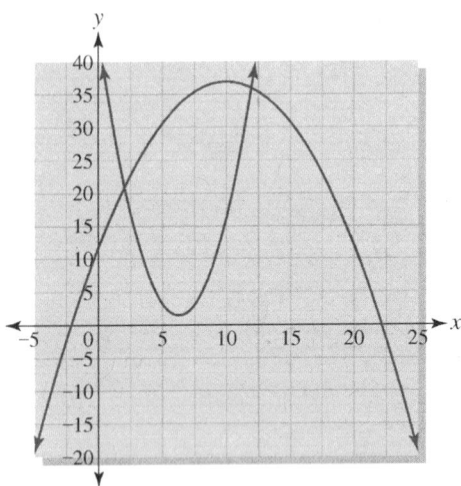

Solution

These graphs appear to cross at about $(2, 21)$ and $(12, 36)$.

EXAMPLE 12

Solve the following system of equations.

$$y = 3x + 5$$
$$y = x^2 + 2x - 1$$

Solution

Using the substitution method, we see that:

$$3x + 5 = x^2 + 2x - 1$$
$$0 = x^2 - x - 6$$
$$0 = (x + 2)(x - 3)$$

$$x = -2 \qquad x = 3$$
$$y = 3(-2) + 5 \qquad y = 3(3) + 5$$
$$y = -1 \qquad y = 14$$

X	Y₁	Y₂
-2	-1	-1
3	14	14

X=

The solutions are $(-2, -1)$ and $(3, 14)$ and are verified in the table.

Section 4.7 Graphing from Standard Form

- When **graphing a quadratic in standard form**, you can use these steps.

  1. Determine whether the graph opens up or down. Find the vertical intercept.
  2. Find the vertex.
  3. Find the horizontal intercepts (if any).
  4. Find the axis of symmetry.
  5. Plot the points, plot their symmetric pairs and sketch the graph. (Find an additional symmetric pair if needed.)

- The x value of the **vertex** can be found in the middle of the x values of any symmetric pair or by using the input value $x = \dfrac{-b}{2a}$.

- To graph a **quadratic inequality,** sketch the graph as if it were an equation, using a dashed line if there is no "equal to" part, a solid line if there is an "equal to" part, and then shade the side of the curve that makes the inequality true. Using a test point that is not on the parabola may help you to determine which side to shade.

EXAMPLE

Graph the function $f(x) = -0.25x^2 + 5x + 12$.

Solution

This graph faces downward because a is negative and is somewhat wide because $|a| < 1$.

The vertical intercept is (0, 12)

The vertex has an input of $x = \dfrac{-5}{2(-0.25)} = 10$ and

$f(10) = 37$, so the vertex is (10, 37).

Because the graph faces downward and has a vertex above the x-axis, the graph will have x-intercepts. We can use the quadratic formula to find them. The solutions are $(-2.17, 0)$ and (22.17, 0).

Another symmetric pair is (5, 30.75) and (15, 30.75). Notice that the x value of the vertex is 10

and is in the middle of the x values of both of these symmetric pairs.

Using all these points we get the following sketch.

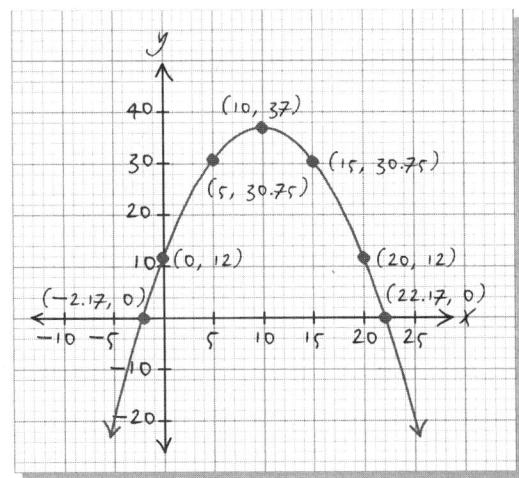

EXAMPLE

Graph the inequality $y < x^2 + 4x - 12$.

Solution

We will sketch the graph with a dashed line since there is no "equal to" part in the inequality symbol. Then we will shade below the curve.

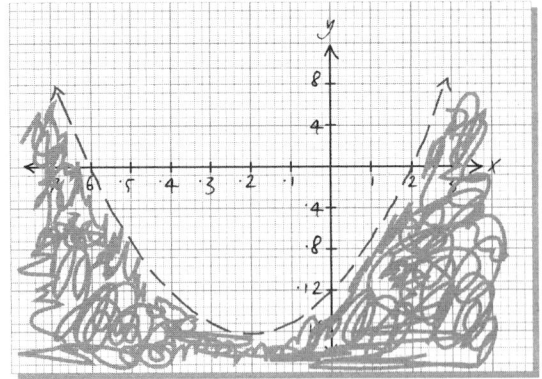

Chapter 4 Review Exercises

CHAPTER

4

1. Use the graph to estimate the following.

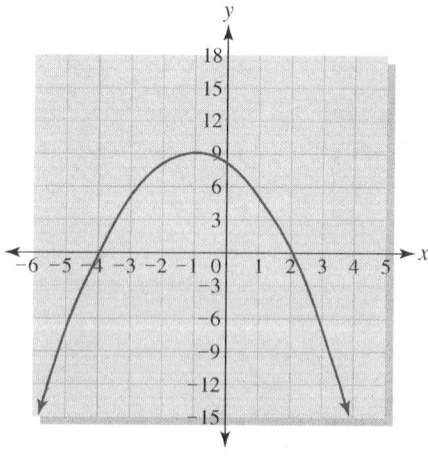

a. Vertex

b. For what x values is the graph increasing?

c. For what x values is the graph decreasing?

d. Horizontal intercept(s)

e. Vertical intercept [4.1]

2. Use the graph from exercise 1 to estimate the following.

a. What is the axis of symmetry?

b. If $f(x)$ is written in vertex form, would a be positive or negative?

c. If $f(x)$ is written in vertex form, what would the values of h and k be?

d. $f(1) = $?

e. What x value(s) will make $f(x) = -12$? [4.2]

3. The population of North Dakota can be modeled by

$$P(t) = -0.89(t - 5.6)^2 + 643$$

where $P(t)$ represents the population in thousands t years since 1990.

Source: Model derived from data in Statistical Abstract 2001.

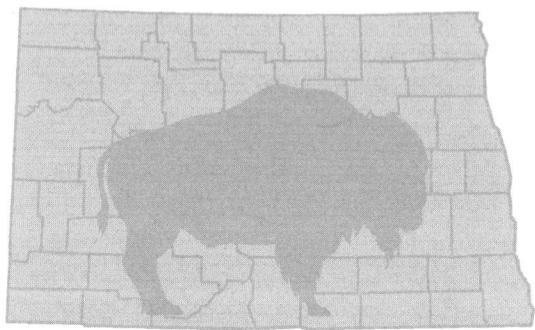

a. Find the population of North Dakota in 1998.

b. Sketch a graph of this model. [4.2]

c. Find the vertex of this model and explain its meaning in this context.

d. When was the population of North Dakota 640 thousand? [4.4]

2. The number of murders, M, in thousands in the United States t years since 1990 can be represented by

$$M(t) = -0.42(t - 2.5)^2 + 23$$

Source: Model derived from data in Statistical Abstract 2001.

a. Estimate the number of murders in the United States in 1996.

b. Find the vertex of this model and explain its meaning in this context.

c. In what year were there 14.5 thousand murders in the United States? [4.4]

3. Sketch the graph and label the vertex, vertical intercept, and the horizontal intercept(s) for the following functions.

a. $f(x) = 3(x - 8)^2 + 4$

b. $g(x) = -0.5(x + 7)^2 - 3$ [4.2]

446

4. The number of juveniles arrested for possession of drugs through the 1990s is given below.   [4.3]

Year	Number of Arrests (in thousands)
1992	47,901
1994	92,185
1995	115,159
1996	116,225
1997	124,683
1998	118,754
1999	112,640

Source: Statistical Abstract 2001.

a. Find an equation for a model of these data.

b. Give a reasonable domain and range for this model.

c. Estimate the number of juveniles arrested for possession of drugs in 2000.

d. If this trend continues, in what year will the number of juvenile arrests be at 50 million again? (50 million = 50,000 thousand).

5. The U.S. Department of Commerce obligations for research and development for several years is given below.   [4.3]

Year	R & D Obligations (in millions of dollars)
1995	1136
1996	1068
1997	1003
1999	990
2000	1041
2001	1127

Source: Statistical Abstract 2001.

a. Find an equation for a model of these data.

b. Give a reasonable domain and range for this model.

c. Estimate the research and development obligations for the Department of Commerce in 1998.

d. Estimate when the Department of Commerce research and development obligations will be $1.5 billion ($1,500 million).

6. The median sales price in thousands of dollars of new houses in the western United States t years since 1990 can be modeled by

$$P(t) = 1.073t^2 - 5.84t + 144.68$$

Source: Model derived from data in Statistical Abstract 2001.

a. Estimate the median price in 1995.

b. When will the median price of a new home in the West reach $250,000?

c. Find the vertex of this model and explain its meaning in this context.   [4.7]

7. Solve the following quadratic equations using the square root property.

a. $3x^2 + 75 = 0$

b. $-0.25(x - 6)^2 + 8 = 0$ [4.4]

8. Solve the following quadratics by completing the square.

a. $x^2 + 26x = 30$

b. $4x^2 - 6x + 20 = 0$ [4.4]

9. Convert the following functions to vertex form.

a. $f(x) = x^2 + 8x + 11$

b. $g(x) = -6x^2 + 20x - 18$ [4.4]

10. Solve the following quadratic equations by factoring.

a. $9x^2 - 24x + 5 = -11$

b. $2x^2 - 8x - 120 = 0$

c. $6x^2 - 8x = 0$

d. $3x^2 - x = 2$ [4.5]

11. Use the graph to find an equation for the quadratic.

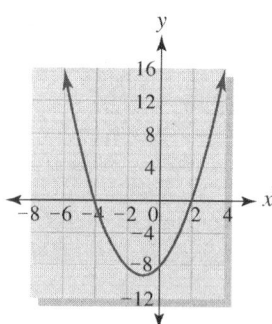

[4.5]

12. Solve the following quadratic equations using the quadratic formula.

a. $3x^2 + 9x - 20 = 14$

b. $-4.5x^2 + 3.5x + 12.5 = 0$

c. $x^2 + 6x + 18 = 0$ [4.6]

13. Sketch the graph. Label the vertex and vertical and horizontal intercept(s).

a. $f(x) = -2.7x^2 + 16.2x - 21.8$

b. $h(n) = 0.3n^2 + 3n + 14.5$ [4.7]

14. Give the domain and range of the following functions.

a. $f(x) = 2(x + 9)^2 - 15$

b. $f(x) = -3x^2 + 12x + 16$ [4.3]

15. Solve the following systems of equations.

a. $y = 2x^2 + 5x + 12$
$y = -2x^2 + 7x + 15$

b. $y = 0.25x^2 + 4x - 7$
$y = 0.5x^2 + 3x - 9$ [4.6]

16. Use the table to estimate the solution(s) to the following system. [4.7]

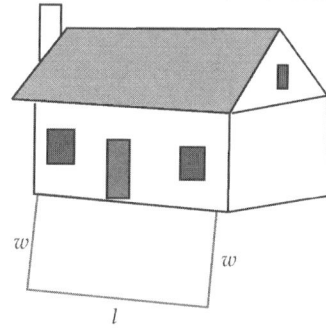

17. Hitomi has 100 ft of lights he wants to put around a rectangular area next to his house. The house will form one side of the area as shown in the figure.

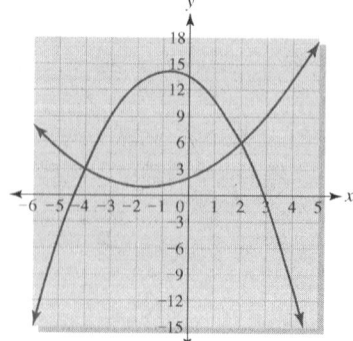

a. What is the largest possible rectangular area?

b. What length and width give the largest possible area? [4.6]

18. Graph the inequalities.

a. $y < 2x^2 + 24x - 10$

b. $y \ge -0.5x^2 - 4x - 6$

c. $y \le 2.5(x - 4)^2 - 18$

d. $y > -0.8(x + 1)^2 - 3$ [4.7]

19. Use the graph to solve the system of equations. [4.7]

Chapter 4 Test

1. The number of violent crimes in the United States during the late 1900s can be modeled by

$$V(t) = -0.044(t - 11)^2 + 14.5$$

where $V(t)$ represents the number of violent crimes in millions t years since 1980.

Source: Model derived from data in Statistical Abstract 2001.

   a. Estimate the number of violent crimes in 1995.

   b. Sketch a graph of this function.

   c. Determine the vertex and explain its meaning in this context.

   d. In what year will there be 10 million violent crimes in the United States?

2. Use the graph to solve the system of equations.

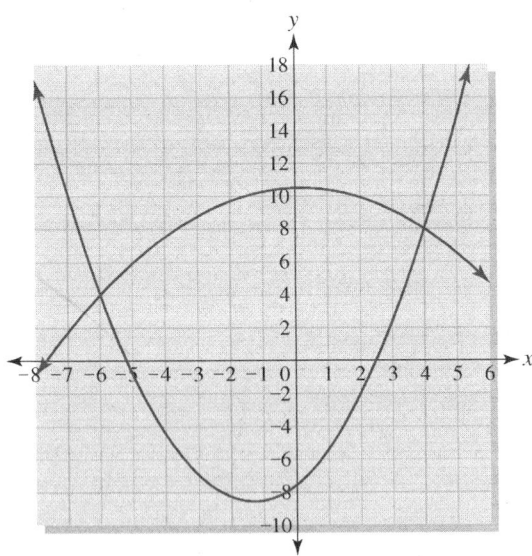

3. Solve the following quadratic equations by factoring.

   a. $3x^2 - 5x + 5 = 33$

   b. $8x^2 - 34x = -35$

4. The total outlays for national defense and veterans benefits by the United States for several years is given below.

Year	Total Outlays (in billions of dollars)
1993	326.8
1995	310
1997	309.8
1999	320.2
2000	337.4

Source: Statistical Abstract 2001.

   a. Find an equation for a model of these data.

   b. Give a reasonable domain and range for this model.

   c. Estimate the total outlays for national defense and veterans benefits in 1998.

   d. When will the total outlays for national defense and veterans benefits reach half a trillion dollars? (Half a trillion = 500 billion).

5. The U.S. commercial space industry revenue for satellite manufacturing for several years is given in the chart.

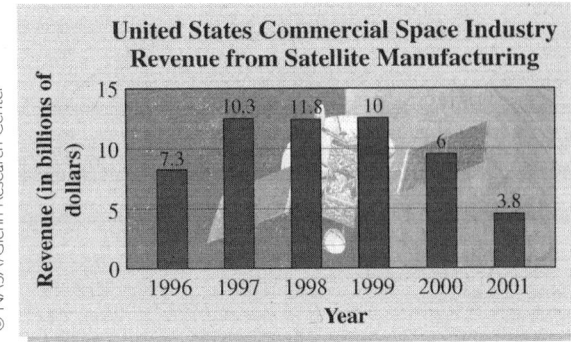

Source: Statistical Abstract 2004–2005.

a. Find an equation for a model of these data.

b. Give a reasonable domain and range for this model.

c. Estimate the commercial space industries revenues for satellite manufacturing in 2000.

d. Give the vertex for your model and explain its meaning in this context.

6. Sketch the graph. Label the vertex and the vertical and horizontal intercepts.

  a. $f(x) = -1.5x^2 + 12x - 20.5$

  b. $g(m) = 0.4m^2 + 1.6m + 11.6$

7. Solve the system of equations

$$y = 2x^2 + 5x - 9$$

$$y = -7x^2 + 3x + 2$$

8. The number of privately owned single unit houses built in the late 1990s can be modeled by

$$H(t) = -52t^2 + 916.2t - 2731.2$$

where $H(t)$ represents the number of privately owned single-unit houses built in thousands t years since 1990.

Source: Model derived from data in the Statistical Abstract 2001.

  a. Estimate the number of single-unit houses built in 2000.

  b. When were there 1000 thousand of these homes built?

9. Jay has a 20 ft ladder that he leaned against the wall of his house. Jay put the base of the ladder 4 ft away from the house. How far up the house is the top of the ladder?

10. Use the table to estimate the solution(s) to the system of equations.

X	Y₁	Y₂
-9	-15.33	24.667
-8	-4	12
-7	4	4
-6	8.6667	.66667
-5	10	2
-4	8	8
-3	2.6667	18.667

X=-3

11. Use the graph to estimate the following:

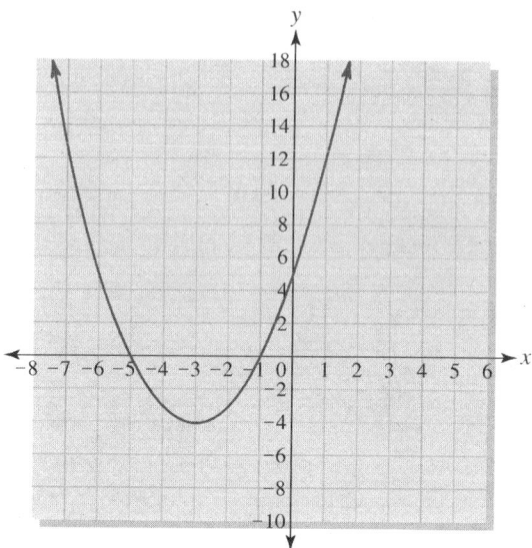

a. The vertex.

b. For what x-values is the graph increasing?

c. For what x-values is the graph decreasing?

d. The axis of symmetry.

e. The vertical intercept.

f. The horizontal intercept(s).

g. Write an equation for the function.

12. Solve the following equations using either the square root property or completing the square.

  a. $x^2 + 12x + 30 = 0$

  b. $4(x - 9)^2 - 20 = 124$

  c. $8x^2 + 15 = 65$

13. Convert the function $f(x) = 3x^2 + 24x - 30$ to vertex form.

14. Sketch the graph of $y < 0.5x^2 - 3x + 9$.

15. Give the domain and range of the following functions.

  a. $f(x) = -0.4(x + 20)^2 - 17$

  b. $g(x) = x^2 + 8x + 15$

Chapter 4 Projects

Group Experiment

Two or more people

What you will need:

- A Texas Instruments CBR unit or a Texas Instruments CBL unit with a Vernier motion detector probe.

- Book or other light object.

CAN YOU MOVE LIKE A PARABOLA?

It will be your job to create data that will best be modeled by a quadratic function. Using the CBR/CBL unit, measure the distance from the unit over time and have someone move a book or other object in front of the motion detector to create data that have the shape of a parabola. It might take several tries before you collect data that are reasonable. Be sure to check your data so that you are satisfied that the observations will be a quadratic function before you move on.

Write up

a. Describe how you collected the data. What was the most important thing to do to get a quadratic model? Did you select only part of the data that were collected to get a good quadratic model?

b. Create a scatterplot using the values that you found in the table on the calculator or computer, and print it out or complete it neatly by hand.

c. Find a model to fit the data.

d. What is the vertex of your model, and what does it represent in this experiment?

e. What is the vertical intercept of your model, and what does it represent in this experiment?

f. Does your parabola face upward or downward? How would you have to change your experiment to get the graph to face the other direction?

g. What are a reasonable domain and range for your model? (Remember to consider any restrictions on the experiment.)

Written Project

One or more people

MAXIMIZE YOUR FRUIT

You are given the task of estimating the number of trees that will maximize the fruit production per acre of a local orchard. By researching other local orchards, you discover that at 400 trees per acre, each tree will produce an average of 30 pounds of fruit. For each tree that is added to the acre, the average production **per tree** goes down 0.05 pounds.

Write up

a. Use the given information to build a table of values for this situation. Let t be the number of trees above 400 planted on an acre and find the total amount of fruit each **acre** will produce. Find at least eight sets of values.

b. Create a scatterplot using the values that you found in the table on the calculator or computer, and print it out or complete it neatly by hand.

c.   Find a function for the total production per acre when t trees over 400 are planted on one acre.

d.   Find the total production per acre when 400 trees are planted.

e.   Estimate how many trees should be planted per acre to produce 12,400 pounds of fruit.

f.   Find the number of trees per acre that will maximize the fruit production.

g.   Find a reasonable domain and range for your model.

WHAT GOES UP MUST COME DOWN?

Group Experiment

Two or more people

What you will need:

•   A Texas Instruments CBR unit or a Texas Instruments CBL unit with a Vernier motion detector probe.

•   A volleyball or other type of ball.

In this project you are going to explore the path of an object that is tossed into the air and its speed and acceleration during that time. Place the CBL/CBR unit on the ground and have someone toss the ball into the air while you record the height of the ball as time goes by. (Be careful that the ball does not land on the CBL/CBR unit.) Do the experiment again, this time recording the velocity (speed) of the ball over time, and a third time, recording the ball's acceleration over time.

Write up

a.   Create tables of the height-versus-time data, the velocity-versus-time data, and the acceleration-versus-time data. (You will have to collect a large number of points for this project.)

b.   Create scatterplots of these three sets of data on your calculator or computer, and print them out or complete them neatly by hand on graph paper.

c.   What kind of model will best fit the height-versus-time data? Create a model for these data.

d.   What kind of model will best fit the velocity-versus-time data? Create a model for these data.

e.   What kind of model will best fit the acceleration-versus-time data? Create a model for these data.

f.   What is the maximum height the ball reached?

g.   What is the maximum velocity of the ball? Was the velocity ever zero? When did these velocities occur?

h.   What was the maximum acceleration of the ball? When did this occur?

i.   Use your models to estimate the height, velocity, and acceleration of the ball after 1.5 seconds.

j.   Explain why you think the graphs of these different values are the shapes they are.

Research Project

One or more people

What you will need:

• Find data for a real-world situation that can be modeled with a quadratic function.

• You might want to use the Internet or library. Statistical abstracts and some journals and scientific articles are good resources for real-world data.

FIND YOUR OWN PARABOLA

In this project you are given the task of finding data for a real-world situation that you can apply a quadratic model to. You may use the problems in this chapter to get ideas of things to investigate, but your data should not be discussed in this textbook.

Write up

a. Describe the data you found and where you found them. Cite any sources that you used.

b. Create a scatterplot of the data on the calculator or computer, and print it out or complete it neatly by hand on graph paper.

c. Find a model to fit the data.

d. What is the vertex of your model, and what does it represent in the context?

e. What is the vertical intercept of your model, and what does it represent in the context?

f. What are a reasonable domain and range for your model?

g. Use your model to estimate an output value of your model for an input value that you did not collect in your original data.